# Circuit
# Troubleshooting
# Handbook

I0031409

# Circuit Troubleshooting Handbook

John D. Lenk

**McGraw-Hill**

New York   San Francisco   Washington, D.C.   Auckland   Bogotá
Caracas   Lisbon   London   Madrid   Mexico City   Milan
Montreal   New Delhi   San Juan   Singapore
Sydney   Tokyo   Toronto

**Library of Congress Cataloging-in-Publication Data**
Lenk, John D.
    Circuit troubleshooting handbook / John D. Lenk.
        p.     cm.
    Includes index.
    ISBN 0-07-038185-2 (hardcover). — ISBN 0-07-038186-0 (softcover)
    1. Electronic apparatus and appliances—Maintenance and repair—
Handbooks, manuals, etc.   2. Electronic circuits—Handbooks,
manuals, etc.
TK7870.2L458     1999                              98-39250
621.3815—dc21                                      CIP

## McGraw-Hill

*A Division of The McGraw·Hill Companies*

1 2 3 4 5 6 7 8 9 0   DOC/DOC   9 0 3 2 1 0 9 8

ISBN 0-07-038185-2 (HC)
ISBN 0-07-038186-0 (SC)

*The sponsoring editor of this book was Steve Chapman. The editing supervisor was
Peggy Lamb, and the production supervisor was Clare Stanley. This book was set in
New Century Schoolbook by Joanne Morbit and Michele Zito of McGraw-Hill's
Professional Book Group composition unit, in Hightstown, N.J.*

*This book is printed on recycled, acid-free paper containing a minimum of
50% recycled, de-inked fiber.*

McGraw-Hill books are available at special quantity discounts to use as premiums
and sales promotions, or for use in corporate training programs. For more information,
please write to the Director of Special Sales, McGraw-Hill, 11 West 19th Street,
New York, NY 10011. Or contact your local bookstore.

# Dedication

*Greetings from the Villa Buttercup!*
*To my wonderful wife, Irene.*
*Thank you for being by my side all these years!*
*To my lovely family, Karen, Tom, Brandon, Justin, and Michael.*
*And to our Lambie and Suzzie, be happy wherever you are!*
*To my special readers; May good fortune find your doorway, bringing you good health and happy things.*
*Thank you for buying my books!*
*Special thanks to Steve Chapman, Stephen Fitzgerald, Leslie Wenger, Ted Nardin, Mike Hays, Lisa Schrager, Patrick Hansard, Peter Mellis, Scott Grillo, Andrew Yoder (best-selling author), and Robert McGraw of McGraw-Hill for making me an international best seller again!*
*This is book number 91*
*Abundance!*

# Contents

# Preface

This book is a "crash course" in circuit troubleshooting, filling the gap between the theory of electronic circuits and the practical how-to of trouble localization. Far more than a simple review of circuits, or an introduction to theoretical troubleshooting, the book is a technician-level guide on how to pinpoint circuit faults. *The information here is of particular value when troubleshooting just-completed experimental circuits, or when a particular circuit fails repeatedly in final form.*

The book is written primarily for field-service engineers and technicians, but will also prove invaluable to the student who is about to face the real world of everyday troubleshooting. Thus, the book is well suited to both the experienced and inexperienced reader. Because the book is written with these two classes of readers in mind, both circuit theory and test/troubleshooting are given for a variety of electronic circuits. Emphasis is placed on how a circuit works because such information is essential to practical troubleshooting.

Each type of circuit is assigned a separate chapter, and the same pattern is maintained throughout all chapters. First, the circuit (usually an integrated circuit or IC) is described in detail, stressing how the circuit functions are related to troubleshooting. This is followed with a test/troubleshooting approach for the circuit or group of circuits. Adjustments and circuit calibration are included where applicable.

*Much emphasis is placed on testing, and the circuit's response to the test.* This is because testing is often the first step in troubleshooting. If you test a circuit and find that it performs as outlined, leave it alone! No further troubleshooting (or tinkering) is required. On the other hand, if the circuit fails to perform one or more functions, or performs the function poorly you have an excellent starting point for trouble localization.

It is assumed that the reader is familiar with electronic test equipment and testing at a level found in the *McGraw-Hill Electronic Testing Handbook*, and with troubleshooting found in the *McGraw-Hill Electronic Troubleshooting Handbook*. However, no reference is made to any other publication. This book stands alone!

*John D. Lenk*

# Acknowledgments

Many professionals have contributed to this book. I gratefully acknowledge the tremendous effort needed to produce this book. Such a comprehensive work is impossible for one person, and I thank all who contributed, both directly and indirectly.

I give special thanks to the following: Alan Haun of Analog Devices, Syd Coppersmith of Dallas Semiconductor, Rosie Hinojosa of EXAR Corporation, Jeff Salter of GEC Plessey, John Allen, Helen Cox, and Linda da Costa of Harris Semiconductors, Ron Denchfield and Bob Scott of Linear Technology Corporation, David Fullagar and William Levin of Maxim Integrated Products, Fred Swymer of Microsemi Corporation, Linda Capcara of Motorola, Inc., Andrew Jenkins and Shantha Natarajan of National Semiconductor, Antonio Ortiz of Optical Electronics Incorporated, Lawrence Fogel of Philips Semiconductors, John Marlow of Raytheon Electronics Semiconductor Division, Anthony Armstrong of Semtech Corporation, Ed Oxner and Robert Decker of Siliconix Incorporated, Amy Sullivan of Texas Instruments, Alan Campbell of Unitrode Corporation, Sally and Barry E. Brown (Broker), and Andrew Yoder (best-selling author).

I also wish to thank Joseph A. Labok of Los Angeles Valley College for help and encouragement throughout the years.

And a very special thanks to Steve Chapman, Stephen Fitzgerald, Leslie Wenger, Patrick Hansard, Peter Mellis, Ted Nardin, Mike Hays, Lisa Schrager, Mary Murray, Carol Wilson, Judy Kessler, Monika Macezinskas, Florence Trimble, Fran Minerva, Jane Stark, Fred Perkins, Robert McGraw, Judith Reiss, Charles Love, Betty Crawford, Jeanne Myers, Peggy Lamb, Thomas Kowalczyk, Clare Stanley, Suzanne Rapcavage, Jaclyn Boone, Kathy Green, Donna Namorato, Regina Frappolli, Sherri Souffrance, Allison Arias, and Midge Haramis of the McGraw-Hill Professional Publishing organization for having that much confidence in me.

And to Irene, my wife and Super Agent, I extend my thanks. Without her help, this book could not have been written.

# About the Author

John D. Lenk has been a technical author specializing in practical electronic design/service guides for over 40 years. He is the best-selling author of more than 90 books on circuit and consumer electronics, which together have sold over 2 million copies in nine languages and 33 countries. His most recent titles include *Lenk's Video Handbook, Lenk's Digital Handbook, Lenk's Audio Handbook, Lenk's Laser Handbook, Lenk's RF Handbook, Lenk's Television Handbook*, the *McGraw-Hill Circuit Encyclopedia*, Volumes 1-4, the *McGraw-Hill Electronic Testing Handbook*, and the *McGraw-Hill Electronic Troubleshooting Handbook*.

# Circuit
# Troubleshooting
# Handbook

Chapter

# 1

# Op-Amp Circuit Troubleshooting

This chapter is devoted to troubleshooting for op amps (operational amplifiers). As promised in the Introduction, the coverage starts with a detailed description of a typical IC op amp, emphasizing how circuit functions are related to troubleshooting. A description of why such circuits fail in either the experimental or final form follows.

## 1.1 The Basic IC Op Amp

IC op amps generally use several differential stages in cascade to provide both common-mode rejection and high gain. (If you are not familiar with such terms as *common-mode rejection,* do not panic. The terms are explained in Section 1.5.) Differential amplifiers require both positive and negative power supplies and have two inputs to provide for in-phase or out-of-phase amplification. The classic op-amp circuit requires that the output be fed back to the input through a resistance or impedance. The output is fed back to the negative or inverting input to produce degenerative feedback (and thus set the desired gain and frequency response).

As in any amplifier, the signal shifts in phase through an op amp when passing from input to output. The amount of phase shift depends on the frequency. If the phase shift approaches 180°, the feedback is in phase with the inverting input (or nearly so), and causes the amplifier to oscillate. The condition of increased phase shift (with increased frequency) limits the bandwidth of the op amp and is a common troubleshooting problem. The condition is often compensated for by adding a phase-shift network (usually an RC circuit, but possibly a single capacitor). Phase compensation is covered in Section 1.3.

## 1.2 IC Op-Amp Circuit

Figure 1-1 shows the block diagram of a typical op amp (the Harris CA3130). Figure 1-2 shows a more detailed schematic. The op amp is a

1

BiMOS, which combines the advantage of both CMOS and bipolar transistors on a single chip. Gate-protected MOSFETs are used at the input to provide very high input impedance and very low input current. Both of these characteristics are quite important in testing troubleshooting, as covered in Sections 1.4 and 1.5.

A CMOS transistor pair, capable of swinging the output voltage to within 10 mV of either supply-voltage terminal (at very high values of load impedance) is used as the output circuit. The op amp can operate at supply voltages that range from 5 to 16 V or ±2.5 to ±8 V, when using split supplies. The IC can be phase compensated with a single external capacitor (Fig. 1-1) and has terminals to adjust the offset voltage (when the external circuit requires offset-null capability). Terminal provisions also permit strobing of the output stage.

The input terminals can be operated down to 0.5 V below the negative supply and the output can swing very close (to about 10 mV) to either supply in many circuits, thus making the op amp well suited to single-supply operation. (The classic op amp requires both positive and negative supplies.)

Three class-A amplifier stages, having the individual gain capability and current consumption shown in Fig. 1-1, provide the total gain. A biasing circuit provides two voltages for common use in the first and second stages. Terminal 8 can be used both for phase compensation and to strobe the output stage into the quiescent condition.

When pin 8 is tied to the negative supply (pin 4), the output potential at pin 6 rises to the approximate positive potential at pin 7. This condition of essentially zero current drain can be achieved only when the load resistance presented to the amplifier is very high (for example, when the op amp is used to drive CMOS digital circuits in comparator applications).

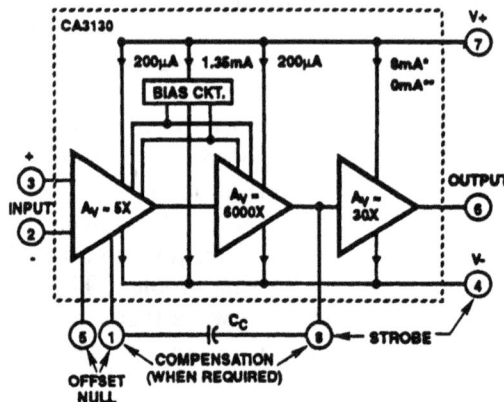

**Figure 1-1**  Block diagram of op amp (Harris Semiconductors, *Linear & Telecom ICs*, 1994, p. 2-112).

TOTAL SUPPLY VOLTAGE (FOR INDICATED VOLTAGE GAINS) = 15V

*WITH INPUT TERMINALS BIASED SO THAT TERM. 6 POTENTIAL IS +7.5V ABOVE TERM. 4.

**WITH OUTPUT TERMINAL DRIVEN TO EITHER SUPPLY RAIL.

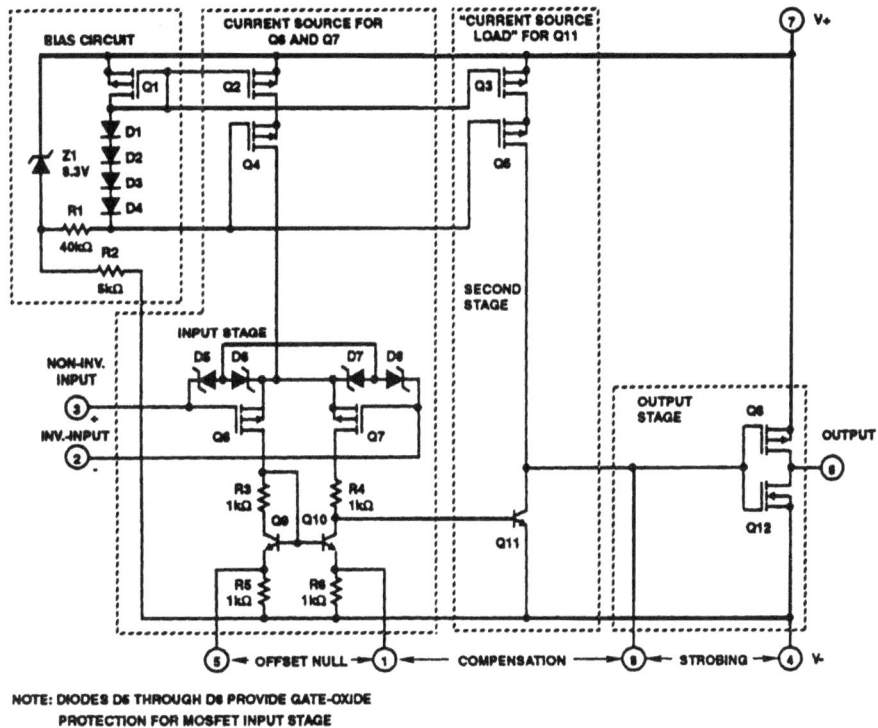

Figure 1-2   Schematic of op amp (Harris Semiconductors, *Linear & Telecom ICs*, 1994, p. 2-111).

## 1.2.1  Input stages

The differential-input stage (Q6/Q7) works into a mirror-pair of bipolar transistors (Q9/Q10) that function as load resistors (together with R3 through R6). Q9 and Q10 also function as a differential-to-single-ended converter to provide base drive to the second-stage bipolar transistor (Q11). Offset nulling (when required) can be provided by connecting a 100-k$\Omega$ pot across pins 1 and 5, with the slider arm at pin 4. The cascade-connected transistors (Q2/Q4) are the constant-current source for the input stage. The small diodes (D5 through D8) provide protection against high-voltage transients, including static electricity, during handling.

## 1.2.2  Second stage

Most of the voltage gain is provided by the second amplifier stage, consisting of Q11 and the cascade-connected load resistance provided by Q3/Q5. Rolloff compensation is provided by connecting a small capacitor between pins 1 and 8. In most circuits, a 47-pF capacitor provides sufficient compensation for this op amp. When testing this or any op amp, always use the recommended compensation scheme. When troubleshooting an op amp that is operative, but not providing the correct frequency or phase response, the external compensation is almost always at fault.

### 1.2.3  Bias circuit

When the supply is above 8.3 V, resistor R2 and Zener Z1 establish a voltage of 8.3 V across the series-connected circuit (consisting of R1, D1 through D4, and Q1). A tap at the junction of R1 and D4 provides a gate bias of about 4.5 V for Q4/Q5, with respect to pin 7. A potential of about 2.2 V is developed across diode-connected Q1, with respect to pin 7. This provides a gate bias for Q2 and Q3. Notice that Q1 is mirror-connected to Q2/Q3. Because Q1, Q2, and Q3 are designed to be identical, the approximately 200-$\mu$A current in Q1 establishes a similar current in Q2/Q3 as constant-current sources for both the first and second amplifier stages.

When the supply voltage is somewhat less than 8.3 V, Z1 becomes nonconducting and the potential (developed across series-connected R1, D1 through D4, and Q1) varies directly with variations in supply voltage. Consequently, the gate bias for Q4/Q5 and Q2/Q3 varies in accordance with supply-voltage variations. This variation results in deterioration of the PSRR ratio at supply voltages below 8.3 V. Operation at supply voltages below about 4.5 V results in seriously degraded performance. If any experimental circuit appears to perform poorly in crucial factors, such as gain, frequency response, etc., be certain that the supply voltage is above the minimum required for that particular IC op amp.

### 1.2.4  Output stage

The output stage consists of a drain-loaded inverting amplifier (Q8/Q12) operating in the class-A mode. When operating into very high-resistance loads, the output can swing within millivolts of either supply. Because the output is drain loaded, the gain depends on load impedance. When you find an op amp that does not produce the correct gain, check the load impedance. What appears to be a fault might actually be a characteristic of the IC (which cannot be altered).

Figure 1-3 shows the transfer characteristics of the output stage for a load returned to the negative supply. The output can drive typical op-amp loads as shown. As a voltage follower, the op amp can achieve 0.01% accuracy levels, including the negative supply.

### 1.2.5  Power-supply problems

As shown in Fig. 1-4, the IC can operate with a dual supply or a single supply. The following summarizes power-supply characteristics from a troubleshooting standpoint.

With dual-supply operation, the currents supplied by both supplies are equal when the output voltage at pin 6 is 0 V. When the gate terminals of Q8/Q12 are driven increasingly positive with respect to ground, current flow through Q12 (from the negative supply) to the load is increased, and current

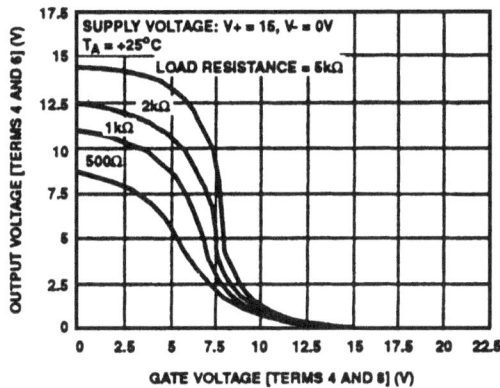

Figure 1-3  Transfer characteristics of output stage (Harris Semiconductors, *Linear & Telecom ICs,* 1994, p. 2-113).

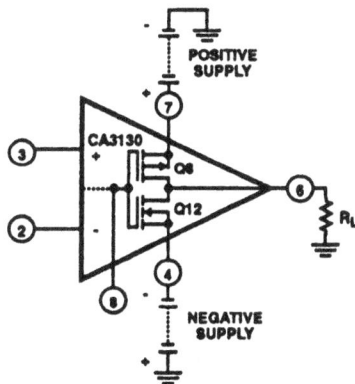

(A) DUAL POWER-SUPPLY OPERATION

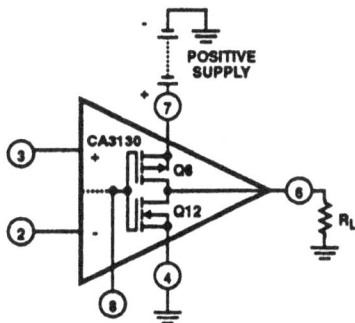

Figure 1-4  Power-supply connections (Harris Semiconductors, *Linear & Telecom ICs,* 1994, p. 2-115).

(B) SINGLE POWER-SUPPLY OPERATION

flow through Q8 (positive supply) decreases by a corresponding amount. When the gate terminals of Q8 and Q12 are driven increasingly negative with respect to ground, current flow through Q8 is increased and current flow through Q12 decreases accordingly.

With single-supply operation, assume that the value of $R_L$ is very high (or disconnected) and that the input terminal bias (pins 2 and 3) is such that the output (pin 6) voltage is at one-half of $V+$ (the voltage drops across Q8 and Q12 are of equal magnitude). Figure 1-5 shows typical quiescent supply current versus supply voltage for the IC operated under these conditions.

Because the output is operating in class A, the supply current will remain constant under dynamic operating conditions as long as the transistors are operated in the linear portion of their voltage-transfer characteristics. If either Q8 or Q12 is driven from their linear regions toward cutoff (a nonlinear region), there will be a corresponding reduction in supply current. In the extreme case (where pin 8 swings to ground potential or is tied to ground), Q12 is completely cut off and the supply current to Q8/Q12 goes (essentially) to zero. However, the two preceding stages continue to draw modest supply current (lower curve of Fig. 1-5), even though the output stage is off. Figure 1-4B shows a dual-supply arrangement for the output stage that can also be strobed off (assuming that $R_L$ is infinite) by pulling the potential of pin 8 down to that of pin 4.

In practical troubleshooting situations, $R_L$ is not infinite, but is at some definite value (such as those shown in Fig. 1-3). Assume that $R_L$ is 2 kΩ and that the input-terminal bias (pins 2/3) is such that the output (pin 6) is at one-half of $V+$. Because Q8 must now supply quiescent current to both $R_L$ and Q12, the supply current increases as an inverse function of $R_L$. Figure 1-6 shows the voltage drop across Q8 as a function of the load current (at several supply voltages).

From this, you can see that the value of load resistance $R_L$ not only affects gain, but has a direct effect on both supply current and linearity. Remember this when troubleshooting an op amp where supply current and/or linearity are incorrect—even though gain is good.

### 1.2.6   Input-current variations

Figure 1-7 shows the relationship of input current to input voltage. Figure 1-8 shows the relationship of input current to ambient temperature. From a design standpoint, the effect of increased input current is a corresponding

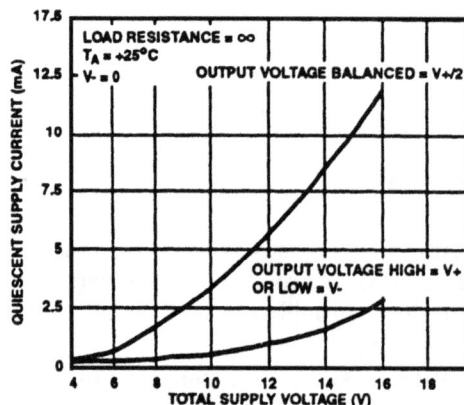

Figure 1-5  Quiescent supply current versus supply voltage (Harris Semiconductors, *Linear & Telecom ICs*, 1994, p. 2-113).

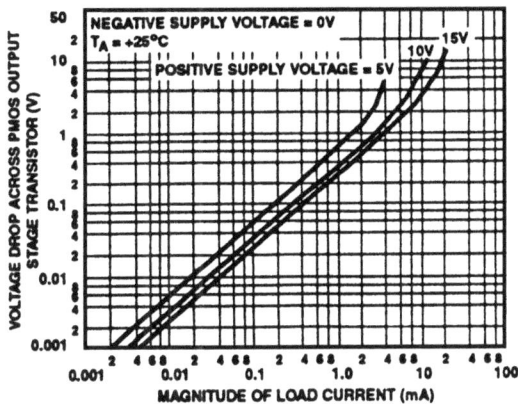

**Figure 1-6** Voltage drop across output transistor versus load current (Harris Semiconductors, *Linear & Telecom ICs*, 1994, p. 2-113).

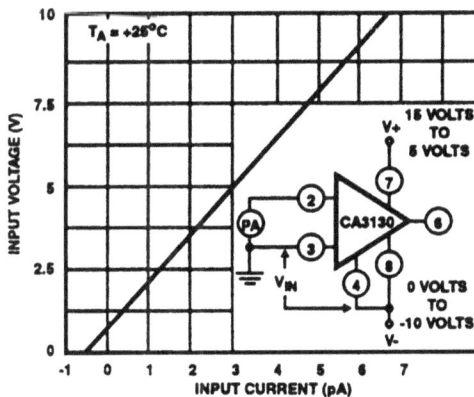

**Figure 1-7** Relationship of input current to common-mode voltage (Harris Semiconductors, *Linear & Telecom ICs*, 1994, p. 2-114).

increase in input-offset voltage across any input resistance. In troubleshooting, the main concern is that input current (and input-offset voltage) increases with temperature. For example, when troubleshooting an experimental circuit that shows excessive input current with anticipated temperatures, a heatsink might be required. In an existing circuit, it is possible that the heatsink is not making proper contact with the IC. Most of the input current is because of leakage current through the gate-protective diodes (D5 through D8) in the input circuit. As a guideline, this leakage current doubles (approximately) for every $+10°C$ increase in temperature.

### 1.2.7 Input-offset voltage variations

The characteristics of any MOSFET device can change slightly when a dc gate-source bias is applied to the device for extended time periods. The magnitude of the change is increased at high temperatures. Figure 1-9 shows typical input-offset shift during lift tests of the IC. Generally, the problems shown in Fig. 1-9 apply only to circuits where the IC must be operated at high temperatures

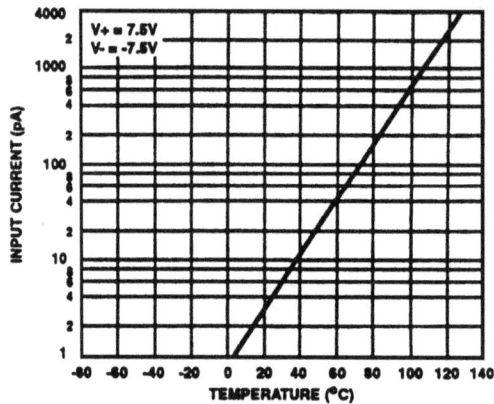

**Figure 1-8** Relationship of input current to ambient temperature (Harris Semiconductors, *Linear & Telecom ICs,* 1994, p. 2-114).

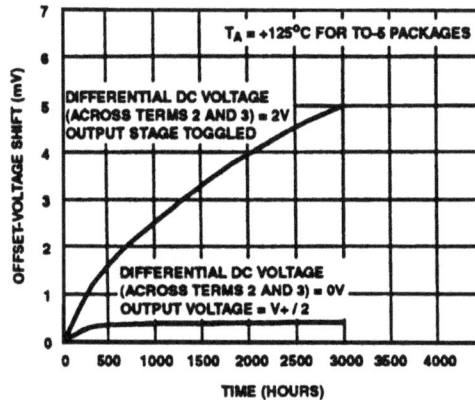

**Figure 1-9** Incremental offset-voltage shift versus operating life (Harris Semiconductors, *Linear & Telecom ICs,* 1994, p. 2-115).

($+125°$C) with large input signals (such as shown in the upper graph). Where the IC is used in a typical linear circuit (such as shown in the lower graph of Fig. 1-9), the shift in offset is small (comparable to that for a bipolar transistor). In troubleshooting, this problem of characteristic changes with temperature often does not show up in experimental circuits, but does occur when the circuit is operated at high temperatures over time. Typically, the only practical solution is a heatsink. Make certain that heatsinks make good contact with the IC.

## 1.3  Frequency-Response and Gain Problems

From a troubleshooting standpoint, the open-loop (without feedback) gain and frequency response of an op amp are characteristics of the basic IC package, but can be modified with *external phase-compensation networks.* The closed-loop (with feedback) gain and frequency response depend (primarily) on *external feedback components.* Except where the IC is provided with built-in compensation, most of the experimental-circuit problems are the result of trade-offs between gain and frequency response or bandwidth). It is therefore essential that you understand the effect of external components on the operation of op amps before attempting to test or troubleshoot. This section starts with the basic feedback configurations.

### 1.3.1  Inverting and noninverting feedback

Figures 1-10 and 1-11 show the two basic op amp configurations, inverting feedback and noninverting feedback, respectively. The equations in Figs. 1-10 and 1-11 are classic guidelines and do not take into account the fact that the open-loop gain is not infinitely high and that the output or load impedance is not infinitely low. As a result, the equations have built-in inaccuracies and are used as guides only.

### 1.3.2  Loop gain

As shown in Fig. 1-10, *loop gain* is defined as the ratio of open-loop gain to closed-loop gain. When loop gain is large, the inaccuracies of the equations in Figs. 1-10 and 1-11 decrease. The relationships in Fig. 1-12 are based on a theoretical op amp (or one with built-in compensation). That is, the open-loop gain rolls off at 6 dB/octave or 20 dB/decade. (The term *6 dB/octave* means that the gain drops by 6 dB each time that the frequency is doubled. This is the same as a 20-dB drop each time that the frequency is increased by a factor of 10.)

If the open-loop gain of an op amp is as shown in Fig. 1-12, any stable, closed-loop gain can be produced by the proper selection of feedback components, provided that the closed-loop gain is less than the open-loop gain. The

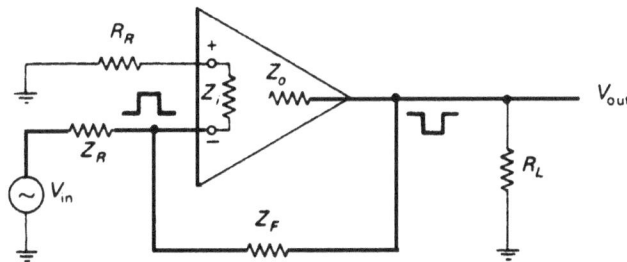

$$V_{out} \approx V_{in} \times \frac{Z_F}{Z_R}$$

$$\text{closed-loop gain} \approx \frac{Z_F}{Z_R}$$

$$\text{loop gain} \approx \frac{\text{open-loop gain}}{\text{closed-loop gain}}$$

$$Z_{in} \approx Z_R, \qquad Z_{out} \approx \frac{Z_o}{\text{loop gain}}$$

**Figure 1-10**  Theoretical inverting feedback op amp.

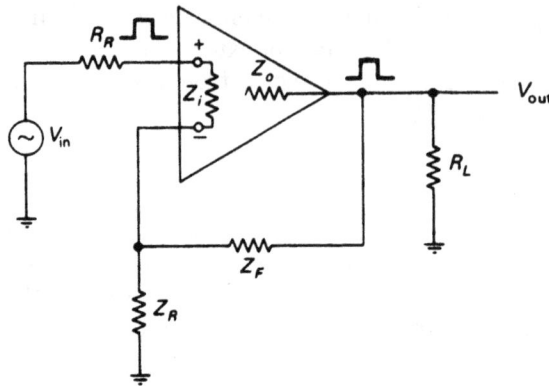

$$V_{out} \approx V_{in} \times \frac{Z_F}{Z_R} + 1$$

closed-loop gain $\approx \dfrac{Z_F}{Z_R} + 1$

loop gain $\approx \dfrac{\text{open-loop gain}}{\text{closed-loop gain}}$

$Z_{in} \approx Z_i + \dfrac{\text{open-loop gain} \times Z_i}{\text{closed-loop gain}}$

$Z_{out} \approx \dfrac{Z_o}{\text{loop gain}}$

**Figure 1-11**    Theoretical noninverting feedback op amp.

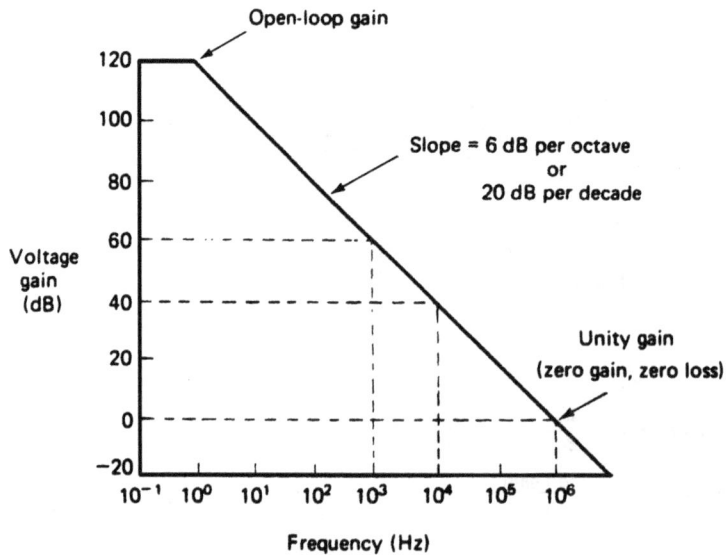

**Figure 1-12**    Frequency-response curve of theoretical op amp.

only concern is a trade-off between gain and frequency response. For example, if a voltage gain of 40 dB ($10^2$) is desired, a feedback resistance $10^2$ times larger than the input is used. The gain is then flat to $10^4$ Hz and rolloff is 6 dB/octave to unity gain at $10^6$ Hz. If a 60-dB gain ($10^3$) is required, the feedback resistance is raised to $10^3$ times the input resistance. This reduces frequency response. The gain is flat to $10^3$ Hz (instead of $10^4$ Hz), followed by a rolloff of 6 dB/octave down to unity gain.

### 1.3.3   Open-loop frequency and phase response

In a practical IC op amp (without compensation), the open-loop frequency response and phase-shift curves more closely resemble those shown in Fig. 1-13 (instead of the theoretical curves of Fig. 1-12). In Fig. 1-13, gain is flat at 60 dB to about 200 kHz. Then it rolls off at 6 dB/octave to 2 MHz. As the frequency increases, rolloff continues at 12 dB/octave (40 dB/decade) to 20 MHz (where gain is about unity or zero), then it rolls off at 18 dB/octave (60 dB/decade).

Some op-amp data sheets provide a curve similar to that shown in Fig. 1-13. If the curves are not available, it is possible to test the op amp and draw actual response curves (as described in Section 1.5).

The sharp rolloff at high frequencies, in itself, is not a problem in op-amp use (unless the op amp must be operated at a frequency very near the high end). However, note that the phase response (phase shift between input and

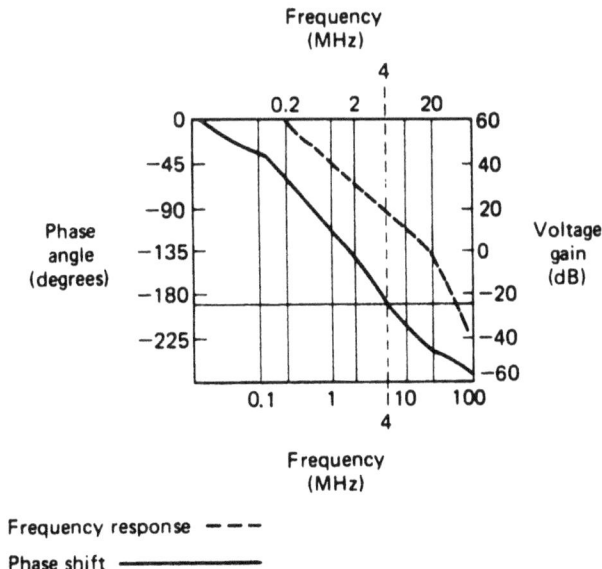

Figure 1-13   Frequency-response and phase-shift characteristics of practical IC op amps.

output) changes with frequency. The phase response of Fig. 1-13 shows that a negative feedback (at low frequencies) can become positive (and cause the op amp to be unstable) at high frequencies (possibly resulting in oscillation). In Fig. 1-13, a 180° phase shift occurs at about 4 MHz. This is the frequency at which open-loop gain is about +20 dB.

As a test/troubleshooting guide, when a selected closed-loop gain is equal to or less than the open-loop gain at the 180° phase-shift point, the op amp is unstable.

As an example, if a closed-loop gain of 20 dB or less is selected, a circuit with the curves of Fig. 1-13 is unstable. (Note the point where the −180° phase angle intersects the phase-shift line. Then draw a vertical line up to cross the open-loop gain line.) The closed-loop gain must be more than the open-loop gain at the frequency where the 180° phase shift occurs, but less than the maximum open-loop gain. Using Fig. 1-13 as an example, the closed-loop gain must be greater than 20 dB, but less than 60 dB.

### 1.3.4 External phase-compensation methods

Several methods of external phase compensation are used in op amps. Each method presents special testing/troubleshooting problems, which are covered here before going on to practical testing and troubleshooting techniques.

*Closed-loop feedback,* through capacitors and/or inductors, can be used to alter closed-loop gain. These components change the pure resistance to an impedance that varies with frequency, thus providing a different amount of feedback at different frequencies and a shift in phase of the feedback signal to offset the undesired through-amplifier phase shift. Such external feedback circuits are rare, and are not generally recommended because of impedance problems at both the high and low ends of the frequency range. However, the circuits do exist and, from a troubleshooting standpoint, must be considered. The main point of concern is that feedback changes with frequency, resulting in a change of both gain and phase shift for each frequency change.

*The open-loop input impedance* can be altered with a resistor and capacitor, as shown in Fig. 1-14. The input impedance of the series C and R decreases as the frequency increases, thus altering through-amplifier (open-loop) gain. As shown in Fig. 1-14, this arrangement causes the rolloff to start at a lower frequency, but produces a stable rolloff, similar to that of the ideal curve in Fig. 1-12. With the circuit properly compensated, the desired closed-loop gain is produced by a selection of external resistors.

In troubleshooting the Fig. 1-14 circuit, where gain and rolloff are not correct, check the values of resistors R1 through R3, and capacitor C1. Also look for leakage in C1. Even a slight leakage will alter the input impedance, and thus change the gain/frequency characteristics of the op amp.

The open-loop gain can be altered by one of several phase-compensation schemes, as shown in Figs. 1-15, 1-16, and 1-17.

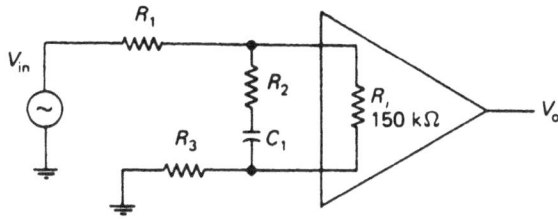

$R_i$ = input impedance of IC

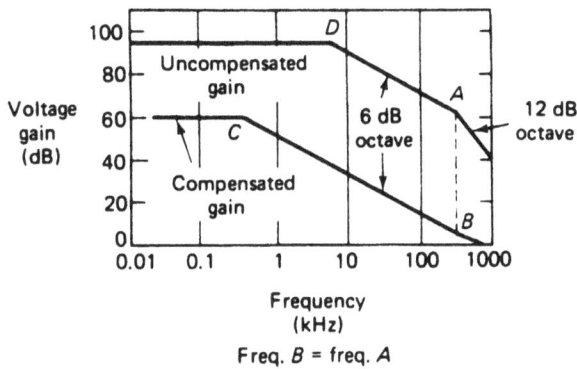

Freq. $B$ = freq. $A$

$R_1 = R_3$

$$R_1 + R_3 = \left( \frac{\text{uncompensated gain (dB)}}{\text{compensated gain (dB)}} - 1 \right) R_i$$

$$R_2 = \frac{R_1 + R_3}{\left( \dfrac{\text{freq. } D}{\text{freq. } C} - 1 \right) \left( 1 + \dfrac{R_1 + R_3}{R_i} \right)}$$

$$C_1 = \frac{1}{6.28 \times \text{freq. } D \times R_2}$$

$$\frac{\text{Compensated}}{\text{gain}} = \frac{\text{uncompensated gain} \times R_i}{R_i + R_1 + R_3}$$

$$\text{frequency } D = \frac{1}{6.28 \times R_2 \times C_1}$$

**Figure 1-14**   Frequency-response compensation by modification of open-loop input impedance.

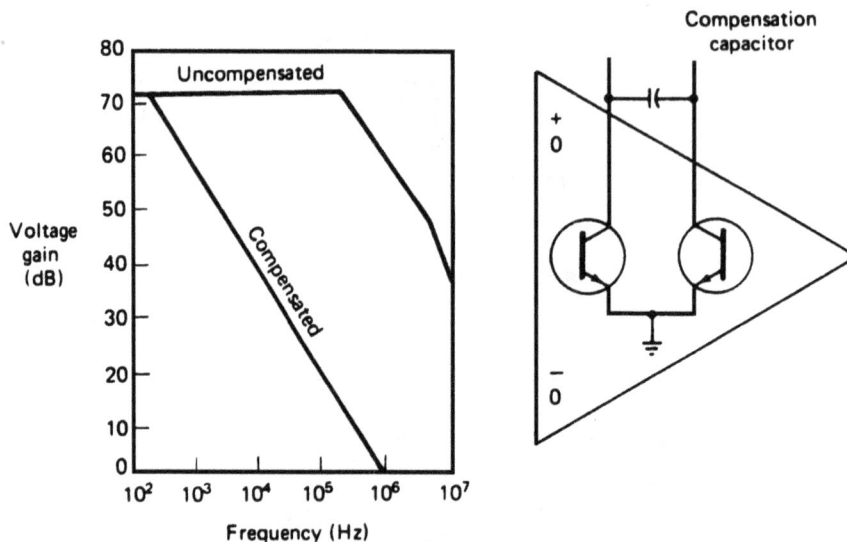

**Figure 1-15**  Phase-lead frequency compensation.

In Fig. 1-15, known as *phase-lead compensation,* the open-loop gain is changed by an external capacitor (often connected between high-gain stages, as shown in Figs. 1-1 and 1-2).

In Fig. 1-16, generally referred to as *RC rolloff, straight rolloff,* or *phase-lag compensation,* the open-loop gain is altered by an external RC network connected across a circuit component. With phase-lag, the rolloff starts at the corner frequency produced by an RC network.

In Fig. 1-17, known as *Miller-effect rolloff,* or *Miller-effect phase-lag compensation,* the open-loop gain is altered by an RC network connected between the input and output of an inverting stage in the IC. (Sometimes, Miller-effect compensation is provided by a capacitor alone, as shown in Figs. 1-1 and 1-2.)

### 1.3.5  Built-in phase compensation

Figure 1-18 shows the characteristic of an IC op amp (the Harris CA3140) with built-in compensation. (A 12-pF capacitor is internally connected between the input and output of the high-gain stage, to provide Miller-effect compensation.) Notice that the rolloff of Fig. 1-18 approaches that of the theoretical op amp shown in Fig. 1-12.

Any closed-loop gain (less than open-loop gain) can be selected by feedback resistances (as shown in Figs. 1-10 and 1-11). Likewise, any combination of gain/frequency response can be selected. For example, if the rolloff is to start at 10 kHz, the closed-loop gain will be flat at about 40 dB to 10 kHz and roll off at 6 dB/octave to unity gain at a frequency between 3 and 4 MHz. If the gain must be increased to 60 dB, then the rolloff will start at about 3 kHz and continue at 6 dB/octave.

When testing a given gain/frequency combination, check the phase shift at the highest frequency. With the IC of Fig. 1-18, the maximum phase shift is about 135° at the unity-gain frequency of 3 to 4 MHz. This still provides a 45° phase margin (Section 1.4.2) so that the IC should be stable even when operated at the maximum frequency.

When troubleshooting, remember that the gain/frequency characteristics of the IC shown in Fig. 1-18 cannot usually be changed. If an entirely different gain/frequency response is required (for example, 70 dB gain at 10 kHz), the IC is not suitable. (However, some IC data sheets show how to alter the built-in compensation, usually with a small capacitor connected externally.)

Also remember that the gain/frequency characteristics shown in Fig. 1-18 are based on a specific load resistance and capacitance. If the load is substantially

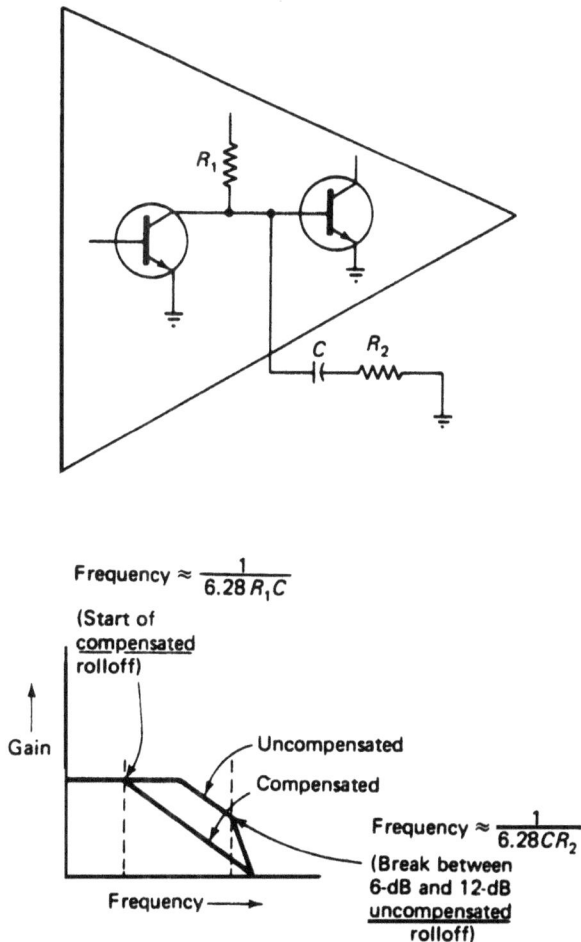

Figure 1-16  RC rolloff or phase-lag compensation.

$$\text{Frequency} \approx \frac{1}{6.28 \left[ (1 + G_m Q_2 R_2) C_x R_1 \right]}$$

(Start of compensated rolloff)

Gain

Uncompensated

Compensated

$$\text{Frequency} \approx \frac{1}{6.28 C_x R_x}$$

(Break between 6-dB and 12-dB uncompensated rolloff)

Frequency ⟶

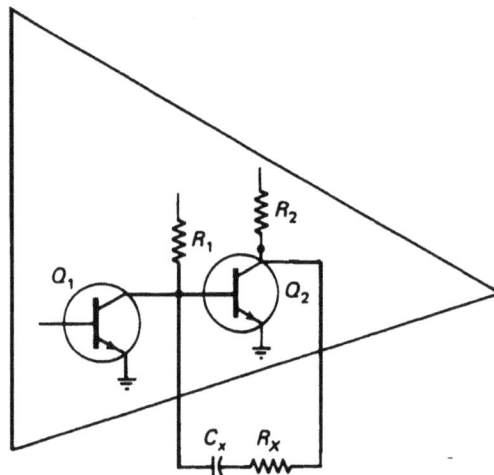

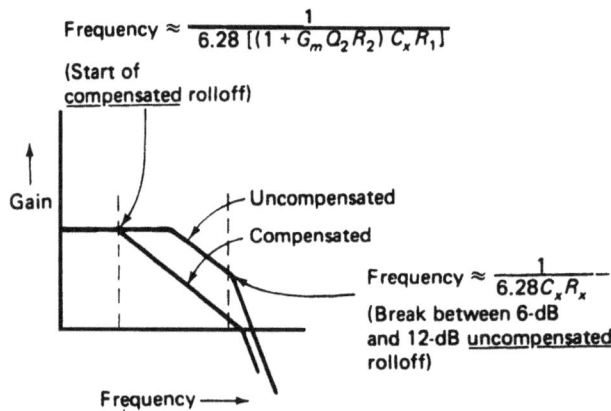

Figure 1-17   Miller-effect phase-lag compensation.

different from that used when the characteristics were measured, the response will be different.

## 1.4   Using Op-Amp Data Sheets in Testing/Troubleshooting

Most of the characteristics required to test and troubleshoot IC op amps can be found in the data sheet. The following paragraphs describe the essential characteristics and cover how they apply to testing/troubleshooting.

### 1.4.1   Open-loop voltage gain

The open-loop voltage gain (AVOL or AOL) is defined as the ratio of a change in output voltage to a change in input voltage. Open-loop gain is always measured without feedback, and sometimes without phase-shift compensation. The AOL shown in Fig. 1-19 includes graphs for four different values of compensation. Figure 1-18 shows the AVOL with the built-in compensation.

Open-loop gain is frequency dependent (gain decreases with increased frequency). Typically, the gain is flat (within ±3 dB) up to a certain frequency (for example, about 10 Hz, as in Fig. 1-18). Then, gain rolls off to unity at another higher frequency (for example, 3 or 4 MHz). AOL is also temperature, supply voltage, and load dependent. The effects of these characteristics on AOL vary

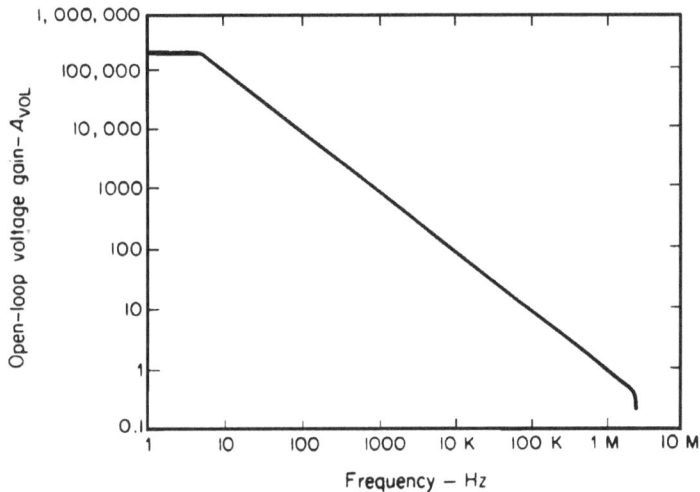

**Figure 1-18**  Characteristics of IC op amp with built-in compensation (Harris Semiconductors, *Linear & Telecom ICs,* 1994, p. 2-133).

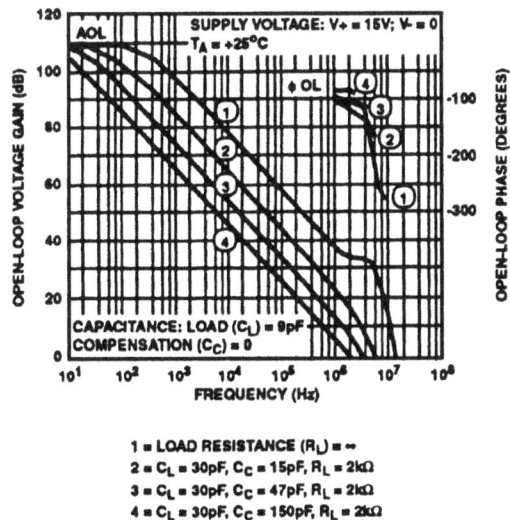

1 = LOAD RESISTANCE (R$_L$) = ∞
2 = C$_L$ = 30pF, C$_C$ = 15pF, R$_L$ = 2kΩ
3 = C$_L$ = 30pF, C$_C$ = 47pF, R$_L$ = 2kΩ
4 = C$_L$ = 30pF, C$_C$ = 150pF, R$_L$ = 2kΩ

**Figure 1-19**  Open-loop voltage gain and phase shift versus frequency (Harris Semiconductors, *Linear & Telecom ICs,* 1994, p. 2-112).

with the IC, so the data sheet should be consulted when the IC must be tested or operated at extreme conditions.

Ideally, open-loop gain should be infinitely high because the primary function of the op amp is to amplify. In general, the higher the gain, the higher the accuracy of the transfer function (the relationship of input signal to output signal). However, there are practical limits to gain (and there are gain levels at which an increase buys little in the way of increased performance). In practical testing/troubleshooting, open-loop gain determines closed-loop accuracy limits, rather than ultimate accuracy.

The important requirement is that the open-loop gain must be greater than the closed-loop gain over the frequency of interest. For example, if a 40-dB op amp and a 60-dB op amp are used in a 20-dB closed-loop-gain circuit and the open-loop gain is decreased by 50% in each case (for example, because of component aging, power-supply variations, etc.), the closed-loop gain of the 40-dB op amp varies by 9% and that of the 60-dB op amp varies by only 1%.

The frequency-rolloff characteristics determine the frequency response of an IC op amp (primarily). An 18-dB/octave rolloff is generally considered the maximum slope that can occur in the active region before phase compensation becomes extremely difficult (or impossible) to get. In addition, because IC op amps have useful applications down to and including unity gain, the active region of the IC can be considered as the entire portion of the frequency characteristics above the 0-dB bandwidth.

A typical IC op-amp data sheet shows the results of compensation by means of graphs (such as those shown in Figs. 1-19 and 1-20). After com-

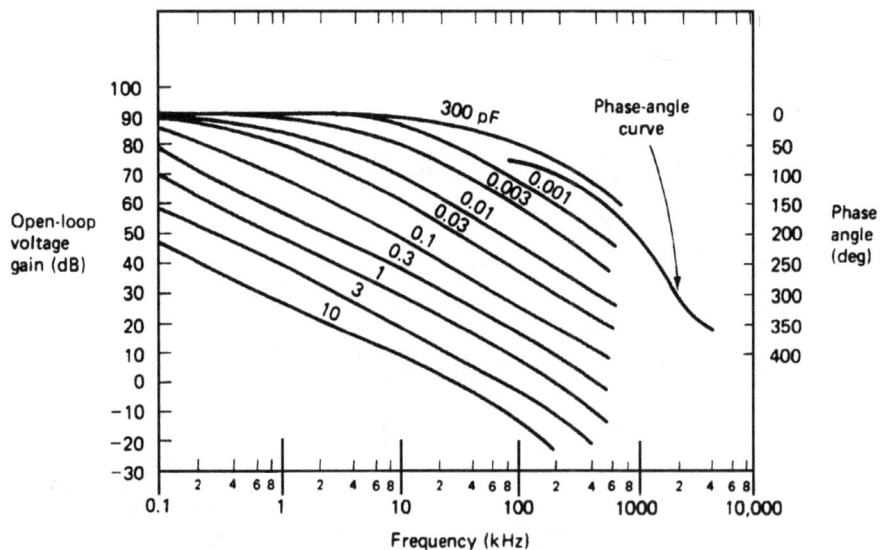

Figure 1-20  Typical phase-compensation characteristics.

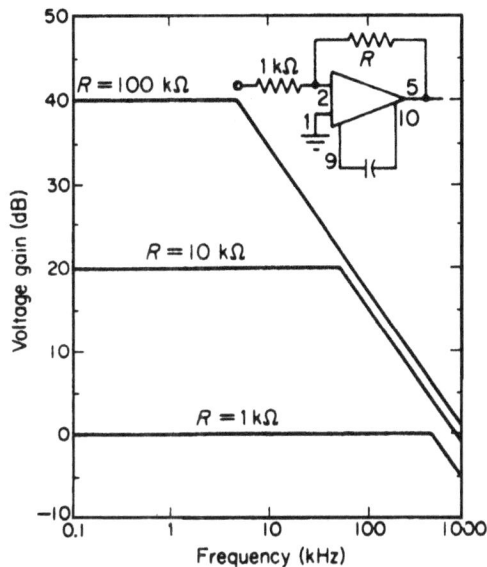

Figure 1-21   Closed-loop voltage gain of IC versus frequency.

pensation is applied, or with built-in compensation, the IC can be connected in the closed-loop configuration. Closed-loop voltage gain is usually not listed as such on the data sheets. In some cases, the data sheet might show some typical gain curves with various ratios of feedback. Figure 1-21 is an example. If available, such curves can be used directly to select values of feedback components.

### 1.4.2   Phase shift and phase margin

One figure of merit commonly used in evaluating the stability of an IC op amp is *phase margin*. Oscillation might occur if the total phase shift around the loop (from input to output and back to input) can reach 360° before the total gain around the loop drops below unity (as the frequency is increased). Because an IC op amp is often used in the inverting mode (Fig. 1-10), 180° of phase shift is available at the beginning. Additional phase shift is developed by the op amp because of internal circuit conditions.

Phase margin represents the difference between 180° and the phase shift produced by the IC at the frequency where loop gain is unity. A value of 45° phase margin is considered quite conservative to provide a guard against production variations, temperature effects, and other stray factors. This means that the IC should not be operated at a frequency where the phase shift exceeds 135° (180°−45°). However, it is possible to operate some IC op amps at frequencies where the phase shift is kept within the 160° to 170° region. Of course, when troubleshooting, the ultimate stability of an IC op amp must be established by the tests described in Section 1.5.

### 1.4.3  Definition of op-amp terms used in testing/troubleshooting

Before getting further into IC op-amp data sheets, here are some basic definitions. Figure 1-22 provides a number of definitions for IC op amps. (These particular definitions are used by Harris Semiconductors to describe their products.) Other manufacturers might define the same characteristics in a different way. To help cut through this confusion, the following definitions apply to the testing and troubleshooting procedures in this book.

### 1.4.4  Bandwidth, slew rate, and output characteristics

From a troubleshooting standpoint, the bandwidth, slew rate, output voltage swing, output current, and output power of an IC op amp are all interrelated. These characteristics depend on frequency and phase compensation. The characteristics also depend on temperature and power supply, but to a lesser extent. Before covering the interrelationship, each of the characteristics are defined (for our purposes). Section 1.5 provides additional definitions for some of the terms.

IC op-amp bandwidth can be expressed in terms of open-loop operation. For example, as shown in Fig. 1-23, an open-loop bandwidth of 800 kHz means that the open-loop gain of the op amp drops to a value 3 dB below the flat or low-frequency level at a frequency of 800 kHz.

The term *frequency range* is sometimes used instead of open-loop bandwidth. The useful frequency range of an IC op amp is similar to the term $f_T$ (total frequency, which is used with transistors). The IC shown in Fig. 1-19 lists a "unity-gain crossover frequency" or $f_T$ of 15 MHz with no compensating capacitor, and 4 MHz with compensation. The IC shown in Fig. 1-18 lists a "gain-bandwidth product" $(f_T)$ of 3.7 MHz (with a built-in capacitor).

From a troubleshooting standpoint, *power bandwidth* is a more useful characteristic because the term represents the bandwidth of the IC in closed-loop operation (connected to a normal load). When available, power bandwidth represents the output capability of the IC (working into a given load) across a band of frequencies.

Some data sheets list the power output of an IC op amp, but the trend is toward listing maximum output voltage and current. For example, the IC shown in Fig. 1-19 lists a maximum output voltage of 13 to 15 V with a maximum output current of 22 mA across a 2-k$\Omega$ load. Usually, *output voltage* is defined as the peak or peak-to-peak output-voltage swing (referred to zero) that can be obtained without clipping. From a testing/troubleshooting standpoint, a symmetrical voltage swing depends on frequency, load current, output impedance, and slew rate. Generally, an increase in frequency decreases the possible output-voltage swing. In some cases, the data sheet shows output-voltage swing versus frequency and phase compensation.

Slew rate is the maximum rate of change in output voltage, with respect to time, that the IC is capable of producing when maintaining linear characteristics

# Operational Amplifiers Glossary of Terms

**AVERAGE INPUT OFFSET CURRENT DRIFT** - The average change in offset current between room ($+25°C$) and high temperature ($+125°C$, $+85°C$ or $+75°C$) or between room temperature and low temperature ($0°C$, $-25°C$ or $-55°C$) divided by the temperature difference.

**AVERAGE OFFSET VOLTAGE DRIFT** - The average change in offset voltage between room ($+25°C$) and high temperature ($+125°C$, $+85°C$ or $+75°C$) or between room temperature and low temperature ($0°C$, $-25°C$ or $-55°C$) divided by the temperature difference.

**CHANNEL SEPARATION** - The ratio of the output of a driven amplifier to the output (referred to input) of an adjacent undriven amplifier.

**COMMON MODE INPUT VOLTAGE ($V_{IC}$)** - The average of the voltages present at the differential input terminals.

**COMMON MODE INPUT VOLTAGE RANGE ($V_{ICR}$)** - The range of voltage that if exceeded at either input terminal will cause the amplifier to cease operating properly.

**COMMON MODE REJECTION RATIO (CMRR)** - The ratio of change in input offset voltage to change in input common mode voltage, expressed in dB.

$$CMRR = 20 \times \log_{10} \left[ \frac{V_{10}}{V_{CM}} \right]$$

**COMMON MODE RESISTANCE ($r_{IC}$)** - The ratio of change to input common mode voltage to the resulting change in input current.

**DIFFERENTIAL INPUT RESISTANCE ($r_{ID}$)** - The ratio of change in input differential voltage (small signal, assumes amplifier operating linearly) to the resulting change in differential input current.

**FULL POWER BANDWIDTH (FPBW)** - The maximum frequency at which a full scale undistorted (THD $<$ 1%) sine wave can be obtained at the output of the amplifier.

**GAIN BANDWIDTH (GBW)** - The open loop gain of an op amp (in V/V at a mid-band, linear region frequency (usually between 1kHz and 10kHz) times that frequency (in Hz). GBW = $[A_{VOL}] \cdot f$.

**INPUT BIAS CURRENT ($I_{BIAS}$)** - The average of the currents flowing into or out of the input terminals when the output is at zero volts.

**INPUT CAPACITANCE ($C_{IB}$)** - The equivalent capacitance seen looking into either input terminal.

**INPUT NOISE CURRENT ($I_N$)** - The input noise current that would reproduce the noise seen at the output if all amplifier noise sources were set to zero and the source impedances were large compared to the optimum source impedance.

**INPUT OFFSET CURRENT ($I_{OS}$)** - The difference in the currents flowing into the two input terminals when the output is at zero volts.

**INPUT OFFSET VOLTAGE ($V_{IO}$)** - The differential DC voltage required to zero the output voltage with no input signal or load. Input offset voltage may also be defined for the case where two equal resistances are inserted in series with the input leads.

**INPUT NOISE VOLTAGE ($e_N$)** - The input noise voltage that would reproduce the noise seen at the output if all the amplifier noise sources and source resistances were set to zero.

**LARGE SIGNAL VOLTAGE GAIN ($A_V$)** - The ratio of the peak to peak output voltage swing (over a specified range) to the change in input voltage required to drive the output.

**OUTPUT CURRENT ($I_{OUT}$)** - The output current available from the amplifier at some specified output voltage.

**OUTPUT RESISTANCE (RO)** - The ratio of the change in output voltage to the change in output current.

**Figure 1-22**  Op-amp definitions (Harris Semiconductors, *Linear & Telecom ICs,* 1994, p. 2-706).

## Operational Amplifiers Glossary of Terms

**OUTPUT SHORT CIRCUIT CURRENT ($I_{SC}$)** - The output current available from the amplifier with the output shorted to ground (or other specified potential).

**OUTPUT VOLTAGE SWING ($V_{OUT}$)** - The maximum output voltage swing, referred to ground, that can be obtained under specified loading conditions.

**OVERSHOOT** - Peak excursion above final value of an output step response.

**POWER SUPPLY REJECTION RATIO (PSRR)** - The ratio of the change in input offset voltage to the change in power supply voltage producing it.

**RISE TIME ($t_R$)** - The time required for an output voltage step to change from 10% to 90% of its final value, when the input in subjected to a small signal voltage pulse.

**SETTLING TIME ($r_{SET}$)** - The time required, after application of a step input signal, for the output voltage to settle and remain within a specified error band around the final value.

**SLEW RATE (SR)** - The rate of change of the output under large signal conditions. Slew rate may be specified separately for both positive and negative going changes.

**SUPPLY CURRENT ($I_S$)** - The current required from the power supply to operate the amplifier with no load and the output at zero volts.

**SUPPLY VOLTAGE RANGE** - The range of power supply voltage over which the amplifier may be safely operated.

**UNITY GAIN BANDWIDTH** - The frequency range from DC to that frequency where the amplifiers open loop gain is unity.

**Figure 1-22** *(Continued)*

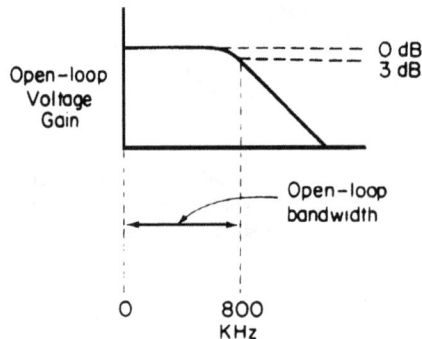

**Figure 1-23** Bandwidth and open-loop gain relationships.

(symmetrical output without clipping). During testing or troubleshooting, remember that slew rate depends on the compensating capacitor, all other factors being the same. This is because the current required to charge and discharge the capacitor can limit available current to the op-amp stages or load, thus producing slower slew rates (particularly at higher frequencies).

In testing/troubleshooting, the major effect of slew rate is on power output. All other factors being the same, a lower slew rate results in lower power output. Slew rate and the term *full-power response* are directly related. Full-power response is the maximum frequency measured in a closed-loop unity-gain configuration for which rated output voltage can be obtained for a sine-wave signal, with a specified load and without distortion because of slew-rate limiting. The slew rate versus full-power response relationship can be shown as:

$$\text{Slew rate (V/s)} = 3.14 \times F_M \times V_{Op\text{-}p}$$

where $F_M$ is the full-power response frequency in Hz and $V_{Op\text{-}p}$ is the peak-to-peak output voltage.

For example, assume that $V_{Op\text{-}p}$ is 26 V at a frequency of 30 kHz. The slew rate is:

$$3.14 \times 30{,}000 \times 26 = 2{,}449{,}200 \text{ V/s} = 2.45 \text{ V/}\mu s, \text{ or simply } 2.45$$

The equation can be turned around to find the full-power response frequency. For example, assume that the rated slew rate is 2.5 (2.5 V/$\mu$s) and the $V_{OP\text{-}p}$ is 20. Find the full-power frequency response ($F_M$) as follows:

$$\frac{2.5 \text{ V/}\mu s}{3.14 \times 20} = \frac{2{,}500{,}000}{62.8} = 39{,}808 \text{ Hz} = 39.8 \text{ kHz}$$

In testing/troubleshooting, remember that the power output depends on output voltage, assuming a constant load. In turn, all other factors being equal, output voltage depends on slew rate. Because slew rate depends on compensation capacitance, output power also depends on compensation.

Some data sheets omit slew rate, but provide a graph similar to that in Fig. 1-24. This graph shows the direct relationship between full-power output frequency and compensation capacitance. For example, with a capacitance of 0.01 $\mu$F and a 500-$\Omega$ load, the IC shown in Fig. 1-24 delivers full-rated output power up to a frequency of about 80 kHz.

### 1.4.5   Output impedance

*Output impedance* is defined as impedance "seen" by a load at the output of the IC, as shown in Fig. 1-25. In troubleshooting, note that excessive output

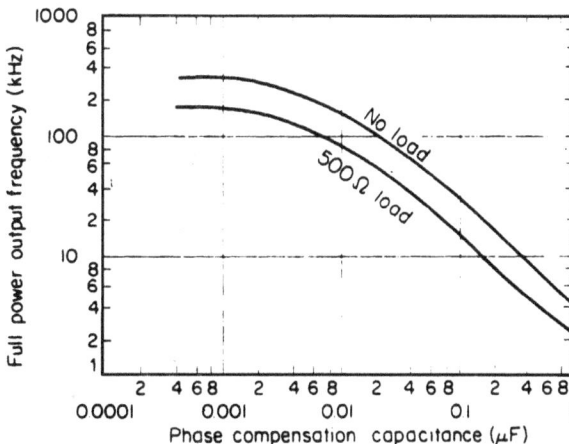

Figure 1-24   Frequency response for full-power output as a function of phase-compensating capacitance.

Open loop

Closed loop

$$\text{Closed loop } Z_{out} \approx \frac{\text{open loop } Z_{out}}{1 + \text{open loop gain} \times \left(\dfrac{R_1}{R_1 + R_2}\right)}$$

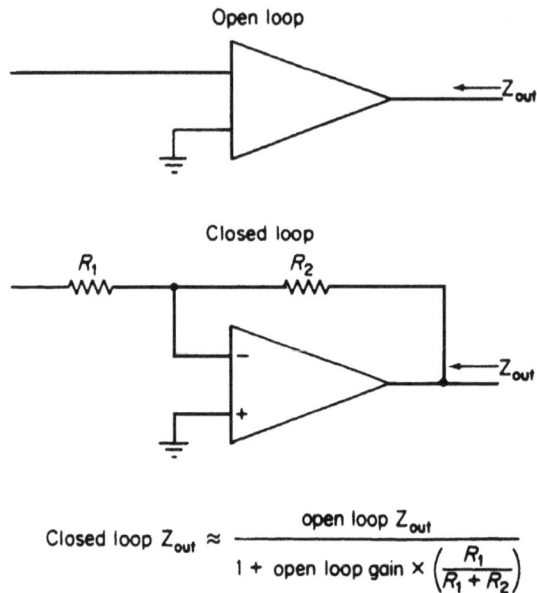

**Figure 1-25**  Open-loop and closed-loop output impedance relationships.

impedance can reduce the gain because, in conjunction with load and feedback resistance, output impedance forms an attenuator network.

The output impedance of a typical IC op amp is less than 200 $\Omega$. When compared with the high input impedance and feedback resistance, output impedance has little effect on gain (especially when the IC is used as a voltage amplifier). In practical troubleshooting, output impedance has a more significant effect on power circuits that must supply large amounts of load current. Also, as shown by the equation in Fig. 1-25, output impedance increases as frequency increases because open-loop gain decreases.

### 1.4.6  Input impedance

*Input impedance* is defined as the impedance seen by a source looking into one input of the IC, with the other input grounded, as shown in Fig. 1-26. In both testing and troubleshooting, the primary effect is to reduce loop gain. Input impedance changes with temperature and frequency. Generally, input impedance is listed on the data sheet at 25°C, possibly at some specific frequency.

The important point to remember is that if input impedance is quite different from the impedance of the device driving the amplifier, there is a loss of input signal because of the mismatch. Of course, in the real world, it is not practical to alter the IC input impedance. If impedance match is crucial, either the IC or the driving source must be changed to effect a match.

### 1.4.7   Input common-mode voltage swing

*Input common-mode voltage swing* ($V_{\text{ICM}}$) is defined as the maximum peak input voltage that can be applied to either input terminal of the IC amplifier without causing abnormal operation or damage, as shown in Fig. 1-27. Some IC data sheets list a similar term, *input common-mode voltage range* ($V_{\text{ICR}}$).

During testing/troubleshooting, $V_{\text{ICM}}$ or $V_{\text{ICR}}$ limits the differential-signal amplitude that can be applied to the input. As long as the input signal does

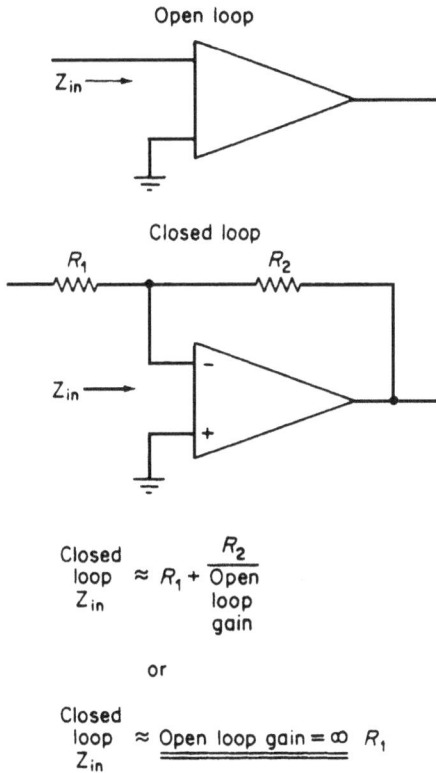

Open loop

Closed loop

$$\text{Closed loop } Z_{\text{in}} \approx R_1 + \frac{R_2}{\text{Open loop gain}}$$

or

$$\text{Closed loop } Z_{\text{in}} \approx \underline{\underline{\text{Open loop gain} = \infty}} \ R_1$$

**Figure 1-26**  Open-loop and closed-loop input-impedance relationships.

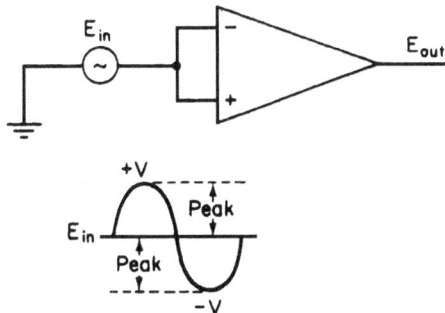

**Figure 1-27**  $V_{\text{ICM}}$ relationships.

not exceed the $V_{ICM}/V_{ICR}$ values, no test problems should be encountered. Notice that some data sheets also list single-ended input voltage signal limits, where the differential input is not to be used.

### 1.4.8    Input-bias current

*Input-bias current* is defined as the average values of the two input-bias currents of the IC differential input stage, and is a function of the input-stage current gain. From a testing/troubleshooting standpoint, the significance of input-bias current is that the current produces a voltage drop across the input resistors. This voltage drop must be overcome by the input signal and can thus restrict the $V_{ICR}$. Input-bias current is sometimes listed as input current.

### 1.4.9    Input-offset voltage

IC *op-amp input-offset voltage* is defined as the voltage that must be applied at the input terminals to get zero output voltage and it indicates the matching tolerance in the differential-amplifier stages. A perfect match requires zero input voltage to produce zero output voltage.

In testing/troubleshooting, the effect of input-offset voltage is that the input signal must overcome the effect before an output is produced. For example, if an op amp has a 1-mV input-offset voltage and a 1-mV signal is applied, there is no output. If the signal is increased to 2 mV, the IC produces only the peaks. Because input-offset voltage is increased by gain, the effect of input-offset voltage is increased by the ratio of feedback resistance to input resistance, plus 1 (unity) in the closed-loop condition. For example, if the ratio is 100:1 (for a gain of 100), the effect on input-offset voltage is increased by 101.

### 1.4.10    Input-offset current

IC *op-amp input-offset current* is defined as the difference in input-bias current into the input terminals. It is an indication of the degree of matching in the input differential stages. In practical testing/troubleshooting, input-offset current can be of greater importance than the input-offset voltage when high impedances are involved. Here is why.

If the input-bias current is different for each input, the voltage drops across the resistors are not equal. If the resistance is large, there is a large imbalance in input voltage. This condition can be minimized by connecting a resistor between the noninverting input and ground, as shown in Fig. 1-28. When making up a practical test circuit, the value of resistor R3 should be equal to the parallel equivalent of the input and feedback resistors R1 and R2, as shown by the equations. Use the equations for the first trail value of R3. Then adjust R3 for minimum difference at both terminals (under normal operating conditions, but with no signal).

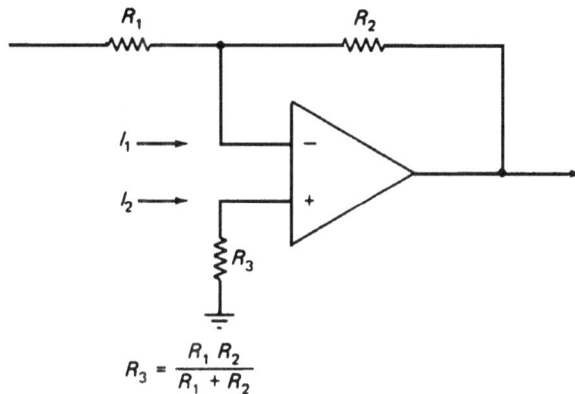

**Figure 1-28** Minimizing input-offset current (and input-offset voltage).

$$R_3 = \frac{R_1 R_2}{R_1 + R_2}$$

### 1.4.11 Input and feedback

Offset voltages and currents are usually referred back to the input because the output values depend on feedback. (That is, data sheets rarely list output-offset characteristics.) Any offset in an IC op amp results from a combination of offset voltage and current. For example, if an IC has a 1-mV input-offset voltage and a 1-μA input-offset current, with the inputs returned to ground through 1-kΩ resistors, the total input offset is either 0 or 2 mV, depending on the phase relationship between the two offset characteristics.

### 1.4.12 Offset null

The offset of an IC op amp is a dc error that should be minimized for many reasons, including the following:

1. The use of an IC op amp as a true dc amplifier is limited to signal levels much greater than the offset.

2. Comparator applications require that the output voltage be zero (within limits) when the two input signals are equal and in phase.

3. Any offset that the input is multiplied by the gain at the output. If the IC drives additional amplifiers, the increased offset at the output is multiplied even further. The gain of the entire system must then be limited to a value that will not cause limiting at the final output stage.

Some IC op amps include provisions to null or neutralize any offset. Typically, an external voltage is applied through a pot to terminals on the IC, such as terminals 1 and 5 of the IC in Fig. 1-2. The voltage is adjusted until the offset at both the input and output is zero. For example, the data sheet for the IC in Fig. 1-2 shows a ±22-mV input-offset voltage adjustment range when a 10-kΩ pot is connected across terminals 1 and 5, with the slider at terminal 4. This is more than sufficient because input-offset voltage is listed as 8 mV (typical) and 15 mV (maximum).

### 1.4.13    Offset-null test circuits

For IC op amps without offset-null provisions, the effects of input offset can be minimized by an external circuit. Figure 1-29 shows two such circuits that can be used during testing/troubleshooting of both inverting and noninverting configurations. The equations in Fig. 1-29 assume that resistance $R_B$ must produce a null range of ±7.5 mV. If this is not sufficient, substitute a wider adjustment-range value (such as ±22 mV) for the ±7.5 mV.

One reason for offset null during testing/troubleshooting is that the input and output dc levels should be equal or nearly equal. This condition is desirable to ensure that the resistive feedback network can be connected between the input and output without upsetting the differential or common-mode dc bias.

### 1.4.14    Average temperature coefficient of input offset

The average temperature coefficient of input-offset voltage, listed on some data sheets as $T_{CV}$, depends on the temperature coefficients for various com-

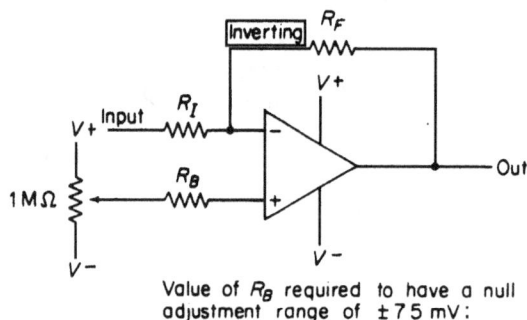

Value of $R_B$ required to have a null adjustment range of ±7.5 mV:

$$R_B \approx \frac{R_I V+}{7.5 \times 10^{-3}} \text{assuming } R_B \gg R_I$$

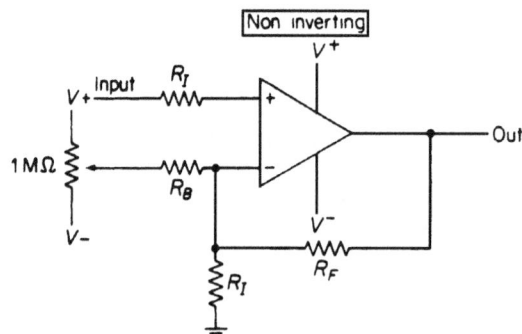

Value of $R_B$ required to have a null adjustment range of ±7.5 mV:

$$R_B \approx \frac{R_I R_F V+}{(R_I + R_F) 7.5 \text{ mV} \times 10^{-3}}$$

**Figure 1-29**  Input-offset minimizing circuits.

ponents within the IC. Temperature changes affect stage gain, match of differential amplifiers, etc., thus changing input-offset voltage. In testing/troubleshooting, $T_{CV}$ needs to be considered only if the parameter is large and the IC must be operated under extreme temperatures, either during test or in final form. For example, if the input-offset voltage doubles with an increase to a temperature that is likely to be found during normal operation, the higher input-offset voltage should be considered as the "normal" value.

### 1.4.15   Power-supply sensitivity of input offset

*Power-supply sensitivity* (sometimes listed as *power-supply rejection ratio, PSRR)* is defined as the ratio of change in input-offset voltage to the change in supply voltage producing the change, with the remaining supply held constant. Some data sheets list the parameter as *input-offset voltage sensitivity*. The parameter is expressed in terms of microvolts per volt ($\mu$V/V), representing the change (in microvolts) of input-offset voltage to a change (in volts) of one power supply. Often, a separate sensitivity parameter is used for each supply, with the opposite power supply assumed to be held constant. For example, the *PSRR* for the IC shown in Fig. 1-19 is a typical 32 and a maximum 320 ($\mu$V/V).

From a troubleshooting standpoint, the effect of PSRR (or input-offset voltage sensitivity) is obvious. If an IC has considerable sensitivity to power-supply variation, the supply regulation must be increased to provide correct operation with the anticipated input-signal levels. In practical terms, you might be trying to troubleshoot an amplifier circuit when you should be troubleshooting the power supply!

### 1.4.16   Noise voltage problems

Both the background noise and signal-to-noise ratio (Sections 1.5.16 and 1.5.17) should be considered when testing and troubleshooting any IC amplifier, but especially when the input signal is small. In typical testing/troubleshooting situations, IC noise becomes a problem only when the noise is large in relation to the input signal. Obviously, an IC amplifier with 10-$\mu$V noise at the input would mask a 10-$\mu$V input signal. If the input is raised to 1 mV with the same IC, the noise is unnoticed. (Note that noise depends on temperature as well as the method of compensation used.)

In practical testing/troubleshooting, if you have any doubt and the input voltage is small in relation to the data-sheet noise voltage, test the IC for both background noise and signal-to-noise ratio, then compare the actual IC noise with the anticipated input signal level.

### 1.4.17   Power dissipation problems

In practical situations, the power dissipation capabilities of ICs become a problem only when circuit failure occurs at high temperatures. In some cases, the problem can be solved by simply adding a heatsink (especially in new or experimental circuits). In those circuits that have been operating for some time and show problems at high temperatures only, the heatsink might be inadequate (or not properly installed). Of course, if repeated circuit failure occurs at high temperatures, the problem is likely one of design. In any event, you should know the effects of power dissipation when testing and troubleshooting any circuit.

One rating found on IC data sheets is *total device dissipation,* which includes the load current. Another rating is the *device dissipation* (often shown as $P_D$ or $P_T$, which is the dc power dissipated by the IC itself (with the output zero and no load). Other data sheets list only one power-dissipation rating, but include *derating factors*.

For example, the IC in Fig. 1-18 shows a typical $P_D$ of 120 mW and a maximum $P_D$ of 180 mW. The IC in Fig. 1-19 shows a dissipation (without a heatsink) of 630 mW at temperatures up to 55°C. Above 55°C, the dissipation must be derated linearly at 6.67 mW/°C. With a heatsink, the dissipation is increased to 1 W at temperatures up to 90°C. Above 90°C, the dissipation must be derated linearly at 16.7 mW/°C.

When both total dissipation and device dissipation are available, the device dissipation must be subtracted from the total dissipation to calculate the available load dissipation. For example, if an IC can dissipate a total of 500 mW (at a given temperature, supply voltage, and with or without a heatsink) and the IC itself dissipates 120 mW, the load cannot exceed 380 mW ($500-120=380$).

## 1.5    Basic Amplifier Tests

This section is devoted to basic tests for amplifier circuits. These procedures can be applied to a complete amplifier device (such as a stereo system) or to specific circuits (such as the IC amplifier circuits described throughout this book). The procedures can also be applied to amplifier circuits at any time during design or experimentation.

As a minimum, the tests should be made when the circuit is first completed in experimental form. If the test results are not as desired, the component values can be changed as necessary to get the desired results. (The test results also provide a good starting point for troubleshooting.) The circuit can then be retested in final form (with all components and the ICs soldered in place). This shows any change in circuit performance because of physical relation of components (a common problem). Additional test procedures for specific types of IC amplifiers are given in the related chapters. Keep in mind that all amplifiers need not be subjected to all tests described in this book. However, if an amplifier circuit produces the desired results for all of the tests, circuit troubleshooting can be avoided. Quit while you are ahead!

### 1.5.1    Testing/troubleshooting equipment for amplifiers

The tests described in this section can be performed using meters, scopes, generators, power supplies, and assorted clips, patch cords, etc. Therefore, if you have a good set of test equipment that is suitable for general electronic work, you can probably get by. A possible exception is a distortion meter (especially if you are interested in audio amplifiers). Consider the following points when selecting and using test equipment.

*Matching test equipment to the circuits*    No matter what test instrument is involved, if pulses, square waves, or complex waves are to be measured (as they are in any IC amplifier test) a peak-to-peak (p-p) meter can possibly provide meaningful indications, but a scope is the logical choice. (Section 1.6 includes a series of notes on matching test equipment to IC amplifiers during testing and troubleshooting.)

*Voltmeters/multimeters*    In addition to making routine voltage and resistance checks during troubleshooting, the main functions for a meter in amplifier work are to measure frequency response and to trace signals from input to output. Many technicians prefers scopes for these procedures. The reasoning is that scopes also show distortion of the waveform during measurement or signal tracing. Other technicians prefer the simplicity of a meter, particularly in such procedures as voltage-gain and power-gain measurements.

It is possible that you can get by with any ac meter (even a basic multimeter, analog or digital) for all amplifier work. However for accurate measurements, use a wideband meter, preferably a dual-channel model. (Obviously, the meter must have a bandwidth greater than the amplifier circuit being tested!) The dual-channel feature makes it possible to monitor both channels of a stereo circuit simultaneously. This is particularly important for stereo frequency-response and crosstalk measurements, but is of no great importance for nonstereo amplifiers.

*Scopes*    If you have a good scope for TV and VCR work, use that scope for all amplifier-circuit measurements (observing the notes in Section 1.6). If you are considering purchasing a new scope, remember that a dual-channel instrument permits you to monitor both channels of a stereo circuit (as is the case with a dual-channel voltmeter). From a troubleshooting standpoint, a dual-channel scope lets you monitor the input and output of a circuit simultaneously. Many examples of this technique are provided throughout this book. A scope also has the advantage over a meter in that the scope can display such common amplifier-circuit conditions as distortion, hum, ripple, overshoot, and oscillation, and is an indispensable tool for measuring such characteristics as settling time, slew rate, and noise. (However, the meter is easier to read if you are concerned only with gain.)

### 1.5.2    Using the decibel scales in testing/troubleshooting

The decibel (dB) is widely used in amplifier work to logarithmically express the ratio between two powers or voltage levels. For example, a typical IC op-amp data sheet lists voltage gain, power gain, and common-mode rejection ratio in decibels. If you are not familiar with dB scales and ranges on meters, read the author's *McGraw-Hill Electronic Testing Handbook* (McGraw-Hill, 1994).

### 1.5.3  Frequency response

Amplifier frequency response can be measured with a generator and a meter or scope. The generator is tuned to various frequencies and the resultant output response is measured at each frequency. The results are then plotted in the form of a graph or response curve. Figure 1-30 shows the test connections to measure open-loop gain ($A_{OL}$) for a typical op amp (Harris CA3100). The term *open-loop gain* applies to the IC amplifier gain without feedback. Figures 1-31 and 1-32 show the frequency response for various power-supply voltages and temperatures, respectively.

As shown, gain (at any given frequency) varies directly with supply voltage and inversely with temperature for this amplifier (and most IC amplifiers). These variations must be considered in troubleshooting. For example, if this particular amplifier is used at 1 MHz, and the maximum available supply voltage is 7 V, the maximum open-loop gain is about 30 dB. If an 18-V supply is available, a gain of 45 dB is possible, all other factors being the same. So, always check the supply voltage against data-sheet characteristics before you condemn an amplifier circuit that appears to be underperforming.

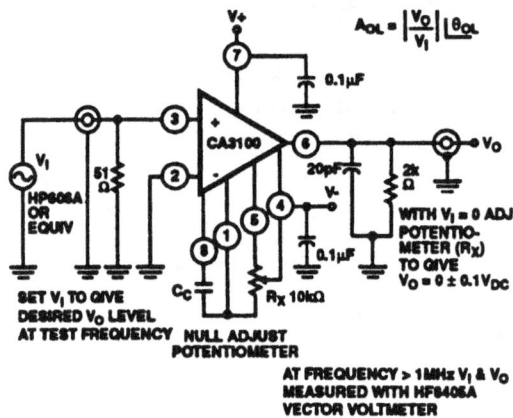

**Figure 1-30**  Open-loop voltage-gain test circuit and offset-adjust circuit (Harris Semiconductors, *Linear & Telecom ICs,* 1994, p. 2-106).

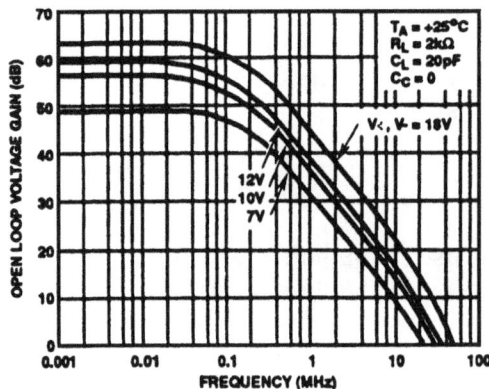

**Figure 1-31**  Open-loop gain versus frequency (Harris Semiconductors, *Linear & Telecom ICs,* 1994, p. 2-104).

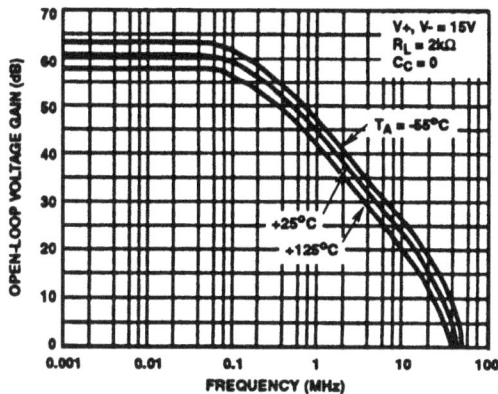

Figure 1-32   Open-loop gain versus frequency at different temperatures (Harris Semiconductors, *Linear & Telecom ICs,* 1994, p. 2-104).

The frequency at which the output begins to drop is called the *rolloff point.* In the amplifier of Figs. 1-31 and 1-32, rolloff starts at a frequency just below 0.1 MHz. The specifications for some IC op amps consider the rolloff point to start when the output drops 3 dB below the flat portion of the curve.

The curves of Figs. 1-31 and 1-32 show the open-loop rolloff and gain-frequency relationship of the IC, without external compensation. As covered, some IC op amps provide for the connection of an external compensation circuit. Such circuits alter both the rolloff point and the gain-frequency relationship. Figure 1-33 shows how the output characteristics are altered when an external compensating capacitor is added to the circuit of Fig. 1-30. Also, many IC op amps have built-in compensation, so the open-loop characteristics cannot be altered. In troubleshooting such circuits, where the characteristics are not as desired, it is the fault of the IC and not the external circuit.

The basic procedure for measuring frequency response is to apply a constant-amplitude signal while monitoring the output. The input signal is varied in frequency (but not in amplitude) across the entire operating range of the amplifier. The voltage output at various frequencies across the range is then plotted on the graph as follows.

1. Connect the equipment as shown in Fig. 1-30. Keep in mind that Fig. 1-30 shows the connections for a specific IC op amp. However, the circuit has all of the basic elements of a typical open-loop gain test circuit. The inverting input is grounded, and the test signal is applied to the noninverting input. Thus, the output should be an amplified, noninverting, version of the input. The input is terminated in an impedance equal to that of the signal source (51 ohms). The output is terminated at some specific test values (2 k$\Omega$ and 30 pF).

   Note that the 20-pF load capacitor is not used for the graph of Fig. 1-32, but is used for both Figs. 1-31 and 1-33. The compensating capacitor is used only for the graph of Fig. 1-33. The output load resistance is used for all three graphs. In troubleshooting, it is important to remember that the IC op-amp characteristics will change (sometimes dramatically) with changes

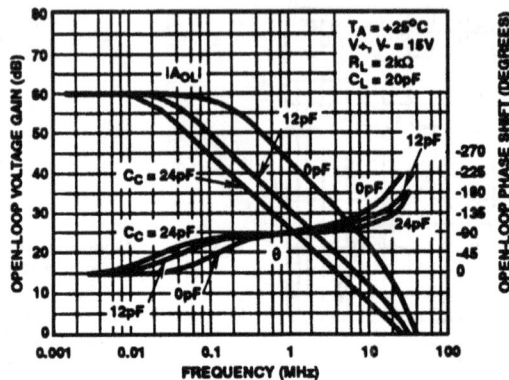

**Figure 1-33** Open-loop gain and phase shift versus frequency (Harris Semiconductors, *Linear & Telecom ICs,* 1994, p. 2-104).

in output load. Therefore, if the data-sheet test values are not close to those of the real load, try testing the IC with real-world values at the output, in addition to the test with data-sheet values. (You might find that the IC will not meet the desired gain-frequency or rolloff requirements when real-world values are used in the external circuit.)

2. Initially, set the generator frequency to the low end of the range (about 1 kHz for this IC amplifier), then set the generator amplitude to the desired input level. In the absence of a realistic test input voltage, set the generator output to an arbitrary value.

   In a practical testing/troubleshooting situation, monitor the circuit output and increase the generator amplitude at the amplifier center frequency (at 1 MHz for this IC) until the amplifier is overdriven. This point is indicated when further increases in generator output do not cause further increases in meter reading (or when the output waveform peaks begin to flatten on the scope display). Set the generator amplitude output just below this point. Then, return the meter or scope to monitor the generator voltage (at the circuit input) and measure the voltage. Keep the generator at this voltage throughout the test.

3. If the circuit is provided with any operating or adjustment controls (volume, loudness, gain treble, bass, balance, etc.), set the controls to some arbitrary point when making the initial frequency-response measurement. The response measurements can then be repeated at different control settings, if desired.

   In troubleshooting op amps, such as shown in Fig. 1-30, it might be necessary to null the IC for internal unbalance. For example, if an unbalance is in the differential input of the IC, or if a level shift is in the stages following the input, the output might not be zero with a zero input. This can be corrected by adjusting $R_x$ as shown in Fig. 1-30. Short pin 3 to pin 2 (or ground), and adjust $R_x$ for zero output, with no signal applied. Note that all IC op amps do not have a null provision.

4. Record the amplifier output voltage on the graph. Without changing the generator amplitude, increase the generator frequency by some fixed

amount and record the new amplifier output voltage. The amount of frequency increase is an arbitrary matter.

For example, when troubleshooting an audio amplifier, use an increase of 10 Hz at the frequencies where rolloff occurs and 1 kHz at the middle frequencies. These values can be increased by 10 for this IC.

5. After the initial frequency-response test, check the effect of operating or adjustment controls (if any). For example, in troubleshooting an audio circuit (Chapter 6), the volume, loudness, and gain controls should have the same effect across the entire frequency range. Treble and bass controls might have some effect on all frequencies. However, a treble control should have the greatest effect at the high end, whereas bass controls should be most effective at the low end.

6. Remember that generator output can vary with changes in frequency (a fact that is sometimes overlooked when making frequency-response tests). In any testing/troubleshooting situation where the generator is used at different frequencies, measure the generator output amplitude after each change in frequency. It is essential that the generator output amplitude remain constant over the entire frequency range of the test.

### 1.5.4   Voltage gain

*Voltage gain* for an amplifier is measured in the same manner as frequency response. The ratio of output voltage $V_O$ to input voltage $V_I$ (at any given frequency or across the entire frequency range) is the voltage gain. In the circuit of Fig. 1-30, open-loop gain (AOL) is measured. Because the input voltage (generator output) is held constant for a frequency-response test, a voltage-gain curve should be identical to a frequency-response curve (such as shown in Figs. 1-31, 1-32, and 1-33).

### 1.5.5   Power output and power gain

The power output of an amplifier is found by noting the output voltage across the load resistance at any frequency across the entire frequency range. For example, in the circuit of Fig. 1-30, if the output voltage ($V_O$) is 10 V, the power is $P=E^2/R=10^2/2000=100/2000=0.05$ W or 50 mW.

To find the power gain, start by finding both the input and output power. Input power is found in the same manner as output power, except that the input impedance must be known (or calculated). Calculating input impedance is not practical in some circuits—especially where input impedance depends on transistor gain (the procedure for finding dynamic input impedance is described in Section 1.5.9). With input power known (or estimated), power gain is the ratio of output power to input power.

In practical testing/troubleshooting, power output and gain depend on both the IC characteristics and the external load. In a new circuit, inadequate output or gain are design problems (unless the IC simply cannot deliver the required power into the load). In an existing circuit, inadequate power could be an IC failure or possibly a change in load. For example, in the circuit of Fig. 1-30, the 20-pF capacitor at pin 6 of the IC could be leaking, thus changing the 2-kΩ load. As a practical matter, never use a wire-wound component or any component that has reactance for the load resistance. Reactance changes with frequency and causes a load to change. Use a composition resistor (fixed or pot) for the load.

### 1.5.6  Input sensitivity

In some amplifier circuits, an input-sensitivity specification is used in place of or in addition to power-output/gain specifications. *Input sensitivity* implies a minimum power output with a given voltage input (such as a 2-W output with a 10-mV input). Input sensitivity usually applies to power amplifiers, and not to op amps. To find input sensitivity, simply apply the specified input and note the actual power output.

### 1.5.7  Bandwidth

Some amplifier specifications require that the circuit deliver a given voltage or power output across a given frequency range. Usually, the voltage bandwidth is not the same as the power bandwidth. For example, an amplifier might produce a full-power output up to 20 kHz—even though the frequency response is flat up to 100 kHz. That is, voltage (without a load) remains constant up to 100 kHz, whereas power output (across a normal load) remains constant only up to 20 kHz. Figure 1-34 shows the test connections and procedures to measure bandwidth at −3-dB points for a typical op amp (Harris CA3020/3020A). Again, from a troubleshooting standpoint, be certain that the external load circuit is correct (pure dc resistance of the desired value) before you condemn the IC.

### 1.5.8  Load sensitivity

Most amplifiers are sensitive to changes in load. This is particularly true of power amplifiers. From a testing/troubleshooting standpoint, an amplifier pro-

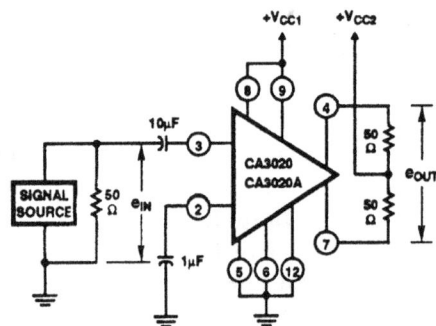

**PROCEDURES:**

1. Apply desired value of $V_{CC1}$ and $V_{CC2}$
2. Apply 1kHz input signal and adjust for $e_{IN}$ = 5mV (rms)
3. Record the resulting value of $e_{OUT}$ in dB (reference value)
4. Vary input-signal frequency, keeping $e_{IN}$ constant at 5mV, and record frequencies above and below 1kHz at which $e_{OUT}$ decreases 3dB below reference value
5. Record bandwidth as frequency range between -3dB points

**Figure 1-34**  Measurement of bandwidth at −3 dB points (Harris Semiconductors, *Linear & Telecom ICs,* 1994, p. 2-50).

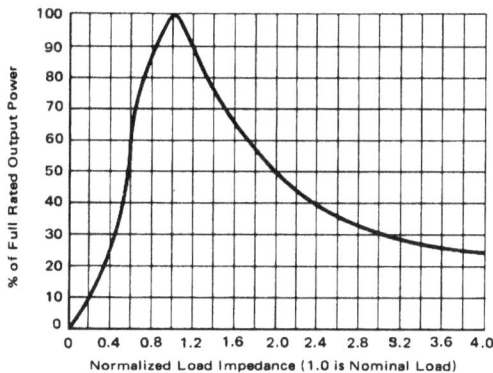

**Figure 1-35**   Typical load-sensitivity response curve.

duces maximum power when the output impedance is the same as the load impedance. The test circuit for load-resistance is the same as for frequency response (Fig. 1-30), except that the load resistance is variable. Again, never use a wire-wound load resistance. The reactance can result in considerable error.

To find *load sensitivity,* measure the power output at various load-impedance and output-impedance ratios. That is, set the load resistance to various values (including a value that is equal to the supposed amplifier-output impedance). Record the voltage and/or power gain at each setting. Repeat the test at various frequencies. Figure 1-35 shows a typical load-sensitivity response curve. Notice that if the load is twice the output impedance (as indicated by a 2:1 ratio, or a normalized load impedance of 2, the output power is reduced to about 50%.

## 1.5.9   Dynamic output impedance or resistance

The load-sensitivity test can be reversed to find the *dynamic output impedance or resistance* of an amplifier. The connections (Fig. 1-30) and procedures are the same, except that the load resistance is varied until maximum power output is found. Power is removed, the load resistance is disconnected from the circuit, and the resistance is measured with an ohmmeter. This resistance is equal to the dynamic output impedance of the circuit (but only at that measurement frequency). This test can be repeated across the entire frequency range, if required.

## 1.5.10   Dynamic input impedance or resistance

Use the circuit and procedures shown in Fig. 1-36 to find the dynamic input impedance or resistance of an amplifier. Notice that the IC shown in Fig. 1-36 has two inputs to be measured. Also notice that the accuracy of this impedance measurement (and the output impedance measurement) depends on the accuracy with which the resistance $(R)$ is measured. Again, a noninductive (not wire-wound) resistance must be used for $R$. The impedance found by this method applies only to the frequency used during the test.

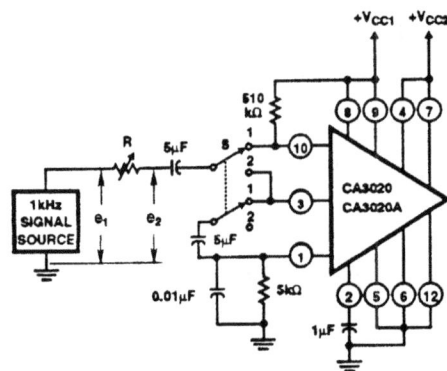

**PROCEDURES:**

Input Resistance Terminal 10 to Ground ($R_{IN10}$)

1. Apply desired value of $V_{CC1}$ and $V_{CC2}$ and set S in Position
2. Adjust 1-kHz input for desired signal level of measurement
3. Adjust R for $e_2 = e_1/2$
4. Record resulting value of R as $R_{IN10}$

Input Resistance Terminal 3 to Ground ($R_{IN3}$)

1. Apply desired value of $V_{CC1}$ and $V_{CC2}$ set S in Position
2. Adjust 1-kHz input for desired signal level of measurement
3. Adjust R for $e_2 = e_1/2$
4. Record resulting value of R as $R_{IN3}$

**Figure 1-36**  Measurement of input resistance (Harris Semiconductors, *Linear & Telecom ICs,* 1994, p. 2-51).

## 1.5.11  Current drain, power output, efficiency, and sensitivity

Figure 1-37 shows a circuit and the procedures suitable for measuring zero-signal dc current drain, maximum-signal dc current drain, maximum power output, circuit efficiency, and transducer power gain. Again, the circuit of Fig. 1-37 applies to a specific IC amplifier (Harris CA3020/CA3020A), but a similar circuit can be used for most IC op amps. The definition of $R_{IN}(10)$ is given in Fig. 1-36.

## 1.5.12  Sine-wave analysis

All amplifiers are subject to distortion. That is, the output signal might not be identical to the input signal. Theoretically, the output should be identical to the input, except for the amplitude. This can be checked by applying a sine wave at the amplifier input (using a circuit similar to Fig. 1-30) and monitoring both the input and output with a scope. If there is no change in the scope display, except for amplitude, there is no distortion.

In practical troubleshooting, analyzing sine waves to pinpoint amplifier problems that produce distortion is a difficult job. Unless distortion is severe, it might pass unnoticed. Sine waves are best used where harmonic-distortion (Section 1.5.14) or intermodulation-distortion (Section 1.5.15) meters are combined with the scope for distortion analysis. If a scope is used alone for distortion analysis, square waves provide the test results. (The reverse is true for frequency response and power measurements.)

## 1.5.13  Square-wave analysis

Distortion analysis is more effective with square waves because of the high odd-harmonic content in square waves (and because it is easier to see a deviation from a straight line with sharp corners than from a curving line). The procedure for checking distortion with square waves is essentially the same as that used with sine waves. Square waves are introduced into the amplifier input, and the output is measured with a scope (Fig. 1-38).

In practical troubleshooting, the primary concern is deviation of the output waveform from the input waveform (which is also monitored on the scope). If the scope has a dual-trace feature, the input and output can be monitored simultaneously. Also, if the scope has an invert function, the output can be inverted from the input for a better comparison of input and output. If there is a change in the waveform, the nature of the change can sometimes reveal the cause of the distortion. Notice that the drawings of Fig. 1-38 are generalized and that the same waveform can be produced by different causes. For example, poor LF (low-frequency) response appears to be the same as HF (high-frequency) emphasis.

Figure 1-39 shows the waveforms that are produced by an actual circuit. Notice that the output (trace B) does a good job of following the input (trace A) at a gain of −1. That is, the output is inverted from the input and there is no gain (unity gain). Also notice that the high-frequency response is reduced in output trace B, but not the exaggerated response shown in Fig. 1-38.

When troubleshooting with square waves, note that the third, fifth, seventh, and ninth harmonics of a clean square wave are emphasized. If an amplifier

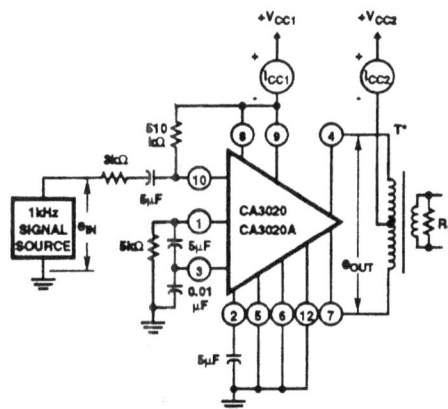

**PROCEDURES:**
Zero-Signal DC Current Drain

1. Apply desired value of $V_{CC1}$ and $V_{CC2}$ and reduce $e_{IN}$ to 0V
2. Record resulting values of $I_{CC1}$ and $I_{CC2}$ in mA as Zero-Signal DC Current Drain

Maximum-Signal DC Current Drain, Maximum Power Output, Circuit Efficiency, Sensitivity, and Transducer Power Gain

1. Apply desired value of $V_{CC1}$ and $V_{CC2}$ and adjust $e_{IN}$ to the value at which the Total Harmonic Distortion in the output of the amplifier = 10%
2. Record resulting value of $I_{CC1}$ and $I_{CC2}$ in mA as Maximum Signal DC Current Drain
3. Determine resulting amplifier power output in watts and record as Maximum Power Output ($P_{OUT}$)
4. Calculate Circuit Efficiency ($\eta$) in % as follows:

$$\eta = 100 \frac{P_{OUT}}{V_{CC1}I_{CC1} + V_{CC2}I_{CC2}}$$

where $P_{OUT}$ is in watts, $V_{CC1}$ and $V_{CC2}$ are in volts, and $I_{CC1}$ and $I_{CC2}$ are in amperes.

5. Record value of $e_{IN}$ in mV (rms) required in Step 1 as Sensitivity ($e_{IN}$)
6. Calculate Transducer Power Gain ($G_p$) in dB as follows:

$$G_p = 10 \log_{10} \frac{P_{OUT}}{P_{IN}}$$

where $P_{IN}$ (in mW) $= \dfrac{e_{IN}^2}{3000 + R_{IN(10)}**}$

**See Figure 10 for definition of $R_{IN(10)}$

*T: PUSH-PULL OUTPUT TRANSFORMER; LOAD RESISTANCE ($R_L$) SHOULD BE SELECTED TO PROVIDE INDICATED COLLECTOR-TO-COLLECTOR LOAD IMPEDANCE ($R_{CC}$)

**Figure 1-37**  Measurement of current drain, power output, efficiency, and sensitivity (Harris Semiconductors, *Linear & Telecom ICs,* 1994, p. 2-50).

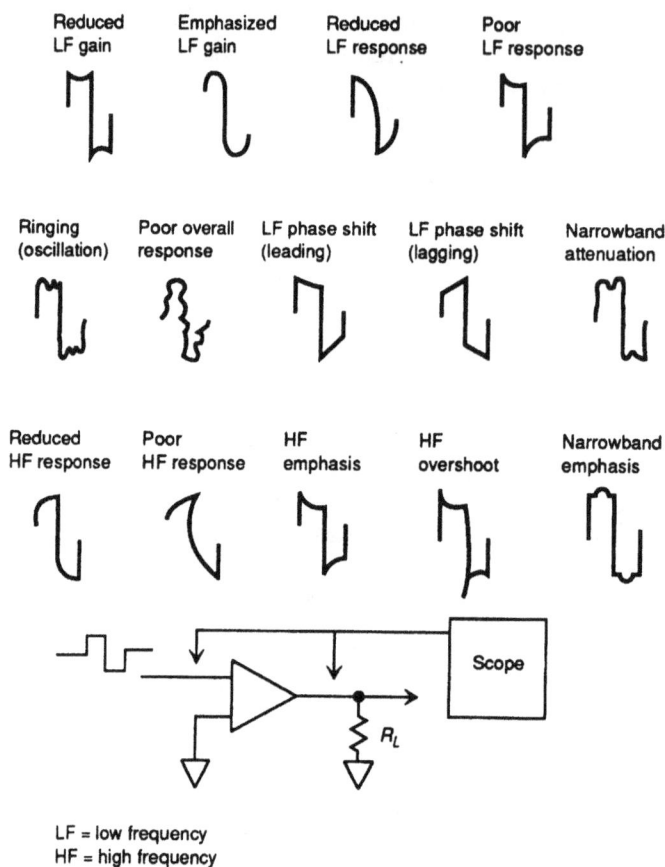

Figure 1-38   Basic square-wave distortion analysis.

passes a given frequency and produces a clean square-wave output, it is reasonable to assume that the frequency response is good up to at least nine times the square-wave frequency. Section 1.6 describes troubleshooting with square-wave displays in greater detail.

## 1.5.14   Harmonic distortion

No matter what amplifier circuit is used or how well the circuit is designed, there is a possibility of odd or even harmonics being present with the fundamental signal. These harmonics combine with the fundamental and produce distortion, as is the case when any two or more signals are combined in a circuit. The effects of second- and third-harmonic distortion are shown in Fig. 1-40.

Harmonic-distortion meters operate on the fundamental-suppression principle. A sine wave is applied to the amplifier input and the output is monitored on a scope or meter. The output is then applied through a filter that suppresses the fundamental frequency. Any output from the filter is the result of harmonics. Figure 1-41 shows typical connections and procedures for measuring harmonic distortion (where a Hewlett-Packard Type 302A, or equivalent, analyzer is used to measure the total harmonic distortion, THD, of a Harris CA3020/3020A). The same circuit is also used for signal-to-noise measurements, as described in Section 1.5.17.

In practical troubleshooting, particularly in audio-amplifier circuits, a scope is combined with a harmonic-distortion meter to find the harmonic frequency. For example, if the input is 1 kHz and the output (after filtering) is 3 kHz, third-harmonic distortion is indicated. (Reduce the scope horizontal sweep down so that you can see one input cycle. If three cycles are at the output for the same time period as the one input cycle, this indicates third-harmonic distortion.)

The percentage of harmonic distortion is also determined by this method. For example, if the output is 100 mV without the filter and 3 mV with the filter, this indicates a 3% harmonic distortion. When troubleshooting a practical circuit, notice that the THD varies with the power output of the amplifier. It is generally necessary to adjust the input voltage for a given power output, as shown in Fig. 1-41. THD also depends on load.

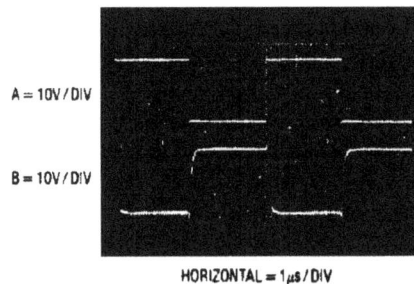

A = 10V / DIV

B = 10V / DIV

HORIZONTAL = 1 μs / DIV

**Figure 1-39**  Amplifier-circuit response to square waves.

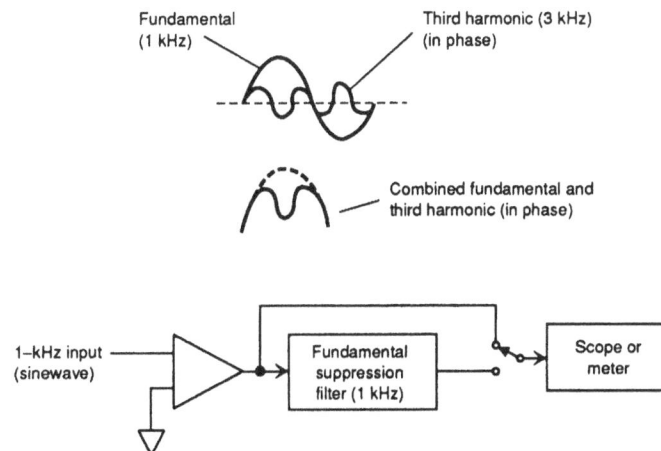

Fundamental (1 kHz)

Third harmonic (3 kHz) (in phase)

Combined fundamental and third harmonic (in phase)

1–kHz input (sinewave)

Fundamental suppression filter (1 kHz)

Scope or meter

**Figure 1-40**  Effects of second- and third-harmonic distortion.

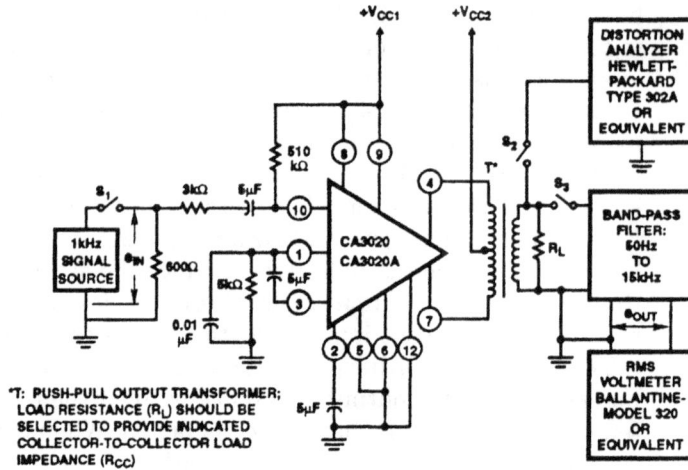

**Figure 1-41**   Measurement of THD and signal-to-noise ratio (Harris Semiconductors, *Linear & Telecom ICs,* 1994, p. 2-51).

PROCEDURES:

Signal-to-Noise Ratio

1. Close $S_1$ and $S_3$; open $S_2$
2. Apply desired values of $V_{CC1}$ and $V_{CC2}$
3. Adjust $e_{IN}$ for an amplifier output of 150mW and resulting value of $E_{OUT}$ in dB as $e_{OUT1}$ (reference value)
4. Open $S_1$ and record resulting value of $e_{OUT}$ in dB as $e_{OUT2}$
5. Signal-to-Noise Ratio   (S/N) = $20\log_{10}\dfrac{e_{OUT1}}{e_{OUT2}}$

Total Harmonic Distortion

1. Close S1 and S2; open S3
2. Apply desired values of $V_{CC1}$ and $V_{CC2}$
3. Adjust $e_{IN}$ for desired level amplifier output power
4. Record Total Harmonic Distortion (THD) in %

## 1.5.15   Intermodulaton distortion

When two signals of different frequencies are mixed in an amplifier, it is possible that the lower-frequency signal will modulate the amplitude of the higher-frequency signal. This produces a form of distortion known as *intermodulation distortion (IMD).* Figure 1-42 shows the basic elements of IMD meters (a signal generator and a high-pass filter). The generator portion produces a higher-frequency signal (usually 7 kHz for standard recording-industry testing) that is modulated by a low-frequency signal (usually 60 Hz).

In practical testing/troubleshooting, the mixed signals are applied to the amplifier input, with the output connected through a high-pass filter to a scope. The high-pass filter removes the low-frequency (60 Hz) signal. The only signal that appears on the scope should be the 7-kHz signal. If any 60-Hz signal is present on the scope, the 60-Hz signal is being passed through as modulation on the 7-kHz signal.

The percentage of IMD can be calculated using the equations shown in Fig. 1-42. For example, if the maximum output (shown on the scope) is 1 mV and the minimum is 0.99 mV, the percentage of IMD is approximately:

$$(1.0-0.99)/(1.0+0.99)=0.005\times100=0.5\%$$

### 1.5.16  Background noise, hum, and oscillation

If a scope is sufficiently sensitive, it can be used to check and measure the background noise level of an amplifier circuit, as well as to check for the presence of hum, oscillation, etc. The scope should be capable of measurable deflection with an input below 1 mV (and considerably less if an IC amplifier is involved).

Figure 1-43 shows the basic procedure for measuring amplifier-circuit background noise. The circuit output is monitored with the volume or gain/loudness controls (if any) at maximum. A meter can be used, but the scope is superior because the frequency and nature of the noise (or other signals) are displayed visually. Scope gain must be increased until there are noise or "hash" indications.

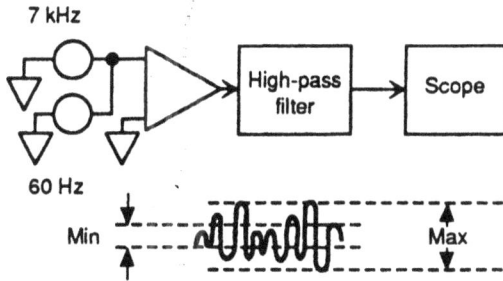

Figure 1-42  Basic IMB analysis.

$$\% \text{ intermodulation distortion} = 100 \times \frac{max - min}{max + min}$$

$R_L$ = normal circuit load impedance

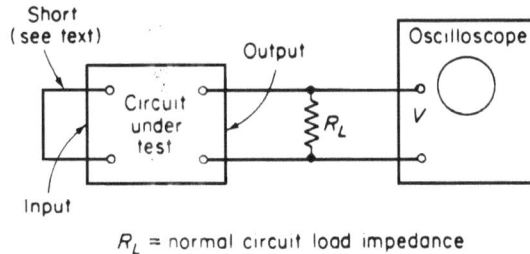

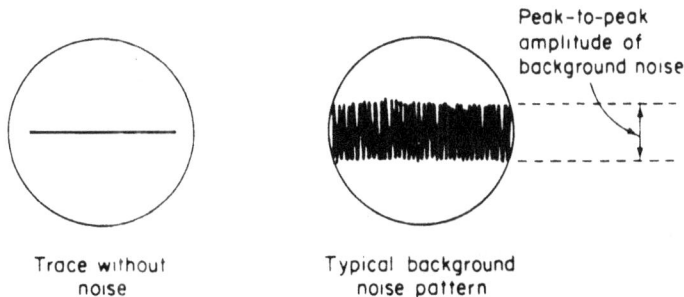

Figure 1-43  Measurement of background noise and hum.

In practical testing/troubleshooting, it is possible that a noise indication can be caused by pickup in the leads between the amplifier and scope. If in doubt, disconnect the leads from the amplifier, but not the scope. If you suspect that 60-Hz power-line hum is present in the amplifier output (picked up from the power supply or other source), set the scope sync controls to the "line" position (if any). If a stationary signal pattern appears, the signal is the result of line hum getting into the circuit. If a signal appears that is not at the line frequency, the signal must be the result of oscillation in the circuit or stray pickup. Short the amplifier input terminals. If the same signal remains, suspect oscillation in the circuit. Section 1.6 covers these problems in greater detail.

Figure 1-44 shows a practical test circuit for background noise. With IC op amps, the internal or background noise is considerably less than 1 mV, and it is impossible to measure directly—even with a sensitive scope. It is necessary to use a circuit that amplifies the output of the IC being tested before the output is applied to the scope. The IC under test (an OP-77) is connected for high gain, as is the following amplifier. This makes it possible to set a typical scope to the X1 position (minimum amplification). It is also possible to monitor (and record) noise on a chart recorder, as shown in Fig. 1-45. In this case, the noise is measured over a 10-s interval, with the peak-to-peak value noted (about 25 nV in the case of Fig. 1-45).

### 1.5.17  Signal-to-noise ratio

Some IC op amps are tested for signal-to-noise ratio instead of (or in addition to) background noise. Figure 1-41 shows the circuit connections and procedures for signal-to-noise measurement of this CA3020/3020A. (This is the same circuit as for THD, except that the distortion analyzer is not connected when the signal-to-noise ratio is measured.)

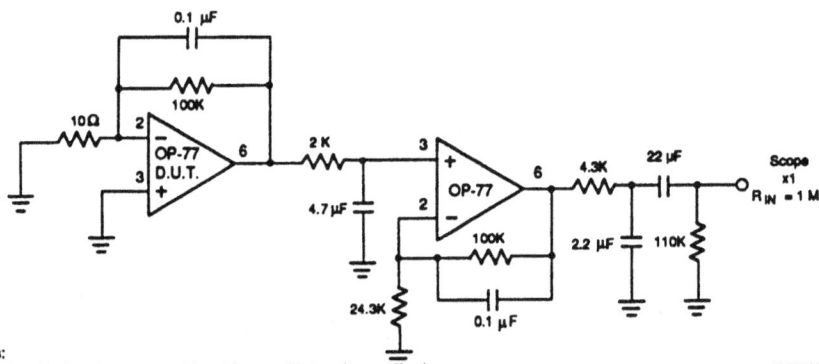

Notes:
1. Peak-to-Peak noise measured in a 10-second interval.
2. The device under test should be warmed up for 3 minutes and shielded from air currents.
3. Voltage Gain = 50,000
4. All capacitor values are for non-polarized capacitors only.
5. Pin numbers shown are for 8-lead packages.

Figure 1-44   A 0.1- to 10-Hz noise test circuit (Raytheon Semiconductors, *Data Book,* 1994, p. 3-531).

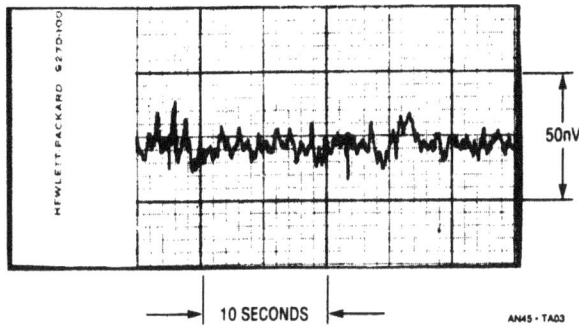

**Figure 1-45** Amplifier background noise as measured on a chart recorder over a 10-second interval.

A signal-to-noise test shows the relationship of background noise to signal amplitude, when the amplifier is operated under specific conditions. For example, in the circuit of Fig. 1-41, the input signal is increased in amplitude until the output is 150 mW and the output voltage is recorded in dB. The input signal is then removed, but the input terminals remain connected together through resistors and capacitors, so the only output is the noise voltage within the IC. This background-noise voltage is also recorded in dB, and the signal-to-noise ratio is calculated as shown.

### 1.5.18    Slew rate (transient response)

Amplifier *slew rate* is the maximum rate of change in output voltage, with respect to time, that the amplifier is capable of producing when maintaining linear characteristics (symmetrical output without clipping). Slew rate is often listed under the heading of transient response in op-amp data sheets. Other transient-response characteristics include rise time, settling time, overshoot, and possibly error band, all of which are covered in Section 1.5.19.

Slew rate is expressed in terms of difference in output voltage divided by difference in time $(d_{V_0}/d_t)$. Usually, slew rate is listed in terms of volts per microsecond. For example, if the output voltage from an op amp is capable of changing 7 V in 1 μs, the slew rate is 7 (which might be listed as 7 V/μs). In practical troubleshooting, the major effect of slew rate is that (all other factors being equal) a higher slew rate results in higher power output.

A simple way to find op-amp slew rate is to measure the slope of the output waveform when a square-wave input is applied, as shown in Fig. 1-46. The input square wave must have a rise time that exceeds the slew-rate capability of the amplifier. As a result, the output does not appear as a square wave, but as an integrated wave. In the example shown, the output voltage rises (and falls) about 40 V in 1 μs. Note that slew rate is usually measured in the closed-loop condition (with negative feedback) and that slew rate increases with higher gain.

Figure 1-47 shows the slew-rate and transient-response test circuit for a typical op amp (the Harris HA-2510). Figure 1-47 also includes some definitions for slew rate, settling time, rise time, overshoot, and error band, all of which are covered next.

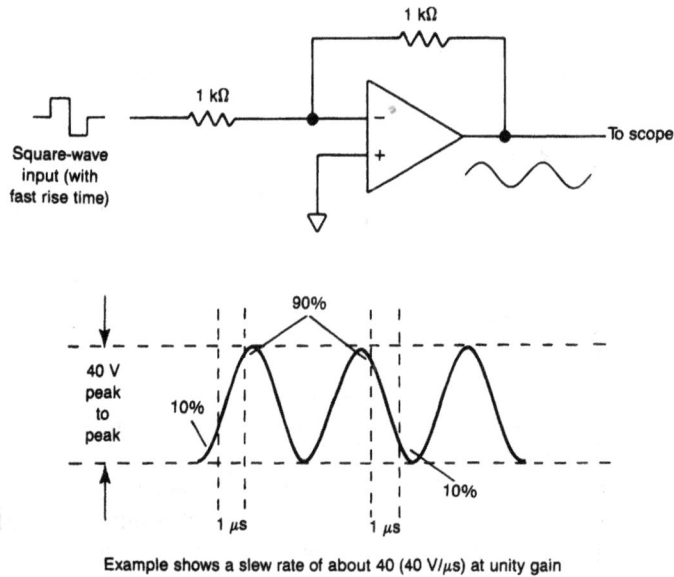

Example shows a slew rate of about 40 (40 V/μs) at unity gain

**Figure 1-46**   Basic slew-rate measurement.

### 1.5.19   Rise time, settling time, and overshoot

Figure 1-48 shows some typical test circuits and scope displays for transient-response measurements.

For this particular op amp (Harris HA-5147), *rise time* is specified with output of 200 mV and a gain of 10. As a result, the small-signal response displays must be used. As shown, the rise time (measured from the 10% point to the 90% point, Fig. 1-47) is about 25 ns. The data sheet specifies a typical rise time of 25 ns and a maximum of 50 ns.

*Settling time* is the total length of time from the input-step application until the output remains within a specified error-band or point around the final value. For the HA-5174, settling time is specified with an output of 10 V and an inverted gain of 10 (−10). Thus, the large-signal response must be used. As shown in Fig. 1-48, settling time is somewhat less than 400 ns (from the start of the input, through the overshoot, and back to where the output levels to 10 V). The data sheet specifies a typical settling time of 400 ns.

From a troubleshooting standpoint, an increase in rise time, settling time, or overshoot lowers the frequency response and bandwidth. If the amplifier is used in pulse applications, excessive rise times and settling times can distort the output pulse.

### 1.5.20   Phase shift problems

The phase shift between input and output of some amplifiers is not a significant problem, but is critical in others, particularly op amps. This is because an op amp generally uses the principle of feeding back output signals to the input.

TRANSIENT RESPONSE

OVERSHOOT

+200mV
INPUT
0mV

+200mV
90%
OUTPUT
10%
0mV

RISE TIME

NOTE: Measured on both positive and negative transitions from 0V to +200mV and 0V to -200mV at the output.

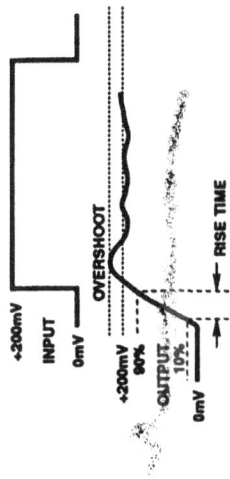

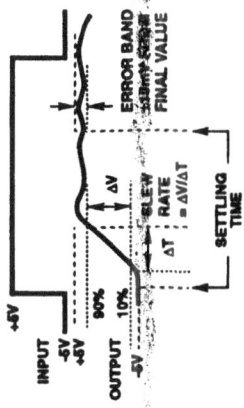

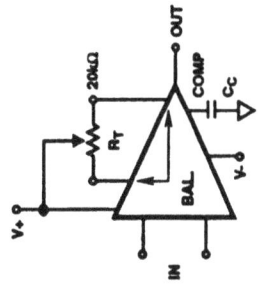

SLEW RATE AND SETTLING TIME

INPUT
+5V
-5V

OUTPUT
+5V
90%
10%
-5V

ERROR BAND

FINAL VALUE

ΔV

SLEW
RATE
= ΔV/ΔT

ΔT

SETTLING
TIME

SUGGESTED $V_{OS}$ ADJUSTMENT AND COMPENSATION HOOK UP

V+
20kΩ
$R_T$
OUT
COMP
$C_C$
BAL
V-
IN

Tested offset adjustment range is $|V_{OS} + 1mV|$ minimum referred to output. Typical ranges are ±6mV with $R_T = 20kΩ$.

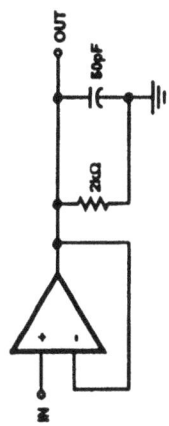

SLEW RATE AND TRANSIENT RESPONSE

IN
2kΩ
50pF
OUT

NOTE: Measured on both positive and negative transitions from 0V to +200mV and 0V to -200mV at the output.

Figure 1-47  Measurement of slew rate, rise time, settling time, and overshoot (Harris Semiconductors, *Linear & Telecom ICs*, 1994, p. 2-298).

47

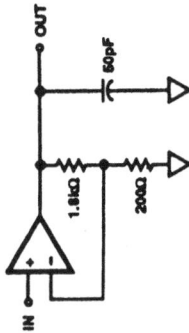

**LARGE AND SMALL SIGNAL RESPONSE TEST CIRCUIT**

IN

OUT

1.8kΩ

200Ω

50pF

**SMALL SIGNAL RESPONSE**

Vertical Scale: (Volts: Input = 10mV/Div)
(Volts: Output = 100mV/Div)
Horizontal Scale: (Time: 100ns/Div)

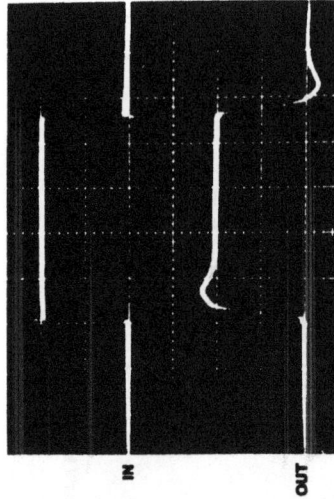

IN

OUT

**LARGE SIGNAL RESPONSE**

Vertical Scale: (Volts: Input = 0.5V/Div.)
(Volts: Output = 5V/Div.)
Horizontal Scale: (Time: 500ns/Div.)

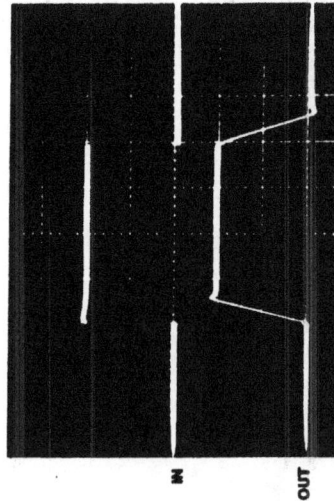

IN

OUT

**Figure 1-48** Large- and small-signal transient-response characteristics (Harris Semiconductors, *Linear & Telecom ICs*, 1994, p. 2-544).

48

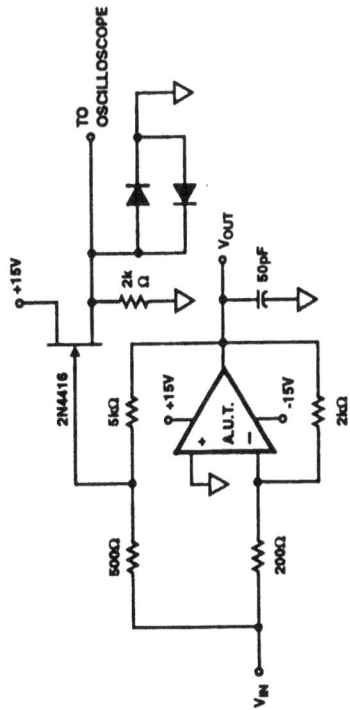

SETTLING TIME TEST CIRCUIT

$A_V = -10$

Feedback and summing resistors should be 0.1% matched.

Clipping diodes are optional. HP5082-2810 recommended.

Figure 1-48 (Continued)

Under ideal open-loop conditions, the output should be exactly 180° out of phase with the inverting input and exactly in phase with the noninverting input. Any substantial deviation from this condition can cause op-amp circuit problems.

For example, in troubleshooting a circuit, assume that an op amp uses the inverting input with the noninverting input grounded, and the circuit output is fed back to the inverting input. If the output is not shifted the full 180° (for example, if the shift is only a few degrees), the circuit might oscillate (because the output being fed back is almost in phase with the input). Even if there is no oscillation, the op-amp gain will not be stabilized and the circuit will not operate properly.

Figure 1-49 shows the basic connections for phase measurement in amplifier circuits. For the most accurate results, the cables that connect the input and output should be of the same length and characteristics. At higher frequencies, a difference in cable length or characteristics can introduce a phase shift.

For simplicity, adjust the scope controls until one cycle of the input signal occupies exactly nine divisions (typically 9 cm horizontally) of the screen. Then find the phase factor of the input signal. For example, if 9 cm represents one complete cycle (360°), 1 cm represents 40° (360/9=40).

With the phase factor established, measure the horizontal distance between corresponding points on the two waveforms (input and output signals). Then multiply the measured distance by the phase factor of 40° to find the phase difference. For example, if the horizontal distance is 0.6 cm with a 40°/cm phase factor, the phase difference is $0.6 \times 40° = 24°$.

If the scope has speed magnification, you can get more accurate results. For example, if the sweep rate is increased 10 times, the magnified phase factor is 40°/cm divided by 10, or 4°/cm. Figure 1-49 shows the same signal with and

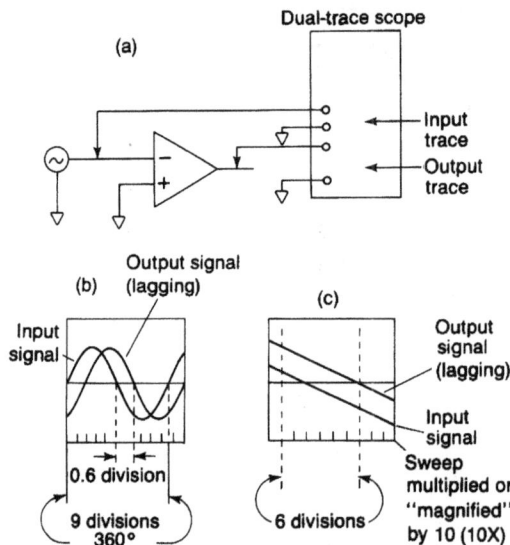

Figure 1-49  Measurement of phase shift.

without sweep magnification. With a $10\times$ magnification, the horizontal distance is 6 cm and the phase difference is $6 \times 4° = 24°$.

### 1.5.21  Feedback measurement

Figure 1-50 shows the basic feedback-measurement connections for an op-amp circuit. Because op-amp circuits usually include feedback, it is sometimes necessary to measure feedback voltage at a given frequency with given operating conditions. Although it is possible to measure the feedback voltage as shown in Fig. 1-50A, a more accurate measurement is made when the feedback lead is terminated in the normal operating impedance, as shown in Fig. 1-50B.

If an input resistance is used in the circuit and this resistance is considerably lower than the IC input resistance, use the circuit-resistance value. If in doubt, measure the input impedance of the IC (Section 1.5.10) and terminate the feedback lead in that value (to measure open-loop feedback voltage). When troubleshooting any op-amp circuit, remember that open-loop voltage gain

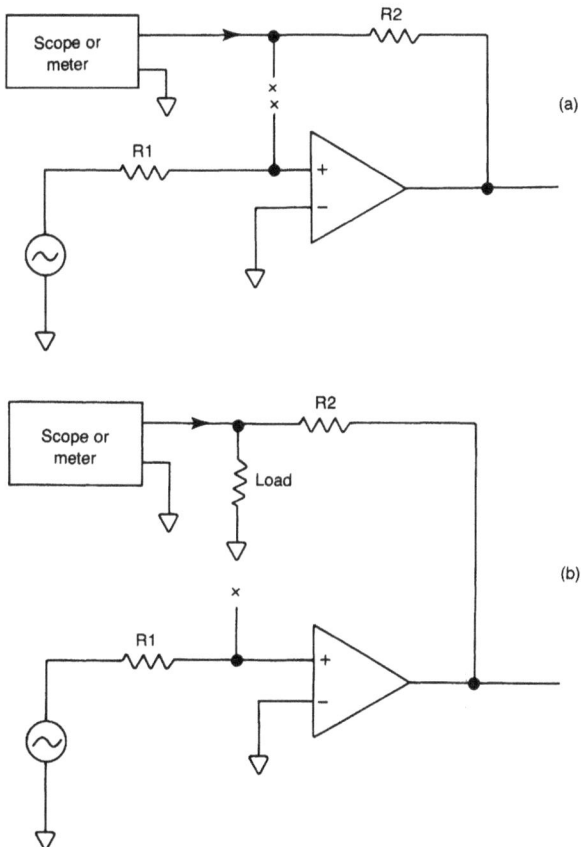

**Figure 1-50**  Measurement of feedback.

must be substantially higher than the closed-loop gain for most circuits to perform properly.

### 1.5.22    Input-bias current

Figure 1-51 shows the basic input-bias measurement connections for an op amp. Any resistance value for $R_1$ and $R_2$ can be used, provided that the value produces a measurable voltage drop and that the resistance values are equal. A value of 1 kΩ (with a tolerance of 1% or better) for both R1 and R2 is realistic for typical op amps. Op-amp input-bias current is the average value of the two input-bias currents of the op-amp differential-input stage.

In troubleshooting, the significance of input-bias current is that the resultant voltage drops across input resistors (such as the resistor at pin 3 of the IC in Fig. 1-30) restrict the input common-mode voltage range at higher impedance levels. The input-bias current produces a voltage drop across the input resistors. This voltage drop must be overcome by the input signal (which can be a problem if the input signal is low and the input resistors are large).

If it is not practical to connect a meter in series with both inputs as shown, measure the voltage drop across R1 and R2, and calculate the input-bias current. For example, if the voltage is 3 mV across 1-kΩ resistors, the input-bias current is 3 μA. Try switching R1 and R2 to see if any difference is the result of difference in resistor values.

In theory, the input-bias currents should be the same for both inputs. In practical troubleshooting, the bias currents should be almost equal. Any great difference in input bias is the result of unbalance in the input differential amplifier of the IC, and it can seriously affect circuit operation (and it usually indicates a defective IC).

### 1.5.23    Input-offset voltage and current

Figure 1-52 shows a circuit for measurement of input-offset voltage and current. Input-offset voltage is the voltage that must be applied at the input ter-

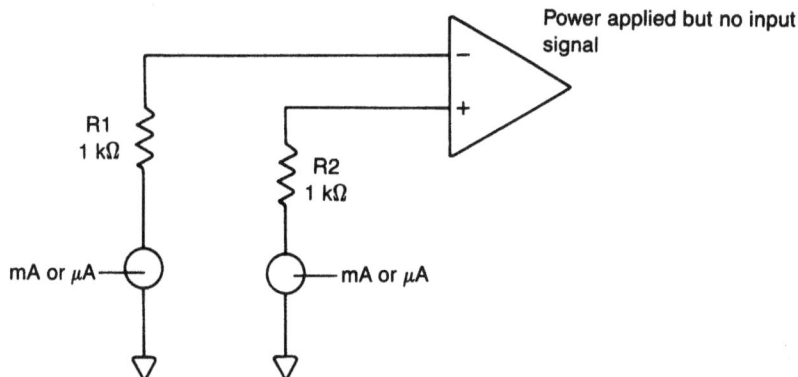

Figure 1-51    Measurement of input-bias current.

minals to get zero output voltage, whereas input-offset current is the difference in input-bias current at the op-amp input. Offset voltage and current are usually referred back to the input because the output voltages depend on feedback.

In troubleshooting, the effect of input-offset is a fixed shift in output level with no input. The input must therefore overcome the offset before an output is produced. This can result in distortion, as shown in Fig. 1-53. In some circuits, the offset (and distortion) can be removed or minimized by an external adjustment circuit. In other circuits, the offset can be reduced by means of an external resistor (R3), as shown in Fig. 1-28, or by a circuit, such as shown in Fig. 1-29.

To use the circuit of Fig. 1-52, measure the output with R3 shorted and with R3 in circuit. Record the two output voltages as $E_1$ (S1 closed, R3 shorted) and $E_2$ (S1 open, R3 in the circuit). With the two output voltages recorded, calculate the input-offset voltage and current using the equation of Fig. 1-52. For example, assume that $R_1$, $R_2$, and $R_3$ are at the values shown, that $E_1$ is 83 mV, and that $E_2$ is 363 mV:

$$\text{Input-offset voltage} = (83 \text{ mV})/(100) = 0.83 \text{ mV}$$

$$\text{Input-offset current} = (280)/(100 \text{ k}\Omega) \times (1 + 5.1 \text{ k}\Omega/51) = 27.7 \text{ nA}$$

### 1.5.24  Common-mode rejection or rejection ratio

Figure 1-54 shows the basic circuit for measurement of common-mode rejection (or common-mode rejection ratio). Many definitions are used for common-mode rejection (Fig. 1-22). No matter what definition is used, the first step to

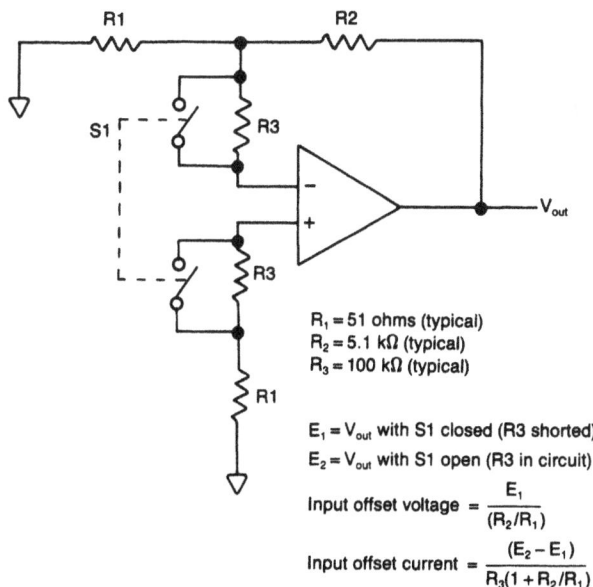

$R_1 = 51$ ohms (typical)
$R_2 = 5.1$ k$\Omega$ (typical)
$R_3 = 100$ k$\Omega$ (typical)

$E_1 = V_{out}$ with S1 closed (R3 shorted)
$E_2 = V_{out}$ with S1 open (R3 in circuit)

$$\text{Input offset voltage} = \frac{E_1}{(R_2/R_1)}$$

$$\text{Input offset current} = \frac{(E_2 - E_1)}{R_3(1 + R_2/R_1)}$$

Figure 1-52  Measurement of input-offset voltage and current.

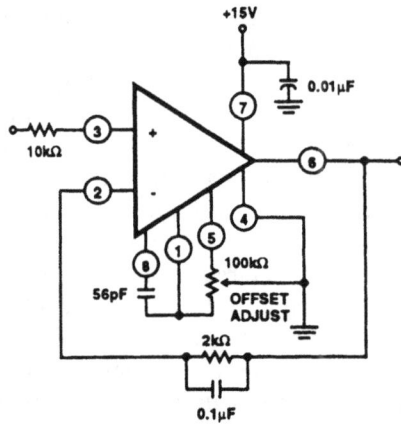

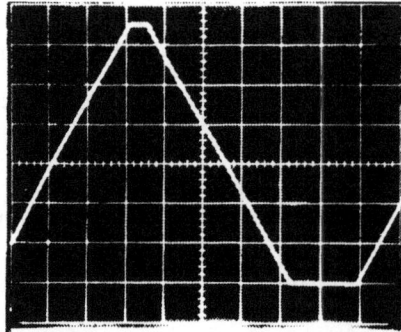

(A) OUTPUT-WAVEFORM WITH INPUT-SIGNAL RAMPING (2V/ DIV. AND 500µs/DIV.)

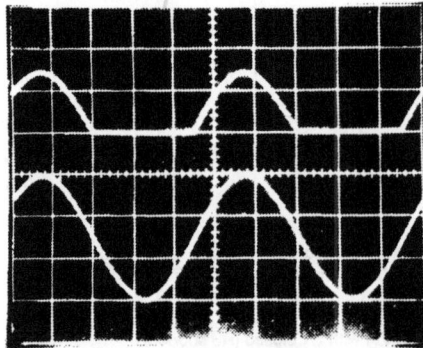

Top Trace:     Output (5V/DIV. and 200µs/DIV.)
Bottom Trace:  Input Signal (5V/DIV. and 200µs/DIV.)

(B) OUTPUT WAVEFORM WITH GROUND-REFERENCE SINE-WAVE INPUT

Figure 1-53  Distortion effects of input offset (Harris Semiconductors, *Linear & Telecom ICs,* 1994, p. 2-117).

measure CMR is to find the open-loop gain of the op amp at the desired operating frequency (Section 1.5.4). Then use the circuit of Fig. 1-54 and increase the common-mode voltage (at the same frequency used for the open-loop gain test) until a measurable output is obtained. Be careful not to exceed the maximum input common-mode voltage specified in the data sheet. If no such value is available, do not exceed the normal input voltage of the IC.

In practical testing/troubleshooting, simplify the calculation by increasing the input voltage until the output is at some exact value, such as the 1 mV shown. Divide this value by the open-loop gain to find the equivalent differential input signal. For example, with an open-loop gain of 100 and an output of 1 mV, the equivalent differential is: 0.001/100 = 0.00001. Now measure the input voltage that produces the 1-mV output, and divide the input by the equivalent differential to find the common-mode rejection ratio. For example, if the output is 1 mV with a 10-V input and a gain of 100, the ratio is 1,000,000 (100 dB).

### 1.5.25    Power-supply sensitivity or rejection ratio

Figure 1-52 shows the basic circuit for measuring *power-supply sensitivity (PSS)* or *rejection ratio (PSRR)*. (This is the same test circuit as for input-offset voltage.) PSS is the ratio of change in input-offset voltage to the change in power-supply voltage that produces the change. On some data sheets, the term is expressed in millivolts or microvolts per volt (mV/V or $\mu$V/V), which represents the change of input-offset voltage (in mV or $\mu$V) to a change (in volts) of the power supply. In other data sheets, PSRR is used instead, and is given in decibels.

The procedure for measuring PSS or PSRR is the same as for measurement of input-offset voltage, except that the supply voltage is changed (in 1-V steps). The amount of change in input-offset voltage for a 1-V change is the PSS or PSRR. (The ratio of change can be converted to dB as required.) The circuit of Fig. 1-52 can also be used when the op amp is operated from two power supplies. One supply voltage is changed (in 1-V steps) while the other supply voltage is held constant.

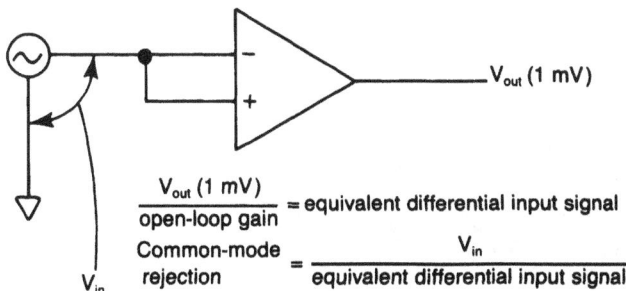

$$\frac{V_{out} (1\ mV)}{open\text{-}loop\ gain} = equivalent\ differential\ input\ signal$$

$$\frac{Common\text{-}mode}{rejection} = \frac{V_{in}}{equivalent\ differential\ input\ signal}$$

**Figure 1-54**   Measurement of common-mode rejection.

## 1.6    Troubleshooting Amplifier Circuits with Scope Displays

As covered in Section 1.5.13, one of the most practical ways to check amplifier performance is to display the amplifier output on a scope with square waves or pulses applied at the input. This procedure checks both the IC amplifier and the external circuit simultaneously. Even though the procedure is simple, many problems occur when using scopes in troubleshooting. This section summarizes the most common problems and provides practical solutions.

Keep in mind that all of the displays covered in this section were made with a good IC amplifier. The poor results were caused by poor layout, inferior scopes and probes, incorrect (or absent) bypassing, and other problems in the external circuit. All of these factors are covered throughout the remainder of this book. However, it is assumed that you have a good scope and probe (both capable of displaying signals at frequencies far beyond those to be used with the amplifier circuit that you are troubleshooting). It is also recommended that you read Application Note 47 by Jim Williams in the *Linear Technology 1993 Linear Applications Handbook, Volume II,* for detailed information on high-speed amplifier testing.

### 1.6.1    Severe ringing

Figure 1-55 shows the display when the generator (square wave or pulse) is unterminated. The result is severe ringing on the pulse edge (caused by reflections), and can be eliminated by terminating the generator cable in its characteristic impedance. If the generator is properly terminated in an existing circuit that worked previously, look for breaks in the PC wiring at the circuit input where the generator is connected.

### 1.6.2    Pulse with abnormal corners

Figure 1-56 shows the display when the generator is terminated, but with a poor-quality termination. The result is a pulse with abnormal corners. On a test bench, the best termination for 50-Ω cable is the BNC coaxial type. For PC-board use, the best termination resistors are carbon or metal-film types,

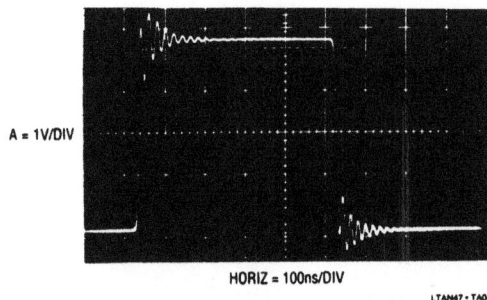

A = 1V/DIV

HORIZ = 100ns/DIV

LTAN47 • TA04

**Figure 1-55**  Display when generator is unterminated (Linear Technology, Application Note 47, p. 7).

with the shortest possible lead lengths. (Never use wire-wound or the so-called "noninductive" types.) The ground end of a terminating resistor should be placed so that the currents flowing from the termination do not disrupt circuit operation. For example, do not return the terminator current to ground near the grounded positive (noninverting) input of an inverting op amp. Typical 5-V pulses through a 50-$\Omega$ terminating resistor produce 100-mA current spikes that could upset the desired 0-volt op-amp reference.

### 1.6.3    Partial ringing

Figure 1-57 shows the display when the ground lead of the scope probe is too long. Keep the probe ground connection as short as possible (preferably less than one inch for higher frequencies). If practical, use scope probes that mate directly to board-mounted coax connectors. When troubleshooting a previously good circuit, check the ground connections, particularly near the amplifier inputs. Look for breaks in the ground plane (if any).

### 1.6.4    Poor probe compensation

Figure 1-58 shows the display when the scope probe is properly grounded, but is not properly compensated (or, even worse, does not match the scope characteristics). Always use the probe recommended by the scope manufacturer and check probe compensation frequently. When troubleshooting a previously good circuit, check for leaking capacitors at the amplifier output.

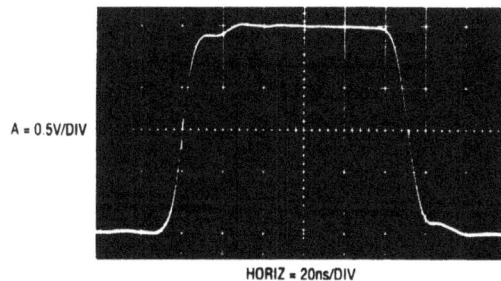

A = 0.5V/DIV

HORIZ = 20ns/DIV

**Figure 1-56**  Display when generator has poor-quality termination (Linear Technology, Application Note 47, p. 8).

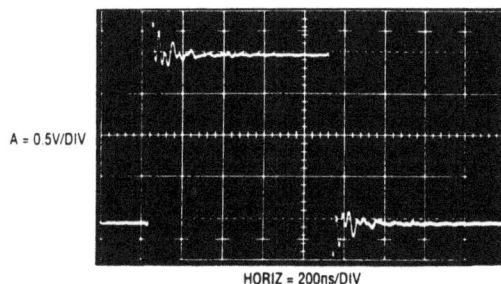

A = 0.5V/DIV

HORIZ = 200ns/DIV

LTAN47 · TA06

**Figure 1-57**  Display when probe ground lead is too long (Linear Technology, Application Note 47, p. 8).

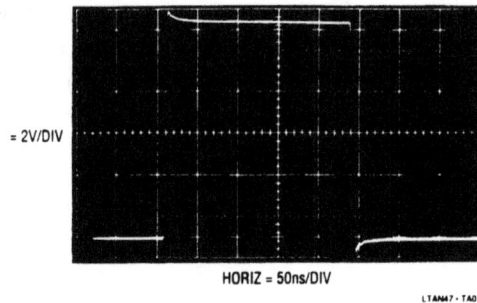

**= 2V/DIV**

**HORIZ = 50ns/DIV**

LTAN47 - TA07

**Figure 1-58**  Display when probe is not properly compensated (Linear Technology, Application Note 47, p. 9).

### 1.6.5  Decreased bandwidth

Figure 1-59 shows the display when the scope probe is heavily compensated (or is simply too slow for the scope), resulting in decreased bandwidth. Be sure that the probe bandwidth is far greater than the measurement frequency. A typical 1× or "straight" probe has a bandwidth of 20 MHz (or less) and produces a large capacitive load. While on the subject of bandwidth, it is assumed that the scope bandwidth is also far greater than the highest frequency used during the test. When troubleshooting a previously good circuit, check for anything that would increase capacitance at the amplifier output.

### 1.6.6  Different delay times

Figure 1-60 shows the display when probes at the input and output of an amplifier have different delay times. This difference in delay is added to the normal amplifier delay, and it produces what appears to be excessive delay. In Fig. 1-60, the delay between the input (trace A, left side) and the output (trace B) is about 12 ns. The amplifier is supposed to have a delay of 6 ns. The apparent excess is because the output probe has a delay of 9 ns, with a 3-ns delay for the input probe. One way to eliminate the problem is to measure the delay of both probes and factor in the difference when interpreting the display. Try connecting both probes to the same point simultaneously. Although it might not shown the true delay of the probe (or the circuit), it will show the difference in delay between the probes. As a point of reference, an active probe (such as an FET probe) and a current probe can have delays of 25 ns. Delay for a fast 10 × or 50-Ω probe is often less than 3 ns. So, do not condemn the circuit or IC as having too much delay until you have checked the probes.

### 1.6.7  Overdriven probes

Figure 1-61 shows the display when an FET probe is overdriven. This causes what appears to be severe distortion in the amplifier. In reality, the common-mode input range of the probe is exceeded, causing the probe to overload. In some cases, an overdriven FET probe will produce excessive delays, but without the obvious distortion. In general, do not use FET probes to monitor signal voltages greater than about ±1 V. Use 10 × and 100 × attenuator probes when

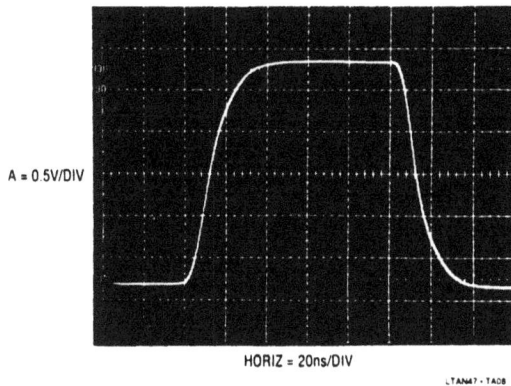

A = 0.5V/DIV

HORIZ = 20ns/DIV

LTAN47 - TA08

**Figure 1-59**   Display when probe
is too heavily compensated or
slow (Linear Technology,
Application Note 47, p. 9).

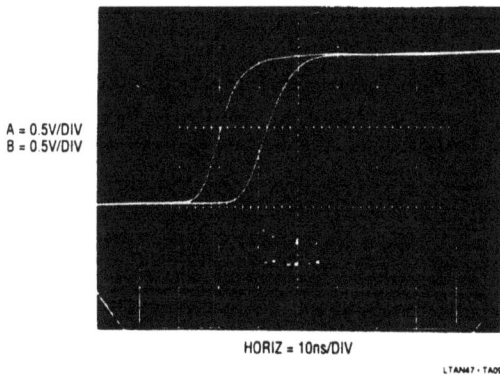

A = 0.5V/DIV
B = 0.5V/DIV

HORIZ = 10ns/DIV

LTAN47 - TA09

**Figure 1-60**   Display when
probes have different delay
times (Linear Technology,
Application Note 47, p. 9).

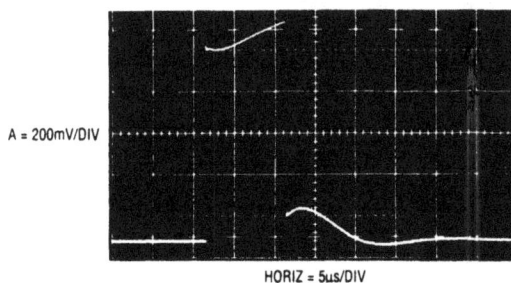

A = 200mV/DIV

HORIZ = 5us/DIV

**Figure 1-61**   Display when FET
probe is overdriven (Linear
Technology, Application Note
47, p. 9).

required. When troubleshooting a previously good circuit with a good probe,
look for a leaking (almost shorted) capacitor somewhere in the circuit.

## 1.6.8   Peaking and ringing

Figure 1-62 shows the display when scope-probe capacitance has caused what
appears to be peaking and ringing. In this case, the probe capacitance is only
10 pF, but the probe is connected to the summing point of an op amp, causing
a lag in feedback action. This forces the amplifier to overshoot and hunt as it

seeks the feedback null point. The problem can be minimized by monitoring with FET probes (or other low-capacitance probes). When troubleshooting a previously good circuit, look for anything that could have caused increased capacitance at the amplifier output.

### 1.6.9   Wideband problems

Figure 1-63 shows the display when the probe is not designed for wideband use. The waveform shown is the final 40 mV of a 2.5-V amplifier output swing. Instead of a sharp corner that settles cleanly, peaking occurs, followed by a long, trailing decay.

### 1.6.10   Overdriven scope

Figure 1-64 shows the display when the scope is overdriven. The waveform shown is the final movement of an amplifier output swing. Set the scope for 1 mV/division; the purpose of the measurement is to view the settling pattern (Fig. 1-47) with high resolution. Any observation that requires off-screen positioning of the display (where only a portion of the waveform is observed) should be approached with caution. It is possible that the scope will be overdriven under extreme conditions (high resolution, time expansion, etc.).

### 1.6.11   Defective ground plane

Figure 1-65 shows the display when the amplifier circuit is without a ground plane. This display is essentially the same as that for a probe with long

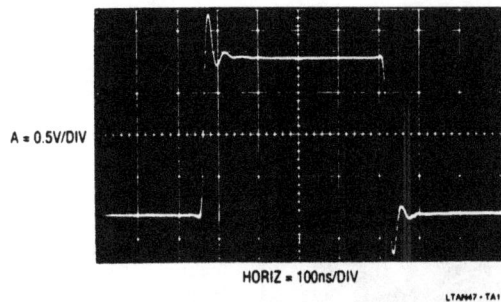

A = 0.5V/DIV

HORIZ = 100ns/DIV

LTAN47 - TA11

**Figure 1-62**  Display when probe capacitance causes peaking and ringing (Linear Technology, Application Note 47, p. 10).

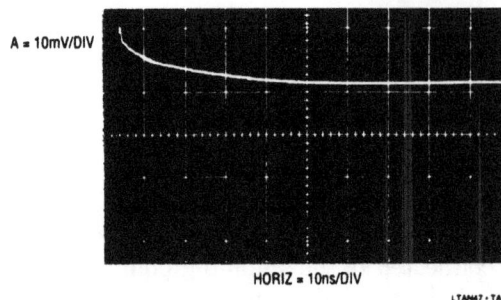

A = 10mV/DIV

HORIZ = 10ns/DIV

LTAN47 - TA12

**Figure 1-63**  Display when probe is not designed for wideband use (Linear Technology, Application Note 47, p. 10).

A = 1mV/DIV

HORIZ = 1µs/DIV

LTAN47 - TA13

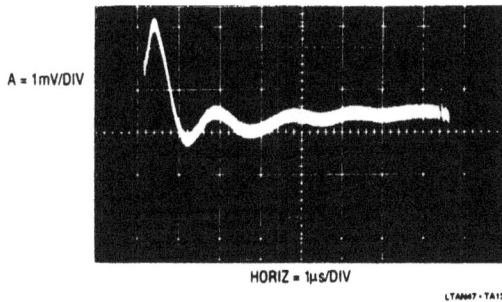

**Figure 1-64**  Display when scope is overdriven (Linear Technology, Application Note 47, p. 10).

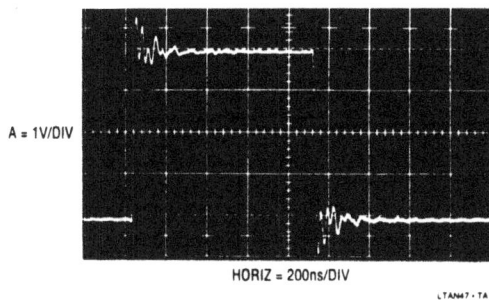

A = 1V/DIV

HORIZ = 200ns/DIV

LTAN47 - TA1

**Figure 1-65**  Display when amplifier circuit has no ground plane (Linear Technology, Application Note 47, p. 11).

ground connection (Fig. 1-57), or with breaks in the ground plane, as covered in Section 1.6.3.

## 1.6.12  Bypass problems

Figures 1-66 through 1-70 show the displays when an IC-amplifier power supply is not properly bypassed. This is a common problem. When troubleshooting any amplifier (either an experimental circuit or one that has operated properly in the past), always check the bypass capacitors. Bypassing is necessary to maintain low supply impedance. Dc resistance and inductance in the supply wires and PC traces can quickly build up to unacceptable levels. This allows the supply line to change when internal current levels of the IC change, resulting in all manner of improper operation!

The display of Fig. 1-66 is that of a completely unbypassed amplifier feeding a 100-Ω load. The supply seen at the IC terminals has high impedance at high frequencies. This impedance forms a voltage divider with the amplifier and load, allowing the supply to change as internal conditions change. This results in local feedback and oscillation.

Figure 1-67 shows the same amplifier, but without the load. This improves the oscillation problem, but still results in unacceptable ringing and overshoot.

Figure 1-68 shows the display where poor-quality bypass capacitors are used or where the capacitors are connected too far from the IC terminals.

Figure 1-69 shows the display where capacitors are connected in parallel to get better bypassing. Although this is acceptable, it must be done carefully.

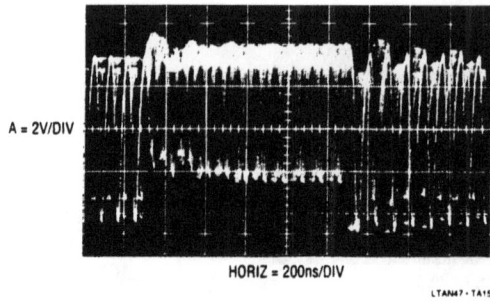

A = 2V/DIV

HORIZ = 200ns/DIV

LTAN47 · TA15

**Figure 1-66**   Display when power supply is not properly bypassed (Linear Technology, Application Note 47, p. 11).

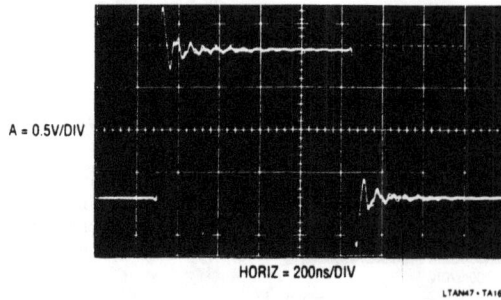

A = 0.5V/DIV

HORIZ = 200ns/DIV

LTAN47 · TA16

**Figure 1-67**   Display when power supply is not properly bypassed, but there is no load (Linear Technology, Application Note 47, p. 11).

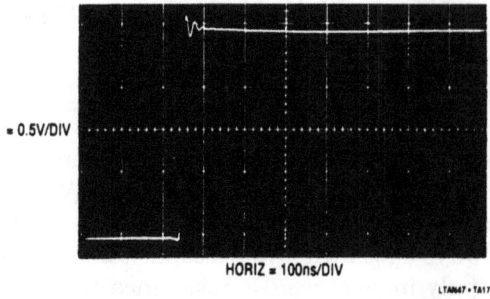

= 0.5V/DIV

HORIZ = 100ns/DIV

LTAN47 · TA17

**Figure 1-68**   Display when poor-quality bypass capacitors are used (Linear Technology, Application Note 47, p. 12).

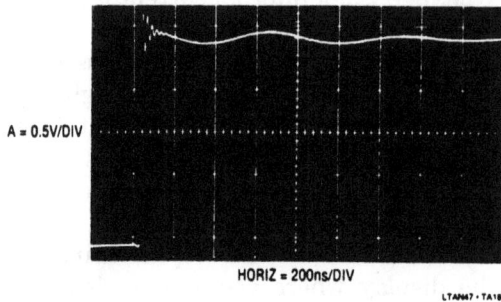

A = 0.5V/DIV

HORIZ = 200ns/DIV

LTAN47 · TA18

**Figure 1-69**   Display when PC traces are too long between parallel capacitors (Linear Technology, Application Note 47, p. 12).

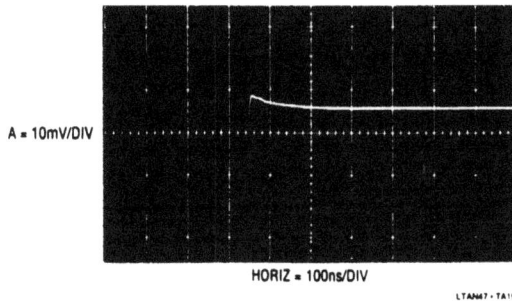

A = 10mV/DIV

HORIZ = 100ns/DIV

LTAN47 - TA19

**Figure 1-70**  Display when bypass capacitance value is not adequate (Linear Technology, Application Note 47, p. 12).

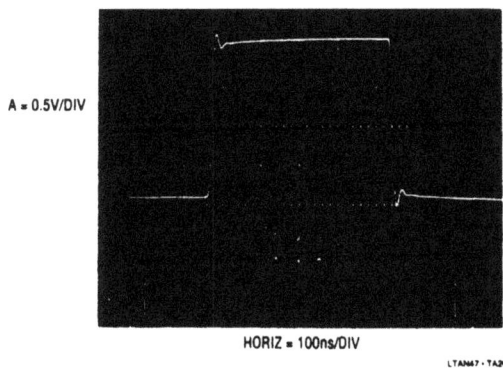

A = 0.5V/DIV

HORIZ = 100ns/DIV

LTAN47 - TA20

**Figure 1-71**  Display when stray capacitance is at the amplifier summing point (Linear Technology, Application Note 47, p. 12).

One common problem is where the PC traces are too long between the parallel capacitors. The capacitors and traces form a resonant circuit and produce the multiple-time-constant ringing shown in Fig. 1-69.

Figure 1-70 shows the display where good bypass capacitors are properly connected at the supply terminals of the IC, but the capacitance value is not adequate. The display of Fig. 1-70 shows the last 40 mV of a 5-V step, almost settling cleanly in 300 ns. The slight overshoot is because of a loaded (500 Ω) amplifier without quite enough bypassing. Increasing the total supply bypassing from 0.1 μF to 1 μF cured the problem. Always use large-value paralleled bypass capacitors when very fast settling is required, particularly if the amplifier is heavily loaded or sees fast load steps.

### 1.6.13  Stray capacitance

Figure 1-71 shows the display when stray capacitance (only 2 pF) at the op-amp summing point causes peaking at both the leading and trailing edges of the output waveform. This can be eliminated with a proper board layout. Minimize the trace area and trace capacitance at crucial points in the circuit. Consider layout as an integral part of the circuit and plan accordingly. Look for layout problems when troubleshooting a circuit that shows repeated failure in the field.

### 1.6.14  Noise problems

Figure 1-72 shows the display when noise is getting into the circuit (for example, from a digital clock or switching regulator), making it appear that the amplifier output is suffering from some form of oscillation. Try to eliminate (or minimize) radiated or conducted noise signals with proper layout and shielding.

### 1.6.15  Stray capacitance and radiated noise

Figure 1-73 shows the combined effects of stray capacitance and radiated noise. The output was taken from a gain-of-10 inverter with a 1-kΩ input resistance. The severe peaking is induced by a 1-pF stray or parasitic capacitance across the resistor terminals—even though the input signal source is terminated in 50 Ω and provides only 20 mV of drive. This problem was cured with a ground-referred shield at a right angle to (and encircling) the 1-kΩ resistor.

### 1.6.16  Compensation problems

Figure 1-74 shows the display when an amplifier is not compensated (no lead or lag compensation) and is operated with low gain (generally to increase speed). The price for the increased speed of an amplifier without compensation is a restriction on minimum allowable gain. Such amplifiers are not stable below some specified minimum gain and can easily break into oscillation, as shown. Always consider gain when troubleshooting an amplifier without compensation.

### 1.6.17  Excessive capacitive loading

Figure 1-75 shows the display when excessive capacitive loading is on an amplifier. Capacitive loading to ground introduces lag in the feedback-signal return path to the input. If enough lag is introduced (with a large capacitive load), the amplifier might oscillate, as shown. Always look for capacitive-loading problems when there is excessive oscillation, as shown. With a new or experimental circuit, check the performance margins of the IC amplifier with capacitive loads.

### 1.6.18  Excessive input-drive signal

Figure 1-76 shows the display when the input drive signal exceeds the common-mode input limit or range. In a new or experimental circuit, always keep the input less than the specified common-mode limits.

### 1.6.19  Feedback booster problems

Figure 1-77 shows the display when an external booster circuit is placed inside the feedback loop of an IC amplifier and problems are in the booster. In this

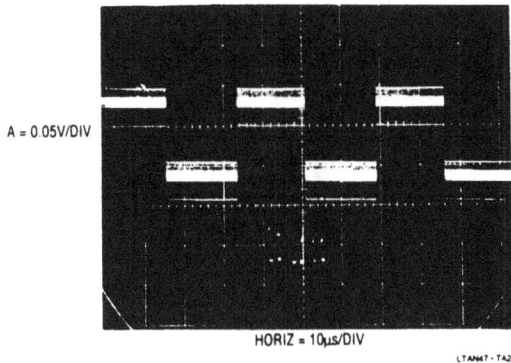

A = 0.05V/DIV

HORIZ = 10μs/DIV

Figure 1-72  Display when noise is getting into the amplifier circuit (Linear Technology, Application Note 47, p. 13).

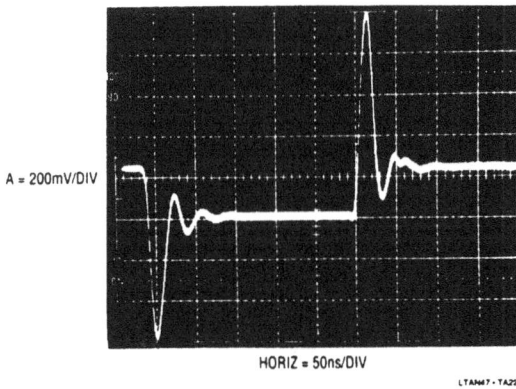

A = 200mV/DIV

HORIZ = 50ns/DIV

Figure 1-73  Display when there is stray capacitance and radiated noise (Linear Technology, Application Note 47, p. 13).

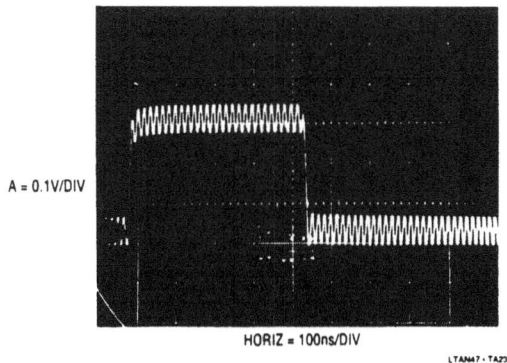

A = 0.1V/DIV

HORIZ = 100ns/DIV

Figure 1-74  Display when amplifier is not compensated (Linear Technology, Application Note 47, p. 13).

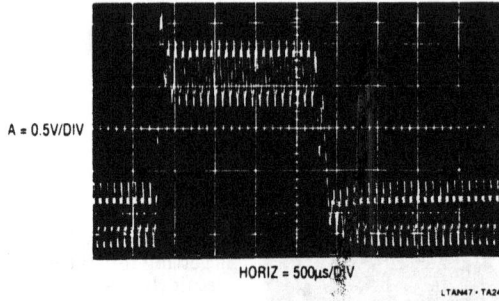

A = 0.5V/DIV

HORIZ = 500µs/DIV

LTAN47 · TA24

**Figure 1-75** Display when excessive capacitive loading is on the amplifier (Linear Technology, Application Note 47, p. 13).

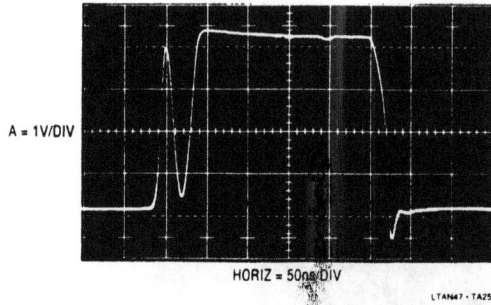

A = 1V/DIV

HORIZ = 50ns/DIV

LTAN47 · TA25

**Figure 1-76** Display when input signal exceeds common-mode limit (Linear Technology, Application Note 47, p. 14).

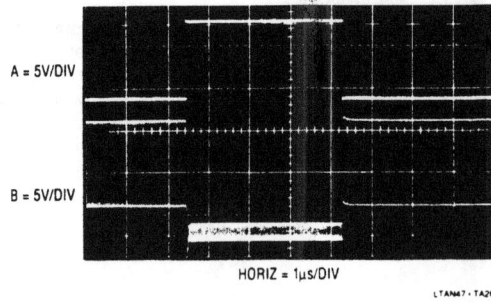

A = 5V/DIV

B = 5V/DIV

HORIZ = 1µs/DIV

LTAN47 · TA26

**Figure 1-77** Display when there are problems in a booster feedback loop (Linear Technology, Application Note 47, p. 14).

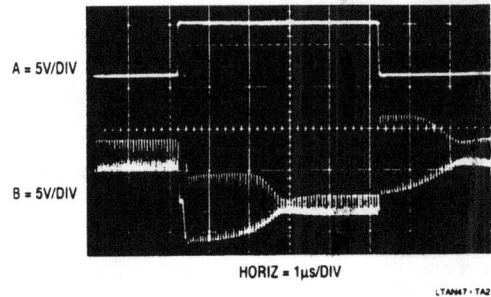

A = 5V/DIV

B = 5V/DIV

HORIZ = 1µs/DIV

LTAN47 · TA27

**Figure 1-78** Display when excessive lag is in a booster feedback loop (Linear Technology, Application Note 47, p. 14).

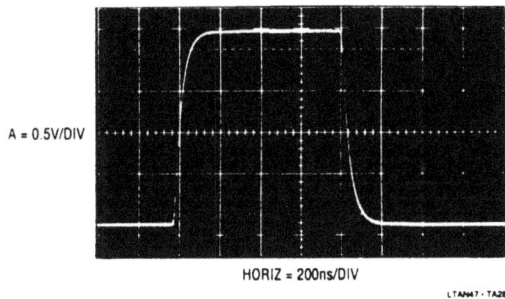

A = 0.5V/DIV

HORIZ = 200ns/DIV

LTAN47 - TA28

**Figure 1-79**  Display when a high source impedance limits bandwidth (Linear Technology, Application Note 47, p. 14).

example, the output of a unity-gain inverter is apparently breaking into oscillation (a rare occurrence). In reality, the oscillation occurs in the external booster and is fed back to the IC amplifier. Be certain that any external booster stages are stable before you condemn the IC amplifier. Wideband booster stages are particularly prone to high-frequency oscillation (usually caused by parasitic capacitance).

### 1.6.20  External-booster lag problems

Figure 1-78 shows another display where oscillation is produced by an external booster. However, in this example, the booster circuit is not oscillating, but is so slow that it introduces enough lag in the feedback to produce oscillation. Be certain that booster stages are fast enough to maintain stability when placed in the feedback loop of the IC amplifier.

### 1.6.21  High source-impedance problems

Figure 1-79 shows the display when a high source impedance combines with amplifier input capacitance to band-limit the circuit. When troubleshooting such a problem, always look for stray capacitance at the input (which can generally be avoided by good circuit layout).

# 2

# OTA Circuit Troubleshooting

This chapter is devoted to troubleshooting for *OTAs (operational transconductance amplifiers)*. Again, we start with a detailed description of a typical OTA, emphasizing how circuit functions are related to troubleshooting. It then describes why such circuits fail in either the experimental or final form. An OTA is similar to the op amps described in Chapter 1. However, OTAs and op amps are not always interchangeable. For that reason, an explanation of unique characteristics found in OTAs is in order.

From a troubleshooting standpoint, the OTA not only includes the differential inputs of an op amp, but also contains an additional control input in the form of an amplifier bias current ($I_{ABC}$). This control increases flexibility of the OTA for use in a wide range of applications, but it also introduces some troubleshooting problems. Another major difference is that the OTA has an extremely high output impedance (unlike the low output impedance of an op amp). When troubleshooting OTA circuits, the output signal is best thought of as a current that is proportional to the difference between the voltages at the two inputs (inverting and noninverting).

## 2.1 Basic OTA Circuit

Figure 2-1 shows a simplified circuit diagram of an OTA (one of three identical circuits in the Harris CA3060). The circuit output signal is a "current" that is proportional to the OTA transconductance (established by the $I_{ABC}$) and the differential input voltage. The OTA can either source or sink current at the output, depending on the polarity of the input signal.

In troubleshooting, the OTA transfer characteristics (or input/output relationship) are best defined in terms of transconductance, rather than voltage gain. Transconductance (often listed as *gm* or *g21*) is the ratio between the difference in current output ($I_{out}$) for a given difference in voltage input ($E_{in}$). Except for the high output impedance and the definition of input/output

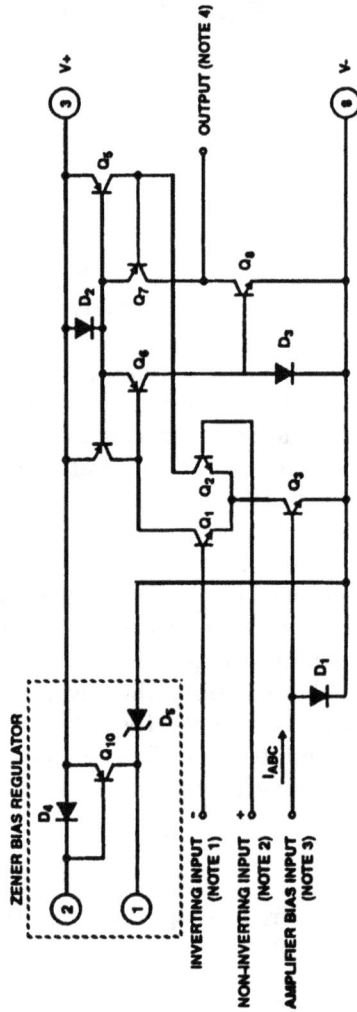

**Figure 2-1.** Simplified diagram of OTA with bias regulator (Harris Semiconductor, *Linear & Telecom ICs*, 1994, p. 2-54).

NOTES:

1. Inverting Input of Amplifiers 1, 2 and 3 is on Terminals 13, 12 and 4, respectively.
2. Non-Inverting Input of Amplifiers 1, 2 and 3 is Terminals 14, 11 and 5, respectively.
3. Amplifier Bias Current of Amplifiers 1, 2 and 3 is on Terminals 15, 10 and 6, respectively.
4. Output of Amplifiers 1, 2 and 3 is on Terminals 16, 9 and 7, respectively.

relationships, the basic troubleshooting approach for OTA circuits is similar to that of a typical op amp.

## 2.2  Definition of OTA Terms

The following is a summary of terms commonly found in OTA literature:

- *Amplifier bias current* $(I_{ABC})$ is the current supplied to the amplifier bias terminal to establish the operating point (such as the $I_{ABC}$ current at the base of Q3 in Fig. 2-1).

- *Amplifier supply current* $(I_A)$ is the current drawn by the amplifier from the positive supply source. The *total supply current* (which includes the sum of the amplifier supply current, the amplifier bias current, and the regulator bias current) is not to be mistaken for the amplifier supply current.

- *Bias regulator current* is the current flowing from the zener bias regulator (such as at terminal 2 of Fig. 2-1) set by an external source, which establishes the operating conditions of the bias regulator.

- *Bias terminal voltage* $(V_{ABC})$ is the voltage existing between the amplifier bias terminal and the negative supply-voltage terminal (such as between the $I_{ABC}$ terminal and terminal 8 of Fig. 2-1).

- *Peak output current* $(I_{OM})$ is the maximum current drawn from a short circuit of the amplifier output (positive output current) or the maximum current delivered into a short-circuit load (negative output current). *Peak-to-peak current swing* is twice the peak output current.

- *Peak output voltage* $(V_{OM})$ is the maximum positive voltage swing $(V_{OM}+)$ or the maximum negative voltage swing $(V_{OM}-)$ for a specific supply voltage and amplifier bias.

- *Power consumption* $(P)$ is the product of the sum of the supply voltage and the supply current $(V+$ plus $V-)$ times $I_A$. This is not the total power consumed by an operating circuit. The power in the regulator must also be included for total power consumed.

- *Zener regulator voltage* $(V_Z)$ is the regulator voltage (such as across terminals 1 and 8 of Fig. 2-1), measured with current flowing in the bias regulator.

## 2.3  Effects of $I_{ABC}$ on Circuit Troubleshooting

One major difference in troubleshooting OTA circuits is that the characteristics of OTAs can be altered by adjusting $I_{ABC}$ (unlike op amps). In effect, many of the OTA characteristics are programmed (by adjusting $I_{ABC}$) to meet specific circuit requirements. The following is a summary of the effects of $I_{ABC}$ on typical OTA circuits. Notice that the characteristics listed here for OTAs are the same as for op amps.

*Input offset current* (Section 1.4.10) is directly affected by $I_{ABC}$, and increases almost in direct proportion with increases in $I_{ABC}$. The same is essentially true for input bias current, amplifier supply current, device dissipation, transconductance, and peak output current.

*Input offset voltage* (Section 1.4.9) is not drastically affected by variations in $I_{ABC}$. A possible exception is when the OTA is operated at high temperatures. The same is essentially true of *peak output voltage,* which is set (primarily) by supply voltage, as is the case with a conventional op amp.

*Input and output capacitance,* as well as *amplifier bias voltage,* increase with $I_{ABC}$ but not in direct proportion. That is, a large increase in $I_{ABC}$ produces a small increase in input/output capacitance and amplifier bias voltage.

*Input and output resistances* both decrease with increases in $I_{ABC}$.

## 2.4   Effect of Circuit Components on OTA Characteristics

Figure 2-2 shows a basic OTA circuit with external components. The following paragraphs describe how the components affect circuit operation. These factors must be considered when troubleshooting an OTA circuit.

The circuit provides a closed-loop gain of 10 (20 dB), the input-offset voltage is adjustable to zero, the supply voltage is ±6 V, the maximum input voltage is ±50 mV, the input resistance is 20 kΩ, and the load resistance is 20 kΩ.

### 2.4.1   OTA transconductance

As in the case of a conventional op amp, closed-loop gain is set by the ratio of feedback resistance $R_F$ to input resistance $R_S$ in an OTA circuit. The circuit input resistance is 20 kΩ, as set by $R_S$. The 200-kΩ value of $R_F$ is 10 times that of $R_S$ to provide a gain of 10. In troubleshooting an OTA circuit where the only problem is incorrect gain, look for an improper $R_S/R_F$ ratio.

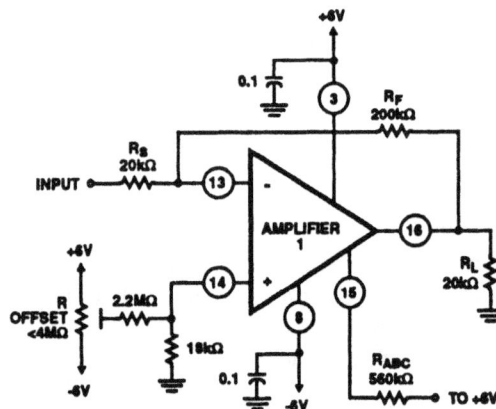

Figure 2-2.   Basic OTA circuit (20-dB amplifier) (Harris Semiconductor, *Linear & Telecom ICs,* 1994, p. 2-58).

If the $R_S/R_F$ ratio is correct, but the gain is low, it is possible that the OTA transconductance is not sufficient to provide the necessary open-loop gain ($A_{OL}$). Typically, the open-loop gain should be at least 10 times the closed-loop gain. In this circuit, with a closed-loop gain of 10, the open-loop gain must be $10 \times 10 = 100$.

Open-loop gain is related directly to load resistance $R_L$ and transconductance. However, the actual load resistance is the parallel combination of $R_L$ and $R_F$, about 18 kΩ for this circuit ($R_L \times R_F/R_L + R_F$). With an $A_{OL}$ of 100 and an actual load of 18 kΩ, the transconductance should be about $100/18,000 = 5.5$ millimho (mmho).

In troubleshooting, keep in mind that the transconductance is set by $I_{ABC}$. With a data-sheet curve similar to that of Fig. 2-3, notice that the minimum $I_{ABC}$ is about 20 μA for a transconductance of 5.5 mmho. $I_{ABC}$ is set by $R_{ABC}$, so, in an experimental circuit, adjust $R_{ABC}$ for the required circuit gain (both closed loop and open loop).

### 2.4.2   Output swing limitations

Before you make any changes to $R_{ABC}$, check that the resulting $I_{ABC}$ will provide the required output swing. For example, with an input of ±50 mV and a gain of 10, the output swing is ±0.5 V. This output appears across the output load of about 18 kΩ. With a 0.5-V swing and an approximate load of 18 kΩ, the total amplifier-current output is about 0.5/18 kΩ=27.7 μA.

In troubleshooting, keep in mind that $I_{ABC}$ also sets the peak output current ($I_{OM}$). With a data-sheet curve similar to that of Fig. 2-4, use the minimum-value curve to check that an $I_{ABC}$ of 20 μA produces an $I_{OM}$ of at least 27.7 μA. As shown in Fig. 2-4, an $I_{ABC}$ of 20 μA produces an $I_{OM}$ of about 40 μA, well above the required 27.7 μA.

### 2.4.3   Approximating $R_{ABC}$ values

In an experimental circuit, a simple method of calculating the value of $R_{ABC}$ will produce the required $I_{ABC}$ (for a given gain, output swing, etc.). However,

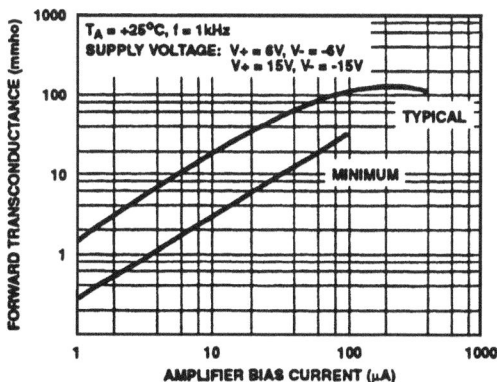

Figure 2-3.  Forward transconductance versus amplifier bias current (Harris Semiconductor, *Linear & Telecom ICs*, 1994, p. 2-56).

you must have a curve similar to that of Fig. 2-5. As shown in Fig. 2-2, $R_{ABC}$ is connected to the +6-V supply. With this arrangement, $R_{ABC}$ and diode D1 are in series between the +V and −V supplies, and a total of 12 V is across the series components (Fig. 2-1). The drop across D1, which is $V_{ABC}$, can be found by reference to Fig. 2-5. An $I_{ABC}$ of 20 μA produces a $V_{ABC}$ of about 0.63 V. The drop across $R_{ABC}$ is 12−0.63 V=11.37 V. For a drop of 11.37 V and an $I_{ABC}$ of 20 μA, the value of $R_{ABC}$ is 11.37/20=568 kΩ. The circuit of Fig. 2-2 uses the next lowest standard resistor of 560 kΩ to ensure that the minimum $I_{ABC}$ is 20 μA.

### 2.4.4   Offset problems (OTA)

The offset pot and fixed resistors at the noninverting input (pin 14) of the OTA in Fig. 2-2 provide for adjustment of any offset. The values for these resistors are selected to provide sufficient offset-adjustment range, but with minimum loading on the power supply. The following is a summary of the guidelines for offset-circuit values.

The value of the resistor between the input and ground should be about equal to the parallel combination of $R_F$ and $R_S$, about 18 kΩ.

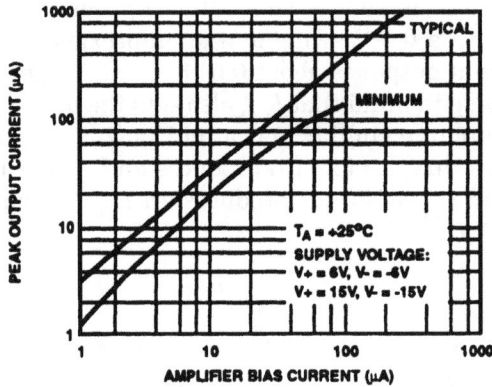

**Figure 2-4.** Peak output current versus amplifier bias current (Harris Semiconductor, *Linear & Telecom ICs,* 1994, p. 2-55).

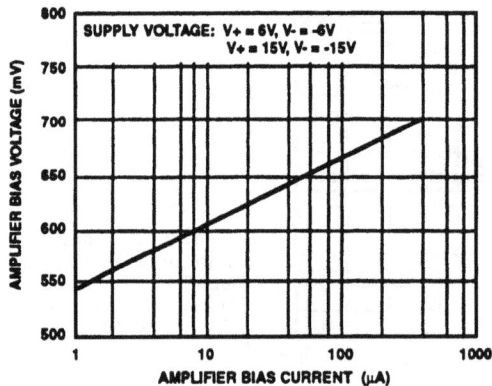

**Figure 2-5.** Amplifier bias voltage versus amplifier bias current (Harris Semiconductor, *Linear & Telecom ICs,* 1994, p. 2-56).

The curves of Fig. 2-6 show that for the $I_{ABC}$ of 20 μA, the input-offset current should be 200 nA. With 200 nA flowing through the input/ground (18 kΩ) resistor, the voltage across the resistor is 200 nA×18 kΩ=3.6 mV. This 3.6 mV must be added to the maximum input-offset voltage possible for the OTA. The data sheet shows a maximum input-offset of 5 mV. Thus, the maximum voltage required at the noninverting input is 5 mV+3.6 mV=8.6 mV.

The current necessary to provide a possible offset voltage of 8.6 mV across the 18-kΩ resistor is about 0.48 μA. This current must flow through the resistor between the input and the offset pot. A possible ±6 V is available from the offset pot to the resistor. However, for a more stable circuit, assume that ±1 V is available to the resistor. With 1 V available and a required current of 0.48 μA, the value of the resistor is about 2 MΩ. The circuit of Fig. 2-2 uses the next larger standard value of 2.2 MΩ.

In most OTA circuits, the value of the offset pot is not crucial, but a larger value draws less current from the supplies. As a guideline, the maximum value should be less than twice the value of the 2.2-MΩ resistor. The circuit of Fig. 2-2 uses 4 MΩ (a standard resistor value).

From a troubleshooting standpoint, the only practical test of the offset circuit is to produce zero output with the input shorted to ground. This should occur somewhere near the midrange of the offset pot. If the pot must be at an extreme to get zero output, the circuit values are not correct (or the IC is defective).

## 2.4.5  Stray capacitance (OTA)

OTA circuits are typically high impedance because OTAs operate at low power. This can create stray-capacitance problems, particularly when the OTA is used in a feedback circuit. For example, a 10-kΩ load with a stray capacitance of 15 pF has a time constant of 1 MHz. Figure 2-7 shows how a 10-kΩ/15-pF load modifies the frequency characteristics of our OTA. With no capacitive loading, the relative gain is −40 dB at 10 MHz. If stray capacitance produces a total load capacitance ($C_L$) of 15 pF, the relative gain drops to −60 dB at 10 MHz. In

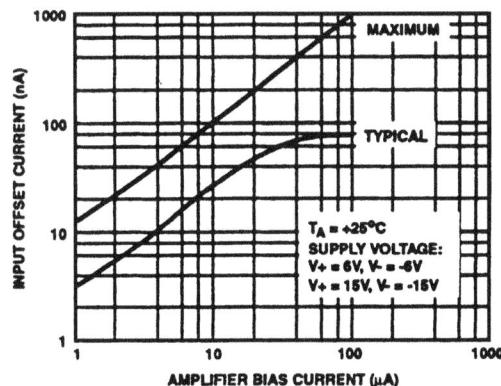

Figure 2-6.  Input offset current versus amplifier bias current (Harris Semiconductor, *Linear & Telecom ICs*, 1994, p. 2-55).

troubleshooting, always look for board layout problems that might increase stray capacitance, as covered in Section 1.6.13 (and shown in Fig. 1-71).

OTA slew rate is also affected by capacitive loading. Because the peak output current is established by the amplifier bias current (Fig. 2-4), the maximum slew rate is limited to the maximum rate at which the capacitance can be charged by the $I_{OM}$. Slew rate equals $dV/dt = I_{OM}/C_L$, where $C_L$ is the total load capacitance, including strays. This relationship is shown graphically in Fig. 2-8.

### 2.4.6  Phase compensation for OTAs

Most OTA circuits do not have external phase compensation. However, compensation can be accomplished with an $R_C$ network at the circuit input, as shown in Fig. 2-9. The values shown provide stable operation for the unity-gain condition, assuming that the capacitive loading on the output is 13 pF or less. (Actually, the circuit of Fig. 2-9 is a slew-rate test circuit for this OTA). From a troubleshooting standpoint, notice that slew rate increases when the resistance values in Fig. 2-9 decrease.

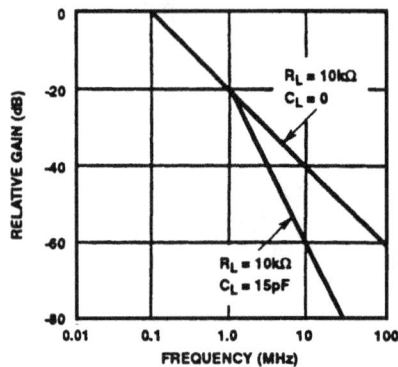

**Figure 2-7.**  Effect of capacitive loading on frequency response (Harris Semiconductor, *Linear & Telecom ICs*, 1994, p. 2-59).

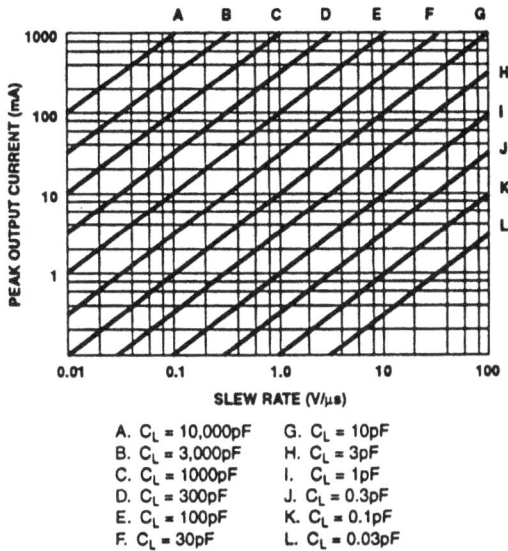

**Figure 2-8.** Effect of load capacitance on slew rate (Harris Semiconductor, *Linear & Telecom ICs,* 1994, p. 2-59).

A. $C_L$ = 10,000pF    G. $C_L$ = 10pF
B. $C_L$ = 3,000pF     H. $C_L$ = 3pF
C. $C_L$ = 1000pF      I. $C_L$ = 1pF
D. $C_L$ = 300pF       J. $C_L$ = 0.3pF
E. $C_L$ = 100pF       K. $C_L$ = 0.1pF
F. $C_L$ = 30pF        L. $C_L$ = 0.03pF

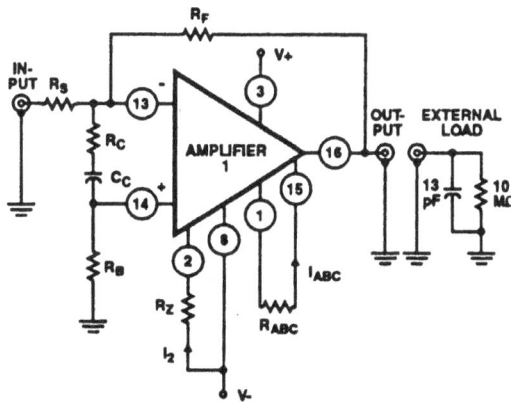

$V_Z$ is measured between Terminal 1 and 8
$V_{ABC}$ is measured between Terminals 15 and 8

$$R_Z = \frac{[\,(V+) - (V-) - 0.7\,]}{I_2}, \quad R_{ABC} = \frac{V_Z - V_{ABC}}{I_{ABC}}$$

Supply Voltage: For both ±6V and ±15V

**Figure 2-9.** Slew rate test circuit input phase compensation (Harris Semiconductor, *Linear & Telecom ICs,* 1994, p. 2-57).

### TYPICAL SLEW RATE TEST CIRCUIT PARAMETERS

| $I_{ABC}$ | SLEW RATE | $I_2$ | $R_{ABC}$ | $R_S$ | $R_F$ | $R_B$ | $R_C$ | $C_C$ |
|---|---|---|---|---|---|---|---|---|
| μA | V/μs | μA | Ω | Ω | Ω | Ω | Ω | μF |
| 100 | 8 | 200 | 62k | 100k | 100k | 51k | 100 | 0.02 |
| 10 | 1 | 200 | 620k | 1M | 1M | 510k | 1k | 0.005 |
| 1 | 0.1 | 2 | 6.2M | 10M | 10M | 5.1M | ∞ | 0 |

# Current-Feedback Amplifier-Circuit Troubleshooting

This chapter is devoted to troubleshooting for current-feedback amplifiers, sometimes called *Norton amplifiers, CFAs,* or *CFBs,* depending on which literature you read. Here, they are called *CFAs* for simplicity. This chapter starts with a detailed description of a typical CFA, emphasizing how circuit functions are related to troubleshooting. Then follows a description of why such circuits fail in either the experimental or final form. CFAs are similar to OTAs (Chapter 2) in that their characteristics are controlled by an external current or voltage. However, the internal circuits and functions of CFAs are quite different from those of OTAs (and from op amps).

When troubleshooting CFAs, keep in mind that there is no separate pin for control current. Any change in amplifier characteristics is set by current that is applied at the $(+)$ noninverting and $(-)$ inverting inputs. Some (but not all) manufacturers use a modified amplifier symbol, such as shown in Fig. 3-1 (which also shows a simplified version of the classic National Semiconductor LM3900). The current arrow between inputs implies a current mode of operation. The symbol also signifies that current is removed from the $(-)$ input and that the $(+)$ input is a current input (which can control amplifier gain). The signal can be applied at either the $(+)$ or $(-)$ inputs.

## 3.1 Basic CFA Circuit

The circuit shown in Fig. 3-1 is for one of four amplifiers, all fabricated on a single IC chip. One common biasing circuit is used for all four amplifiers. The bias reference for the PNP current source $(V_p)$, which biases Q1, is designed to cause the upper current source $(200 \ \mu A)$ to change with temperature and thus provide compensation for beta variation in NPN output transistor Q3. The bias reference for the NPN pull-down current sink $(V_h$, which biases Q7) is designed to stabilize this current (1.3 mA) to reduce the variation when the temperature changes. This provides a more constant pull-down capability for the amplifier over the temperature range. Transistor Q4 provides the class-B action that exists under large-signal operating conditions.

## 3.2 Typical CFA Characteristics

The following is a summary of performance characteristics for the CFA of Fig. 3-1. Figure 3-2 shows open-loop gain characteristics of the CFA (compared with those of the classic 741 op amp).

| | |
|---|---|
| Bias current drain per amplifier stage | 1.3 mA dc |
| Power-supply voltage range | 4 to 36 Vdc or ±2 to ±18 Vdc |
| Open loop: | |
|    Voltage gain ($R_L = 10$ k$\Omega$) | 70 dB |
|    Unity-gain frequency | 2.5 MHz |
|    Phase margin | 40° |
|    Input resistance | 1 M$\Omega$ |
|    Output resistance | 8 k$\Omega$ |
| Output voltage swing | $(V_{CC} - 1)\ V_{pp}$ |
| Input bias current | 30 nA dc |
| Slew rate | 0.5 V/$\mu$s |

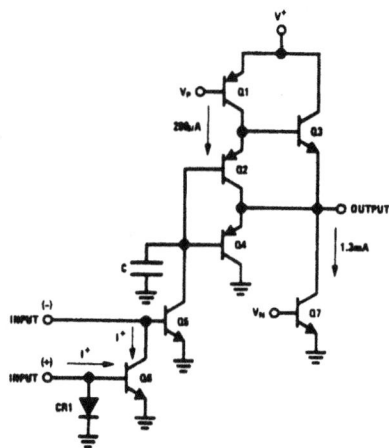

TL/H/7383-4

(a) Circuit Schematic

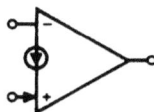

TL/H/7383-5

(b) New "NORTON" Amplifier Symbol

Figure 3.1. Basic current-feedback (Norton) amplifier (National Semiconductor, *Linear Applications Handbook,* 1994, p. 171).

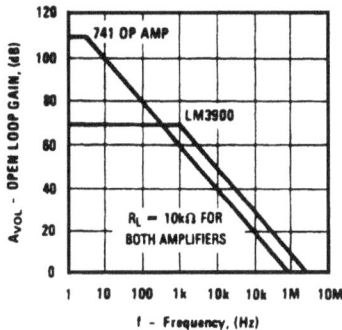

**Figure 3.2.** Open-loop gain characteristics (National Semiconductor, *Linear Applications Handbook,* 1994, p. 172).

TL/H/7383-6

Because the bias currents are all taken from diode forward-voltage drops, only a small change in bias-current magnitude occurs when the power-supply voltage is varied. The open-loop gain changes only slightly over the supply-voltage range and is essentially independent of temperature changes. As shown in Fig. 3-2, the LM3900 provides an additional 10-dB gain for all frequencies greater than 1 kHz (when compared with the 741 op amp).

## 3.3  Basic CFA Testing and Troubleshooting

When troubleshooting CFAs, two input parameters are of particular importance: the input bias current, $I_{BIAS}$, and the mirror gain constant, $A_I$. (The mirror gain is especially important when a CFA is used as a voltage follower.) These parameters cannot be measured in the same way as for op amps or OTAs (there is no equivalent to $A_I$ in op amps and OTAs). Therefore, before covering CFA troubleshooting, the two parameters are examined to see how they are tested and why they are important in troubleshooting.

Figures 3-3 and 3-4 show simplified versions of two classic CFAs (the National Semiconductor LM3900 and LM359, respectively). Notice that the manufacturer describes these ICs as Norton amplifiers.

The circuit of Fig. 3-3 is essentially a common-emitter amplifier (Q3) with an emitter-follower output stage. Added to the base of Q3 is a current mirror (Q1/Q2). If a fixed current is injected into the (+) input and the output is fed back to the (−) input, the output rises until the current in Q2 matches that flowing in Q1. The currents at the input terminals are not equal because some current ($I_{BIAS}$) flows into the base of Q3, and there might be some mismatch in Q1 and Q2. This is quite noticeable when the mirror current is in the 1- to 10-$\mu$A range. The degree of matching is called *mirror gain* $A_I$ and is ideally equal to 1.

The circuit of Fig. 3-4 (LM359) differs from the LM3900 in that Q3 of the LM359 is a cascade stage and Q4 is a Darlington follower. In addition, the internal biasing is variable (the set current, $I_{SET}$, is determined by an external resistor). The gain-bandwidth product, slew rate, input noise, output drive current, input bias current and supply current all vary with $I_{SET}$.

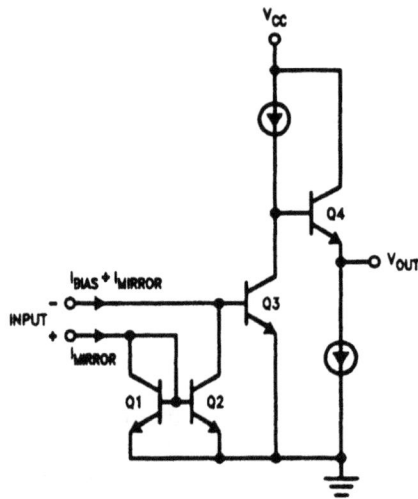

**Figure 3.3.** Simplified schematic of LM3900 (National Semiconductor, *Linear Applications Handbook,* 1994, p. 1138).

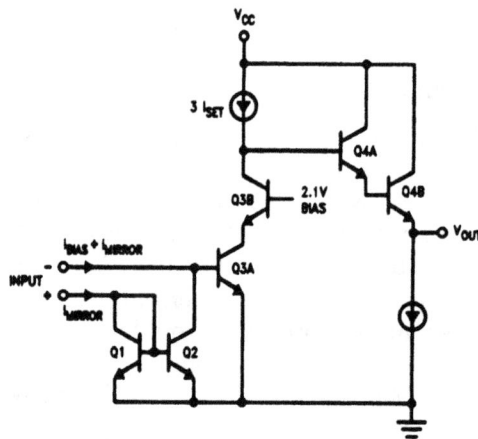

**Figure 3.4.** Simplified schematic of LM359 (National Semiconductor, *Linear Applications Handbook,* 1994, p. 1138).

### 3.3.1  $I_{BIAS}$ test circuits

Figures 3-5 and 3-6 show test circuits for measurement of $I_{BIAS}$ in the LM3900 and LM359, respectively. In the circuit of Fig. 3-5, two voltage measurements are made at the output, one with S1 closed and one with S1 open. The output-voltage increase is equal to the voltage appearing across the 1-M$\Omega$ resistor, multiplied by the output gain, $A_V$. For the values shown, the output-voltage increase multiplied by 200 gives the bias current in nA, or:

$$I_{BIAS} \, (nA) = 200 \, \Delta V_{OUT} = \frac{(10^{-9})}{A_V \times 1 \, M\Omega} \Delta V_{OUT}$$

The circuit of Fig. 3-6 is essentially the same, except that $R_{SET}$ is added (to provide an $I_{SET}$ of 5 $\mu$A), and a 27-pF capacitor is added for circuit stability.

### 3.3.2   Mirror-gain test circuits

Figures 3-7 and 3-8 show test circuits to measure mirror gain $A_I$ in the LM3900 and LM359, respectively. In the circuit of Fig. 3-7, resistors R are selected to provide the desired mirror current ($I_{MIRROR}$) using the values shown. The voltage across each R is measured and the ratio of the two voltages is equal to $A_I$. Mirror gain is affected by $I_{BIAS}$. Where $I_{BIAS}$ is a significant part of the mirror current, the equation becomes:

$$A_I = \frac{(V_2) - R \times I_{BIAS}}{V_1}$$

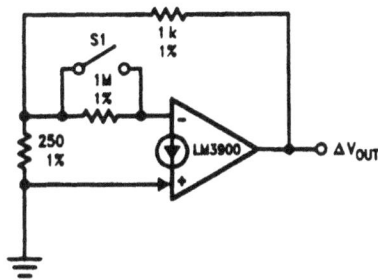

Figure 3.5.   Test circuit for LM3900 $I_{BIAS}$ (National Semiconductor, *Linear Applications Handbook,* 1994, p. 1138).

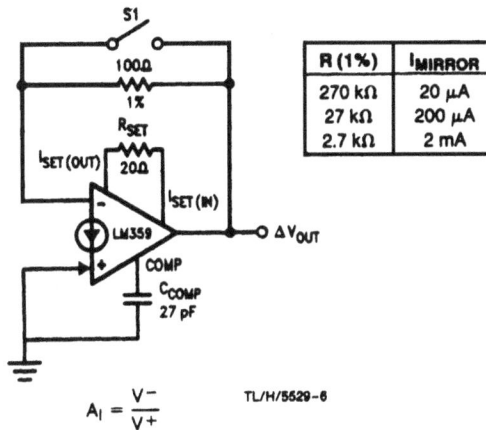

| R (1%) | $I_{MIRROR}$ |
|--------|--------------|
| 270 kΩ | 20 μA |
| 27 kΩ | 200 μA |
| 2.7 kΩ | 2 mA |

$$A_I = \frac{V-}{V+}$$

TL/H/5529-6

Figure 3.6.   Test circuit for LM359 $I_{BIAS}$ (National Semiconductor, *Linear Applications Handbook,* 1994, p. 1139).

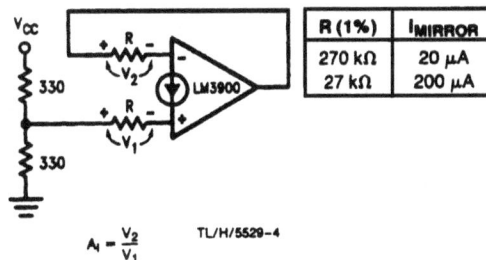

| R (1%) | $I_{MIRROR}$ |
|--------|--------------|
| 270 kΩ | 20 μA |
| 27 kΩ | 200 μA |

$$A_I = \frac{V_2}{V_1}$$

TL/H/5529-4

Figure 3.7.   Test circuit for LM3900 mirror gain, $A_I$ (National Semiconductor, *Linear Applications Handbook,* 1994, p. 1139).

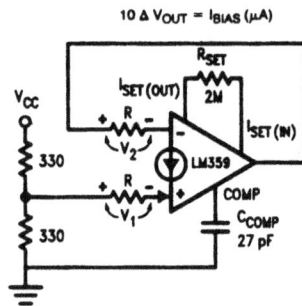

Figure 3.8.  Test circuit for
LM359 mirror gain, $A_I$
(National Semiconductor,
*Linear Applications Handbook,*
1994, p. 1139).

Again, the major difference between the circuits of Figs. 3-7 and 3-8 is the
addition of $R_{SET}$ and the compensating capacitor.

### 3.3.3  Test conditions for CFAs

Many of the LM359 data-sheet parameters, including $I_{BIAS}$, are measured with
an $I_{SET}$ of 0.5 mA. Three times the current flows in the collector of Q3A, making its bias current about 15 µA. The LM3900 has a corresponding Q3 collector current of about 3 µA and an $I_{BIAS}$ of 30 nA. (However, the LM3900 does not
have a 400-MHz gain-bandwidth product, as does the LM359!) The LM3900
mirror gain is measured with an $I_{SET}$ of 5 µA, making $I_{BIAS}$ so small that it has
little effect on the measurement.

In troubleshooting, $I_{BIAS}$ might be a significant part of the mirror current. This
makes the dc bias of CFA circuits a possible source of trouble. In some circuits,
the problem is minimized by making the mirror current at least one-third of $I_{SET}$.

In all of the test circuits, it is assumed that the $V_{CC}$ is 12 V. Test accuracy is
only as good as the resistors and meter used (which is generally the case!).
Matching is very important for the two resistors (R) in Figs. 3-7 and 3-8. A 1%
tolerance is recommended (although you can probably get by with 5% resistors, if they are sorted carefully for a close match). The feedback resistors in
Figs. 3-5 and 3-6 should also be 1%. In either testing or troubleshooting, standard $3^1/_2$-digit DVMs should have sufficient accuracy for voltage measurements. The meter input impedance should be at least 10 MΩ to prevent circuit
loading in the mirror-gain tests.

## 3.4  Inverting AC Amplifier Troubleshooting

Figure 3-9 shows a basic inverting ac amplifier circuit using a CFA with single-
supply biasing. In such circuits, $R_1$ and $R_2$ are selected to provide the desired
voltage gain ($A_V$), as if the IC is a conventional op amp. The third resistor at
the (+) input is equal in value to twice that of the feedback resistance, $R_2$ (in
most circuits).

When noise must be kept to a minimum, the (+) input is often grounded. In
troubleshooting such circuits, remember that the output will bias at a voltage
equal to the base-emitter voltage of the input transistor (typically about 0.5 V).

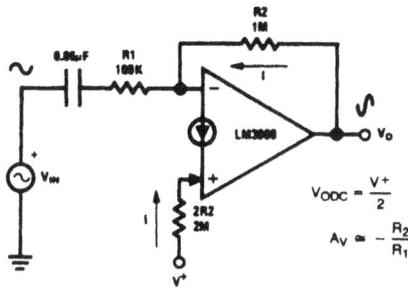

**Figure 3.9.** Inverting ac amplifier with single supply (National Semiconductor, *Linear Applications Handbook,* 1994, p. 175).

In the circuit of Fig. 3-9, the CFA is biased from the same power supply used to operate the IC amplifier. This can be a problem if ripple is present on the $V+$ supply line. Ripple can be coupled into the output with a "gain" of one-half. In troubleshooting such a problem, be certain that supply ripple is at a minimum (with proper bypassing, supply filtering, etc.). Also, if practical, use one source of ripple-filtered voltage for the IC and any other amplifiers in the system, as covered next in Section 3.5.

## 3.5   Noninverting AC Amplifier Troubleshooting

Figure 3-10 shows a basic noninverting ac amplifier circuit using a CFA with an alternate method of biasing. Again, the ac gain is set by the ratio of the feedback resistance to input resistance. The small-signal impedance of the diode at the $(+)$ input (Fig. 3-1) is added to the value of $R_1$ when setting gain, as shown by the equations in Fig. 3-10. By making $R_2$ equal to $R_3$, the output will be biased to the reference voltage applied to R2. The filtered $V+/2$ reference shown can also be used for other amplifiers.

Troubleshooting for this circuit is similar to that for the circuit of Fig. 3-9. However, if there is excessive ripple, look for an open capacitor (C) on the $V+$ line. If bias appears to be incorrect (with the correct values for all resistors),

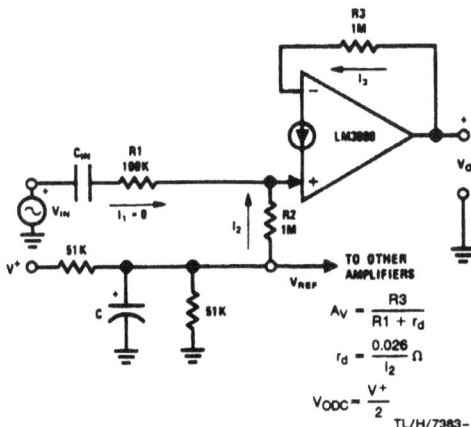

**Figure 3.10.** Noninverting ac amplifier with voltage-reference biasing (National Semiconductor, *Linear Applications Handbook,* 1994, p. 175).

TL/H/7363–15

look for a leaking capacitor (C). (If C is completely shorted, the $V+$ line voltages will change drastically.

## 3.6  Inverting AC Amplifier (NVBE Bias) Troubleshooting

Figure 3-11 shows a basic inverting ac amplifier circuit with what the manufacturer describes as "NVBE" biasing. The technique is most effective for inverting ac amplifier configurations. The input bias voltage, $V_{BE}$, at the $(-)$ input established a current through resistor R3 to ground. Because this current comes from the amplifier output, $V_O$ must rise to a level that causes the same current to flow through R2. The bias voltage ($V_{ODC}$) is calculated from the ratio of $R_2$ and $R_3$, as shown by the equations. When NVBE biasing is used, $R_1$ and $R_2$ are established first, then resistor R3 is added to provide the desired dc output voltage.

For example, assume that the desired input impedance is 1 MΩ, with a gain of 10 and a dc output bias of 7.5 V. By making $R_1=1$ MΩ, the input impedance is 1 MΩ. By making $R_2=10$ MΩ, the gain is 10. To bias the output voltage at 7.5 Vdc, $R_3$ is found by:

$$R_3 = \left( \frac{R_2}{\dfrac{V_O}{V_{BE}} - 1} \right) = \left( \frac{10 \text{ M}\Omega}{\dfrac{7.5}{0.5} - 1} \right) = \text{about } 714 \text{ k}\Omega$$

The circuit of Fig. 3-11 shows a trial value of 680 kΩ.

When troubleshooting this circuit, remember that ac gain is (or should be) independent of biasing, and is set by $R_1/R_2$. However, the dc level at the output is set by $R_3$.

## 3.7  Inverting AC Amplifier (Negative Supply) Troubleshooting

Figure 3-12 shows a basic inverting ac amplifier circuit with a negative supply. Notice that this circuit is very similar to that of Fig. 3-11, except that the dc

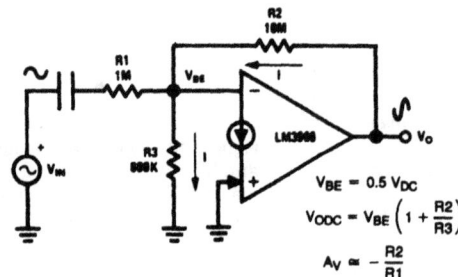

**Figure 3.11.**  Inverting ac amplifier with NVBE biasing (National Semiconductor, *Linear Applications Handbook,* 1994, p. 175).

biasing current ($I$) is established by the negative-supply voltage through resistor R3. This configuration provides a very stable output quiescent ($Q$) point for the amplifier. Also, $V_{BE}$ at the ($-$) can be ignored when calculating the output bias voltage ($V_{ODC}$).

Troubleshooting for this circuit is similar to that for the circuit of Fig. 3-11. Again, ac gain is set by $R_1/R_2$, and the dc output level is set by $R_3$.

## 3.8   High Input-Impedance and Gain

Figure 3-13 shows a basic inverting ac amplifier circuit with high input-impedance ($Z_{IN}$) and high gain ($A_V$). The circuits presented thus far in this chapter provide either high gain or high input impedance, but not both. If the input resistance is increased to get a high input-impedance value, the feedback resistance must be increased by a corresponding amount, for a given gain. This requires that the output voltage be set at a higher value to get the required feedback current.

In the circuit of Fig. 3-13, when the gain from the input is unity (when $R_1=R_3$), the gain of the complete stage is set by the voltage-divider network composed of R4, R5, and C2. When $R_5$ is decreased, the gain of the stage approaches the ac open-loop limit of the IC (about 70 dB with a 10-k$\Omega$ load). The insertion of capacitor C2 allows the dc bias to be controlled by the series combination of R3 and R4, with no effect from R5. As a result, R2 can be select-

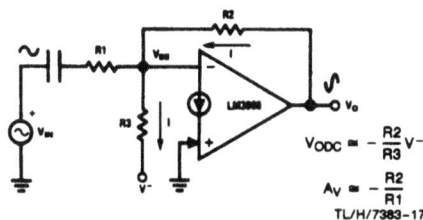

**Figure 3.12.**   Inverting ac amplifier with negative supply (National Semiconductor, *Linear Applications Handbook,* 1994, p. 176).

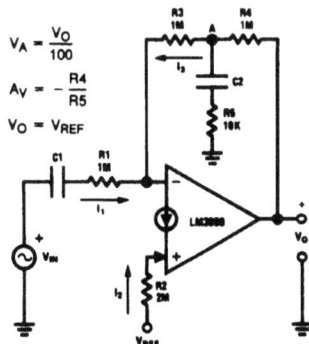

**Figure 3.13.**   Inverting ac amplifier with high input impedance (National Semiconductor, *Linear Applications Handbook,* 1994, p. 176).

ed to get the desired output biasing (using any of the methods described thus far). For reference, the circuit of Fig. 3-13 has a gain of 100 and an input impedance of 1 MΩ.

The main source of trouble in this circuit is capacitor C2. If C2 is leaking, even slightly, the gain will be incorrect, as will the input impedance. If C2 is shorted, open, or leaking badly, the circuit will be inoperative.

## 3.9    Gain-Control Circuit Troubleshooting

Figure 3-14 shows a basic circuit for a noninverting ac amplifier with a dc gain control. The amplifier output is kept from being driven into saturation when the dc gain control is varied by providing a minimum biasing current through R3. For maximum gain, CR2 is off, and both the currents through R2 and R3 enter the (+) input. This causes the amplifier output to bias at about 0.6 V+. For minimum gain, CR2 is on and only the R3 current enters the (+) input, thus biasing the output at about 0.3 V+. The dc gain-control input ranges from 0 Vdc for minimum gain to less than 10 Vdc for maximum gain. The proper output bias for large-output signals is provided for the maximum-gain situation.

The main sources of trouble in this circuit are diodes CR1 and CR2. If either is leaking (both forward leakage and reverse leakage), gain-control operation will be impaired. Of course, if either diode is shorted, open, or leaking badly, the circuit will be inoperative.

## 3.10    DC-Amplifier Circuit Problems

The operation of dc CFAs is more complex than ac amplifier circuits using CFAs. The major problem is when the CFA must operate from a single supply and yet provide an output voltage which goes to 0 V dc, but will also accept inputs of 0 V dc. To overcome this problem, the inputs must be biased into the linear region ($+V_{BE}$) with dc input signals of 0 V and the output must be modified if operation to actual ground (and not $V_{SAT}$) is required. The bias network

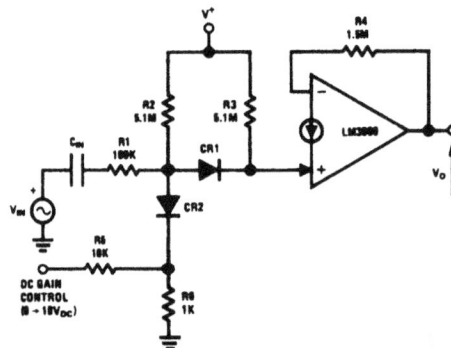

**Figure 3.14.**  Ac amplifier with dc gain control troubleshooting (National Semiconductor, *Linear Applications Handbook,* 1994, p. 176).

TL/H/7383–19

must provide an output voltage $V_O$ approximately equal to zero when the input voltage ($V_{IN}$) is zero. This is done by *common-mode biasing,* which is covered next.

## 3.11    Common-Mode Bias Circuit Troubleshooting

Figure 3-15 shows a basic circuit for a noninverting dc amplifier using a CFA with common-mode biasing. This circuit has two problems. First, the resistors must be closely matched so that $V_O$ will equal $V_{IN}$. Second, and of greater importance, the output $V_O$ cannot go below the saturation voltage ($V_{SAT}$) of the output transistor. Typically, $V_{SAT}$ is about 100 mV.

Figure 3-16 shows a similar circuit where $V_O$ can go down to about 5 mV (without adjustment) and can be adjusted by offset pot R1 so that $V_O$ is 0 V when $V_{IN}$ is 0 V. This is done by adding diode CR1 between the output of the IC and the load, as shown. The diode provides a dc level shift that allows $V_O$ to go to ground. With the load ($R_L$) connected, $V_O$ becomes a function of the voltage divider formed by R4 and $R_L$.

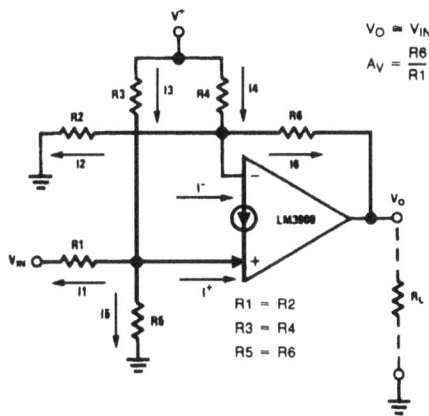

**Figure 3.15.**   Dc amplifier with common-mode biasing (National Semiconductor, *Linear Applications Handbook,* 1994, p. 177).

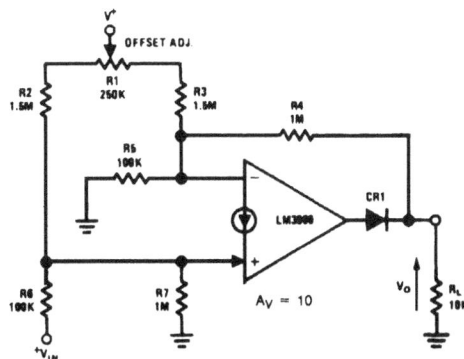

**Figure 3.16.**   Noninverting dc amplifier (National Semiconductor, *Linear Applications Handbook,* 1994, p. 177).

Figure 3-17 shows the voltage-transfer functions for the circuit of Fig. 3-16, both with and without the diode CR1. Although the diode greatly improves operation around 0 V, the voltage drop across the diode reduces the peak output-voltage swing by about 0.5 V. Notice that the load impedance should be large enough to avoid excessive loading of the IC. The value of $R_L$ can be significantly reduced by replacing the diode with an NPN transistor.

In troubleshooting either of the common-mode bias circuits, look for improperly matched resistors (or resistors that have drifted out of tolerance because of temperature changes) when $V_O$ does not match $V_{IN}$. If the gain is incorrect, check the ratio of feedback and input resistors (R6/R1 for Fig. 3-15, and R4/R6 for Fig. 3-16). When all of the resistors are properly matched and the feedback/input ratio is correct, but the circuit output of Fig. 3-16 does not follow the input, look for a defective (leaking) CR1.

## 3.12   Ground-Referenced Circuit Troubleshooting

Figure 3-18 shows a basic circuit for a dc amplifier using a CFA with a ground-referenced differential input. With this circuit, the output ($V_O$) is equal to the differential voltage ($V_R$) across the input resistance ($R$). The resistors are kept large to minimize loading. With the 10-M$\Omega$ resistors shown, an error exists for small values of $V_1$ because of the input-bias current at the ($-$) input. The per-

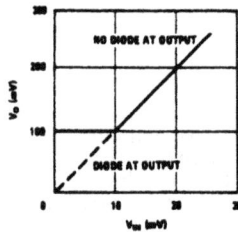

**Figure 3.17.**  Voltage-transfer functions for dc amplifiers (National Semiconductor, *Linear Applications Handbook,* 1994, p. 178).

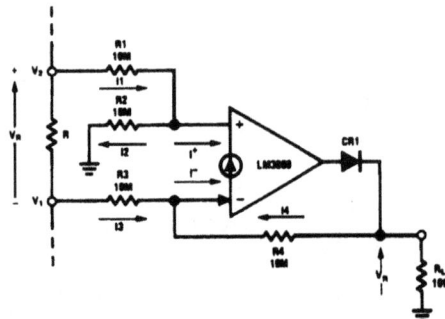

**Figure 3.18.**  Dc amplifier with ground-referenced differential inputs (National Semiconductor, *Linear Applications Handbook,* 1994, p. 178).

centage of error can be reduced by lowering the resistor values from 10 MΩ so it might be necessary to trade off between IC loading and error.

In troubleshooting the circuit of Fig. 3-18, remember that the input voltage ($V_2$) must be greater than 1 V and the differential input voltage ($V_R$) must be limited to within the output dynamic-voltage range of the IC. For example, if $V_2$ is 1 V, the input voltage ($V_1$) can vary from 1 V to $-13$ V when operating from a 15-V supply. Figure 3-19 shows a similar circuit where common-mode biasing is added so that both $V_1$ and $V_2$ can be negative.

### 3.13　Unity-Gain Amplifier Circuit Troubleshooting

Figure 3-20 shows a basic circuit for a unity-gain dc buffer-amplifier using a CFA. In this simple circuit, the voltage applied at the input is reproduced at the output. However, the input voltage must be greater than 1 $V_{BE}$ (about 0.5 V), but less than the maximum output swing ($V_{CC}-1$). In troubleshooting, if the output is not a true 1:1 ratio (unity gain), the problem is almost always a mismatch of the two 1-MΩ resistors (or the IC is defective, which is not likely).

### 3.14　CFA with Output Clamping

Figures 3-21 and 3-22 show the simplified circuit and evaluation-board connections, respectively, for a CFA (the Harris HFA1130) with programmable output clamps. This IC is designed for high-frequency applications that

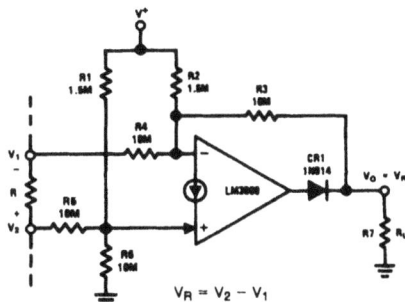

Figure 3.19.　Dc amplifier with both inputs negative (National Semiconductor, *Linear Applications Handbook,* 1994, p. 178).

TL/H/7383-27

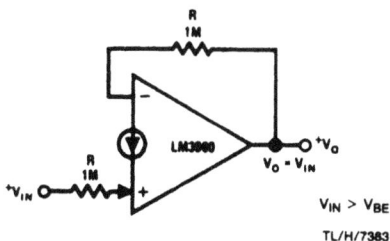

Figure 3.20.　Unity-gain dc buffer-amplifier (National Semiconductor, *Linear Applications Handbook,* 1994, p. 179).

TL/H/7383-28

require output limiting—especially applications that require ultrafast over-drive-recovery times. The output-clamp function permits the maximum positive and negative output levels to be set or programmed, thus protecting later stages from damage or input saturation. The subnanosecond recovery time quickly returns the IC to linear operation following an overdrive condition.

Figure 3-23 shows some typical performance characteristics for the IC. As shown, the gain drops as the frequency or bandwidth increases, as is the case with op amps and OTAs. In turn, bandwidth is set by external feedback resistance $R_F$, as shown in Fig. 3-21B. The external $R_F$, in conjunction with an internal capacitor (at point Z), sets the dominant pole of the frequency response shown in Fig. 3-23. The IC bandwidth is proportional to $R_F$.

In troubleshooting, note that the IC is optimized for a 510-$\Omega$ $R_F$ at a gain of +1. Decreasing $R_F$ in a unity-gain application decreases stability, leading to excessive peaking and overshoot. The IC is more stable at higher gains, so $R_F$ can be decreased in a tradeoff of bandwidth versus stability. The table in Fig. 3-21 shows the recommended $R_F$ values for various gains and the expected bandwidth.

### 3.14.1  PC-board layout problems

Because of the high frequencies involved, the frequency performance of the IC depends largely on the amount of care taken with the PC board. Figure 3-22

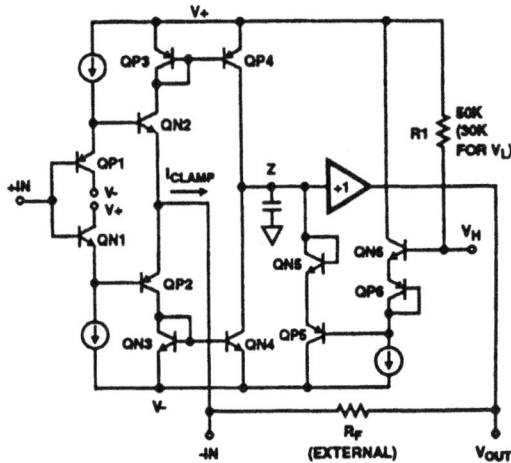

| $A_{CL}$ | $R_F$ ($\Omega$) | BW (MHz) |
|----------|------------------|----------|
| +1 | 510 | 850 |
| -1 | 430 | 580 |
| +2 | 360 | 670 |
| +5 | 150 | 520 |
| +10 | 180 | 240 |
| +19 | 270 | 125 |

**Figure 3.21.** Simplified circuit for CFA with output clamps (Harris Semiconductor, *Linear & Telecom ICs,* 1994, p. 2-662).

BOARD SCHEMATIC

TOP LAYOUT

BOTTOM LAYOUT

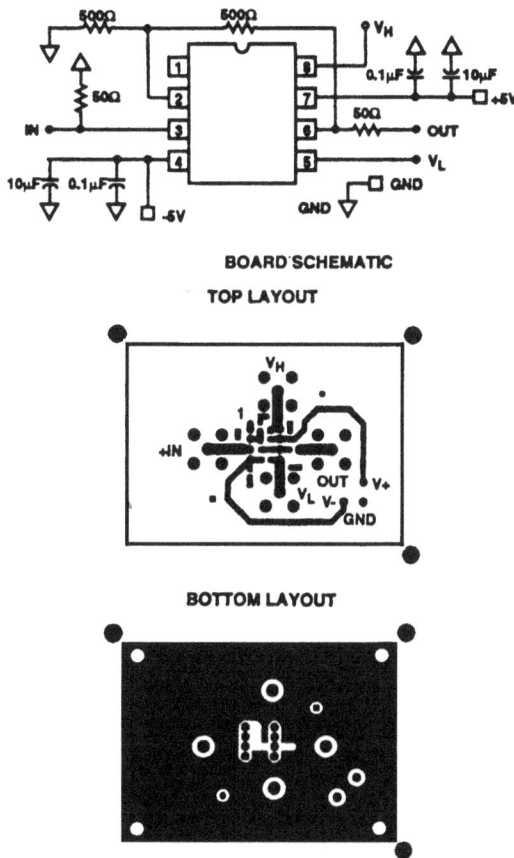

Figure 3.22.  Evaluation-board connections for CFA (Harris Semiconductor, *Linear & Telecom ICs,* 1994, p. 2-663).

shows an example of a good high-frequency layout for the evaluation board. The use of low-inductance components (such as chip resistors and chip capacitors) is strongly recommended. A solid ground plane is a must! As shown in Fig. 3-22, the power supplies must be decoupled. A large-value (10 μF) tantalum in parallel with a small-value (0.01 μF) chip capacitor is recommended by the manufacturer. Terminated microstrip signal lines are recommended at the input and output of the IC.

In troubleshooting this circuit, notice that excessive output capacitance (such as from an improperly terminated transmission line) will degrade the frequency response and might cause oscillation. In most cases, oscillation can be avoided, or cured, by placing a resistor in series with the output, as shown. Also, take care to minimize the capacitance to ground seen by the ($-$) input (pin 2). The larger this capacitance, the worse the gain peaking, which results in pulse overshoot and possible instability. The manufacturer recommends that the ground plane be removed on the traces connected to pin 2, and all connections to pin 2 be kept as short as possible.

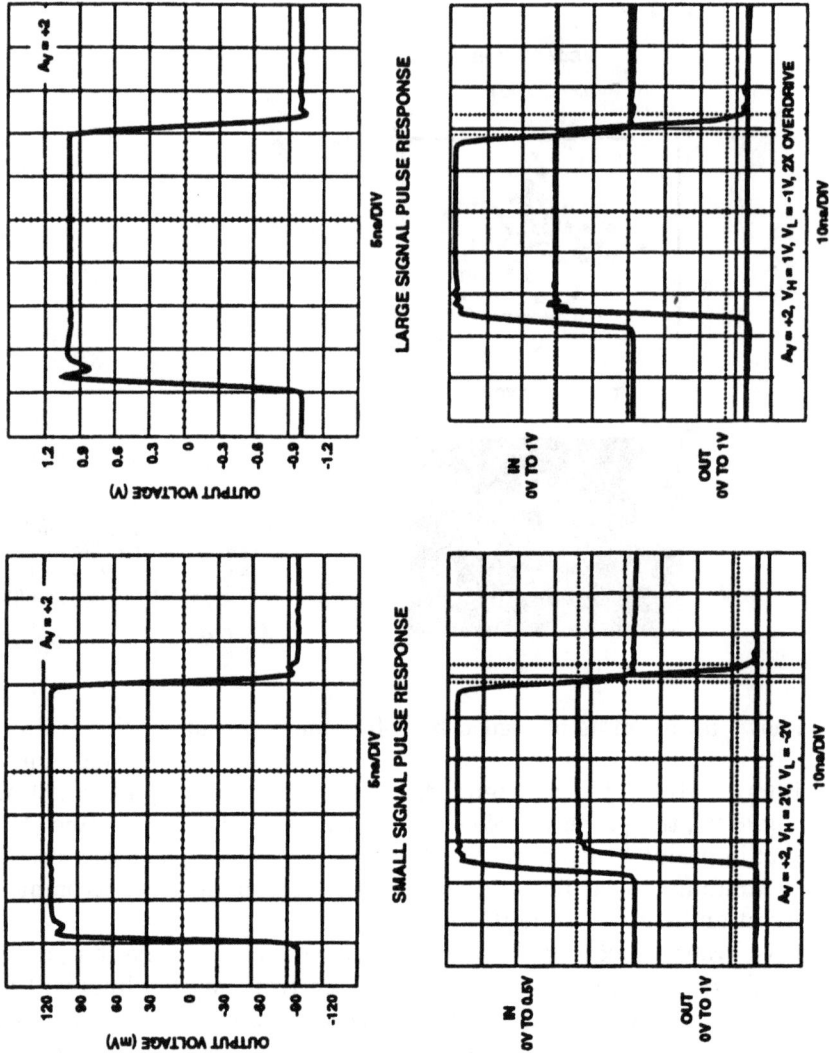

**Figure 3.23.** Typical performance characteristics for CFA (Harris Semiconductor, *Linear & Telecom ICs*, 1994, p. 2-665).

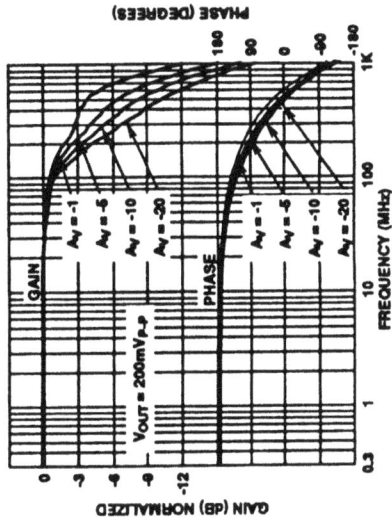

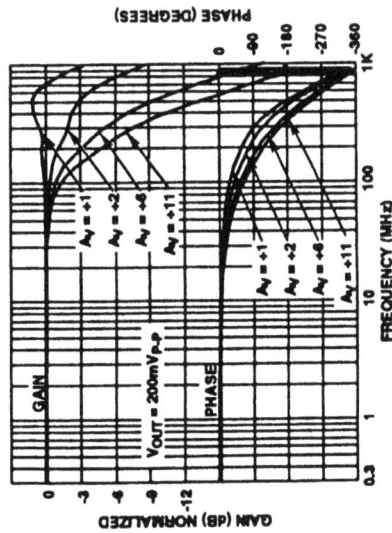

Figure 3.23. (*Continued*)

95

### 3.14.2  Clamp circuit troubleshooting

Clamping action is obtained by applying voltages to the $V_H$ and $V_L$ terminals (pins 8 and 5) of the IC. (Figure 3-21 shows the $V_H$ clamp circuit, as well as the input circuit, of the IC. The $V_L$ clamp circuit is similar, except for some component values.) The voltage at $V_H$ sets the upper output limit and $V_L$ sets the lower clamp level. If the IC tries to drive the output above $V_H$ or below $V_L$, the clamp circuits limit the output voltage at $V_H$ or $V_L$ (± the clamp accuracy). The low input-bias currents at the clamp pins allow the clamp inputs to be driven by simple resistive dividers or active elements, such as other amplifiers or DACs (Chapter 10).

As shown in Fig. 3-21, a unity-gain buffer is between the positive and negative inputs. The buffer forces −IN to track +IN and set up a slewing current of $(V_{IN}-V_{OUT})/R_F$. This current is mirrored into the high-impedance point (Z) and converted to a voltage. The resultant voltage is fed to the output through another unity-gain buffer. If the clamping function is not used (no voltage applied to $V_H$ or $V_L$), the voltage at Z can swing between the limits defined by QP4 and QN4. When the output reaches the quiescent value, the current flowing through −IN is reduced to only that small current ($I_{BIAS}$) required to keep the output at the final voltage.

For troubleshooting purposes, operation of the clamp can be illustrated by tracing from $V_H$ to point Z. $V_H$ decreases by 2 $V_{BE}$ (QN6 and QP6) to set up the base voltage on QP5. When Z reaches a voltage equal to the QP5 base plus 2 $V_{BE}$ (QP5 and QN5), QP5 begins to conduct. Thus, QP5 clamps Z whenever Z reaches $V_H$. Resistor R1 provides a fixed pullup function to keep the circuit operating if the clamp input is floating (no voltage applied to $V_H$ or $V_L$).

When the output is clamped, the (−) input continues to source a slewing current ($I_{CLAMP}$) in an attempt to force the output to the quiescent voltage defined by the input. QP5 must sink this current while clamping, because the −IN current is always mirrored into the high-impedance point (Z).

To calculate the clamping current when troubleshooting the circuit, use $(V_{IN}-V_{OUT})/R_F$. For example, a unity-gain circuit with a $V_{IN}$ of 2 V, a $V_H$ of 1 V, and $R_F$ of 510 ohms would have an $I_{CLAMP}$ of $(2-1)/510\ \Omega = 1.96$ mA. Notice that the IC current drawn from the supply increases by an amount equal to $I_{CLAMP}$ when the output is clamp limited.

### 3.14.3  Clamp-accuracy problems

When troubleshooting a circuit where the clamp accuracy appears to be incorrect, notice that the clamped output voltage is not exactly equal to the voltage applied at $V_H$ or $V_L$. The result in a clamp-accuracy characteristic (usually found on the data sheet). The important point to remember is that clamp accuracy degrades as overdrive increases or as $R_F$ decreases.

For example, the data sheet specifies a clamp accuracy of ±60 mV for a 2 × overdrive, with an $R_F$ of 510 $\Omega$. This means that if the input is 2 V and the output is clamped at 1 V, the output will be 1 V ±60 mV. If the $R_F$ is decreased to 240 $\Omega$ with the same 2 × overdrive, the clamp accuracy degrades to ±220 mV.

If the overdrive is increased to 3 $\times$ but $R_F$ is left at 510 $\Omega$, the accuracy degrades to $\pm 250$ mV.

### 3.14.4  Clamp-range limitations

Both $V_H$ and $V_L$ have usable ranges that cross 0 V. Although $V_H$ must be more positive than $V_L$, both can be positive or negative (within the range restrictions indicated by the specifications). For example, the IC could be limited to digital ECL (emitter-coupled logic) levels by setting $V_H$ at $-0.8$ V and $V_L$ at $-1.8$ V. $V_H$ and $V_L$ can be connected to the same voltage (ground, for example), but the result is not a dc output voltage from an ac input signal. A 150- to 200-mV ac signal will still be present at the output. In troubleshooting, remember that $V_H$ must be more positive than $V_I$—even if both are negative.

### 3.14.5  Overdrive-recovery problems

One of the advantages of a CFA with clamped outputs is the ability to recover from overdrive signals. The output voltage remains at the clamp level as long as the overdrive condition remains. When the input voltage drops below the overdrive level, the amplifier returns to linear operation.

In troubleshooting a CFA where the output appears not to return to linear operation after overdrive, notice the time delay (known as the *overdrive recovery time*), which is required before a return to linear is complete. The plots of "unclamped performance" and "clamped performance" shown in Fig. 3-23 show the recovery time. The difference between the clamped and clamped delays is the overdrive recovery time.

For example, as shown in Fig. 3-23, the delays are 4.0 ns for the unclamped pulse and 4.8 ns for the clamped (with 2$\times$ overdrive) pulse. This produces an overdrive recovery time of 800 ps. The measurement uses the 90% point of the input transition to ensure that the linear operation has resumed. (The manufacturer states that the unclamped/clamped measurements shown in Fig. 3-23 depend on the test fixtures and that the true overdrive recovery time is closer to 500 ps.)

### 3.14.6  Linearity and clamp voltages

In troubleshooting CFA circuits where linearity is crucial, notice that the clamp voltages have a definite effect on linearity. Figure 3-24 shows the effect of several clamp voltages or levels on amplifier linearity.

## 3.15   CFA Troubleshooting Tips

The following is a summary of troubleshooting tips for CFAs. The summary is based on a comparison of a CFA (the Linear Technology LT1223) and a conventional op amp (the Linear Technology LT1220). Because CFAs act very much the same as op amps, the summary concentrates on the differences.

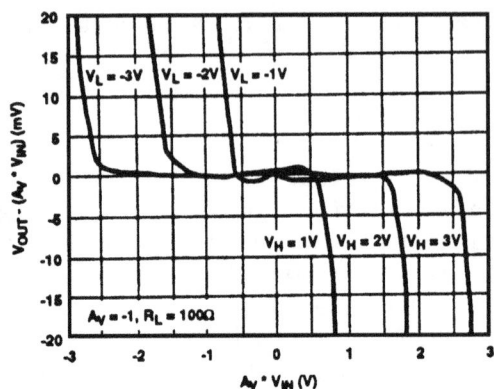

**Figure 3.24.** Effect of clamp voltages on linearity of CFA (Harris Semiconductor, *Linear & Telecom ICs,* 1994, p. 2-669).

That is, the comparison shows how the same circuit can be implemented with an op amp and a CFA, and how the differences affect circuit troubleshooting.

Before getting into troubleshooting for specific CFA circuits, remember that the impedance at the (−) inverting input of a CFA sets the bandwidth and thus the stability of the amplifier. The (−) input impedance must be resistive in all circuits, never capacitive. To slow the amplifier down, increase the resistance driving the (−) input. If the amplifier peaks too much because of capacitive loading (or anything else), increase the value of the feedback resistors.

### 3.15.1  Adjustable-gain CFA amplifier troubleshooting

Figure 3-25 shows a comparison of an op amp and a CFA connected as an adjustable-gain amplifier. The gain of a conventional op amp can be varied by either $R_g$ or $R_f$. The only real restriction is the loading effect on the resistors at the amplifier output (as covered in Chapter 1). With a CFA, the value of $R_f$ should not be varied. For example, if $R_f$ is a pot, then the bandwidth will be reduced at minimum gain and the circuit will oscillate if $R_f$ is very small.

### 3.15.2  Bandwidth-limiting CFA amplifier troubleshooting

Figure 3-26 shows a comparison of an op amp and a CFA connected to provide bandwidth limiting. It is quite common to limit the bandwidth of an op amp with a small capacitor in parallel with $R_f$. This usually works with most op amps, provided that they are stable at unity gain.

Do not connect a small capacitor from the inverting (−) input of a CFA to anything, especially not to the output! The capacitor on the (−) input will cause peaking or oscillation (or both). If the bandwidth of a CFA must be limited, use a resistor and capacitor (such as R1 and C1) at the noninverting (+) input. This technique also tends to cancel the peaking caused by stray capacitance at the (−) input. Unfortunately, the technique does not limit output noise in CFAs, as is the case with op amps.

### 3.15.3 CFA integrator troubleshooting

Figure 3-27 shows a comparison of an op amp and a CFA connected to provide integration (a classic op-amp function). With the CFA, a resistor (1 kΩ, in this case) must be added at the (−) input. This produces a new summing node where the capacitive feedback can be added. With the extra resistor at the (−) input, the CFA integrator does the same job as the op-amp integrator, often with better large-signal capability and accurate phase shift at high frequencies. In troubleshooting any integrator circuit, the most likely cause of inaccurate integration is the feedback capacitor.

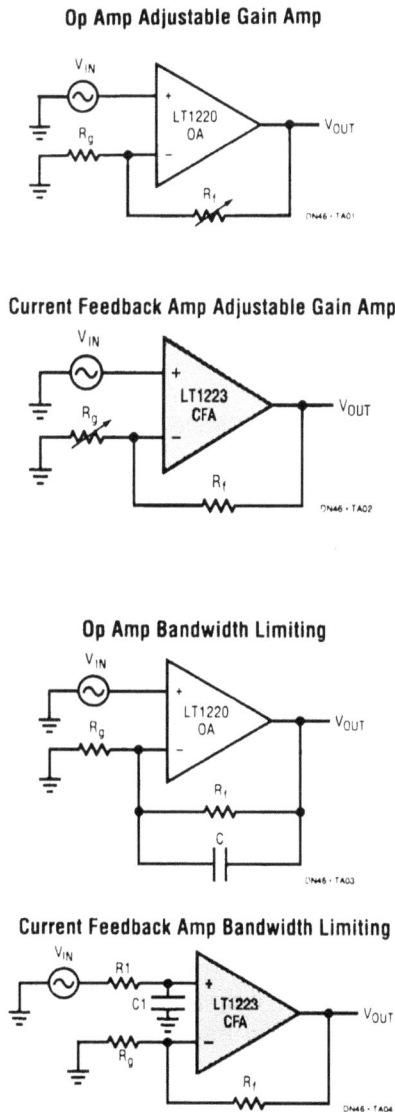

**Op Amp Adjustable Gain Amp**

**Current Feedback Amp Adjustable Gain Amp**

**Figure 3.25.** Comparison of adjustable-gain amplifiers (Linear Technology, Design Note 46, p. 1).

**Op Amp Bandwidth Limiting**

**Current Feedback Amp Bandwidth Limiting**

**Figure 3.26.** Comparison of bandwidth-limiting amplifiers (Linear Technology, Design Note 46, p. 1).

**Op Amp Integrator**

**Current Feedback Amplifier Integrator**

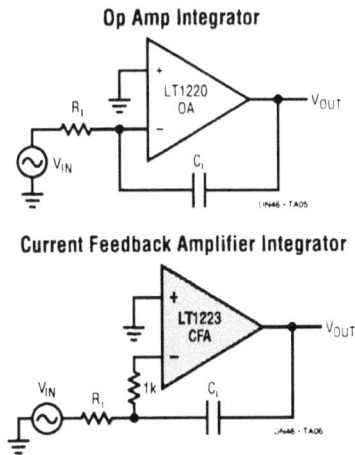

**Figure 3.27.** Comparison of amplifier integrators (Linear Technology, Design Note 46, p. 2).

### 3.15.4 CFA dc-accurate summing amplifier troubleshooting

Figure 3-28 shows a CFA connected as a dc-accurate summing amplifier. No $I_{OS}$ (input offset current) specifications are used for CFAs because there is no correlation between the two input-bias currents. Thus, the circuit is not improved by adding an extra resistor at the $(-)$ input (as is shown in Fig. 3-27). In troubleshooting any summing amplifier with inaccurate outputs, the most likely cause of trouble is feedback resistance.

### 3.15.5 Instrumentation amplifier troubleshooting

Figure 3-29 shows two CFAs connected in the classic op-amp instrumentation-amplifier configuration. The design equations for both CFAs and op amps are given. The only significant difference is to make the feedback resistor of each amplifier the same value, thus making the gain-setting resistors different values. This creates a difference in circuit troubleshooting. For CFAs, the bandwidth of both amplifiers is the same and the common-mode rejection at high frequencies is better than that of the op-amp circuit. When op amps are used in the same circuit, one amplifier has maximum bandwidth because it operates at about unity gain. The other amplifier is limited to the IC gain-bandwidth product divided by the gain.

### 3.15.6 Cable driver troubleshooting

Figure 3-30 shows a CFA connected as a cable driver. This circuit could be used with a conventional op amp. However, most op amps do not have enough output current to drive heavy cable loads. The important consideration in troubleshooting this CFA circuit is that both ends of the cable are properly terminated. This is especially true when even modest high-frequency performance is required. The additional advantage of such termination is that it isolates the capacitive load (presented by the cable) from the amplifier. This permits the CFA to operate at maximum bandwidth. Always look for incorrect terminated-resistor values or poor connections when bandwidth is not correct.

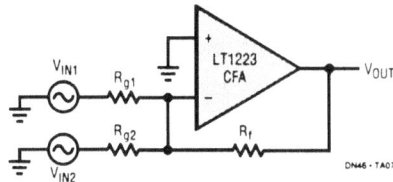

**Figure 3.28.**  CFA dc-accurate summing amplifier (Linear Technology, Design Note 46, p. 2).

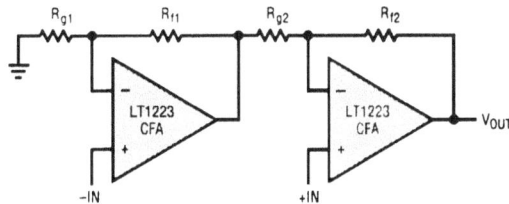

TRIM $R_{g2}$ FOR GAIN, THEN TRIM $R_{g1}$ FOR CMRR. VOLTAGE GAIN, G, IS $V_{OUT}$ DIVIDED BY DIFFERENCE BETWEEN +IN AND −IN.

OP AMP DESIGN EQUATIONS:
$R_{f1} = R_{g2}$;  $R_{f2} = (G-1) R_{g2}$;  $R_{g1} = R_{f2}$

CURRENT FEEDBACK AMP DESIGN EQUATIONS:
$R_{f1} = R_{f2}$;  $R_{g1} = (G-1) R_{f2}$;  $R_{g2} = \frac{R_{f2}}{G-1}$

**Figure 3.29.**  CFA instrumentation amplifier (Linear Technology, Design Note 46, p. 2).

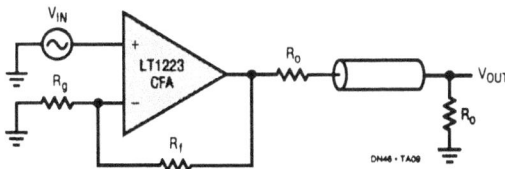

**Figure 3.30.**  CFA cable driver (Linear Technology, Design Note 46, p. 2).

# Chopper-Stabilized-Amplifier Circuit Troubleshooting

This chapter is devoted to troubleshooting for chopper-stabilized amplifiers. The chapter begins with a section concerning why chopper stabilization is required in direct-coupled amplifiers. With this theory out of the way, a detailed description of a typical chopper-stabilized IC amplifier follows, emphasizing how circuit functions are related to troubleshooting. Then follows a description of why such circuits fail in either the experimental or final form. Chopper stabilization is often used where stability is essential over time and with variations in temperature and supply voltage.

## 4.1 Chopper Stabilization

The major problem with any direct-coupled amplifier (any amplifier where there is no coupling capacitor between stages, which is typical for most IC amplifiers) is that the circuit responds the same way to a change in power-supply voltage as to a change in the dc signal level. This effect can be minimized by feedback, temperature-compensating diodes, and complementary circuits. However, none of these methods can compensate for a constantly changing or drifting power supply (from whatever cause) when the signal is also direct current.

Figure 4-1 shows a method for overcoming the voltage-change problems of direct-coupled amplifiers. In this circuit, the dc signal is inverted to an equivalent ac signal (through modulation). The ac signal is amplified in a gain-stabilized ac amplifier and then reconverted to a direct current (through demodulation). During amplification, the signal is not affected by drift.

In this theoretical circuit, the amplifier input is switched alternately to both sides of a transformer. This periodically inverts the polarity of the signal applied to the amplifier, thus converting dc to ac. Although mechanical switches are

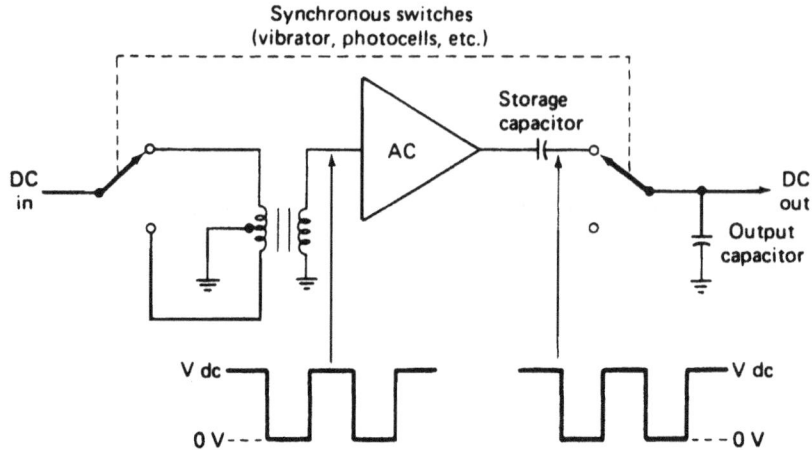

**Figure 4.1.**  Basic chopper stabilization technique.

shown in Fig. 4-1, the switching system is solid state for present-day IC amplifiers (and there is no transformer!).

In the circuit of Fig. 4-1, a second pair of contacts at the output establishes
the ground level for a storage capacitor in series with the output. The output
storage capacitor becomes charged to a level that corresponds to the amplitude
of the output square wave. Synchronous detection preserves the polarity of the
input voltage and recovers both positive and negative voltages with the correct
polarity. The synchronous modulation and demodulation is known as *chopping*
or *chopper* stabilization. A direct-coupled amplifier with a chopper circuit
offers drift-free amplification of low-level signals in the microvolt region.

## 4.2  Frequency Response in Chopper-Stabilized
Amplifiers

One problem with chopper stabilization is frequency response. If the input is
pure direct current, there is no problem. However, if the input is very low-
frequency alternating current, the chopper frequency must be higher than the
signal frequency. If not, the input signal waveform can be distorted or completely lost. For example, if both the input signal and chopping frequency are 100 Hz,
and if the amplifier input is shorted at the same instant as the positive swing of
the input signal, the amplifier sees only the negative portion of the input.

Figure 4-2 shows a method of overcoming the frequency-response problem.
Here, a chopper-modulated amplifier is used to correct the direct-coupled
amplifier for a voltage change or drift. Notice that the input-signal direct current is amplified through a conventional dc amplifier. A portion of the amplified
output is tapped off from a divider network and compared with the original
input signal. The divider network reduces the output by the same amount that
is amplified by the dc amplifier. Thus, the divider output should be equal to the
input level at the summing point. Any difference at the summing point (caused

by voltage change, drift, etc.) is amplified through the modulated amplifier and then applied to the main-channel dc amplifier as negative feedback to cancel the drift.

## 4.3    Basic Chopper-Stabilized-Amplifier Circuit

Figures 4-3 and 4-4 show the internal functions and pinouts, respectively, of a chopper-stabilized op amp (the Harris ICL7650S). The IC is a direct replacement for the industry-standard ICL7650, but with improved input-offset voltage, a lower input-offset temperature coefficient, reduced input-bias current, and wider common-mode voltage range.

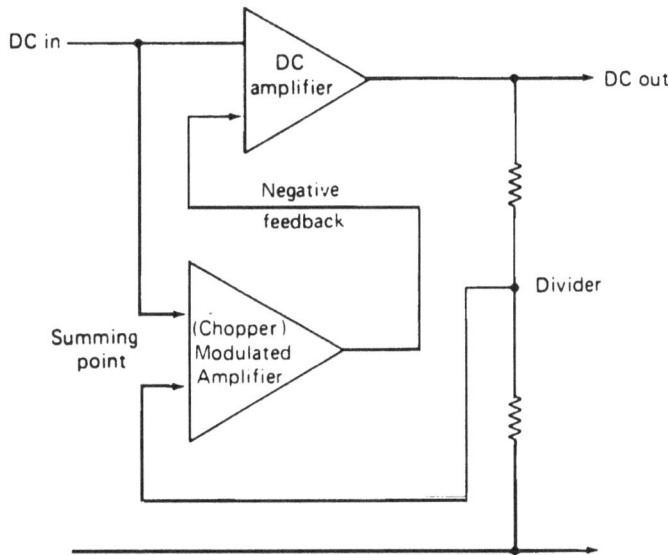

**Figure 4.2.**   Chopper-modulated technique.

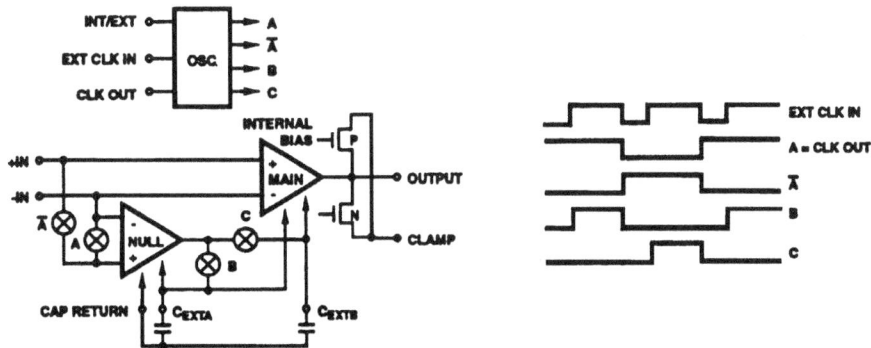

**Figure 4.3.**   Internal functions of chopper-stabilized amplifiers (Harris Semiconductor, *Linear & Telecom ICs,* 1994, p. 2-695).

ICL7650S
(PDIP, SOIC)
TOP VIEW

$C_{EXTA}$  1    8  $C_{EXTB}$
-IN  2    7  V+
+IN  3    6  OUTPUT
V-  4    5  $C_{RETN}$

ICL7650S
(TO-99 CAN)
TOP VIEW

$C_{EXTB}$  8
$C_{EXTA}$  1    7  V+/CASE
-IN  2    6  OUTPUT
+IN  3    5  $C_{RETN}$
4
V-

ICL7650S
(PDIP, CDIP, SOIC)
TOP VIEW

$C_{EXTB}$  1    14  INT/EXT
$C_{EXTA}$  2    13  EXT CLK IN
NC (GUARD)  3    12  INT CLK OUT
-IN  4    11  V+
+IN  5    10  OUTPUT
NC (GUARD)  6    9  OUT CLAMP
V-  7    8  $C_{RETN}$

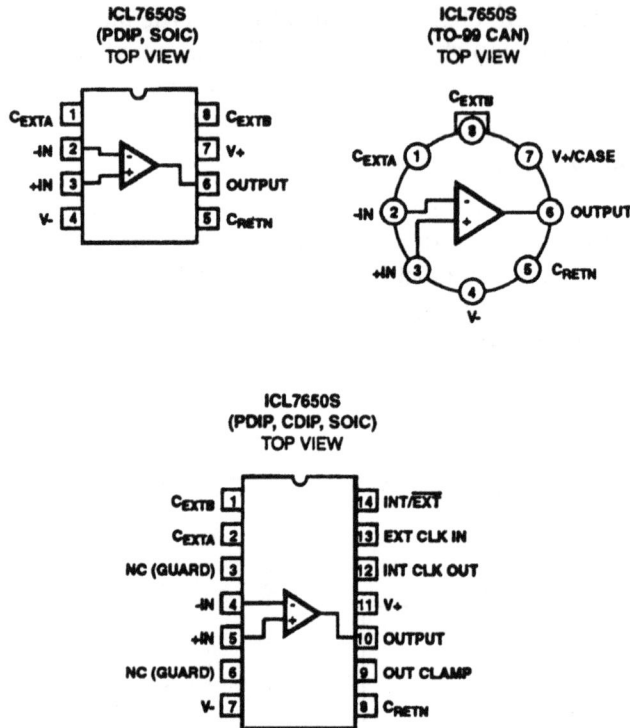

**Figure 4.4.** Pinout of a chopper-stabilized amplifier (Harris Semiconductor, *Linear & Telecom ICs,* 1994, p. 2-694).

The IC provides the low offset by comparing the inverting and noninverting input voltages in a nulling amplifier operated by alternate clock phases. Two external capacitors are required to store the correcting potentials on the two amplifier nulling inputs. The clock oscillator and all other control circuits are self-contained. However, the 14-pin version (Fig. 4-4) includes a provision for an external clock if required. The IC is internally compensated for unity-gain operation. The chopper-amplifier circuits are user-transparent, eliminating the common chopper-amplifier problems of intermodulation effects, chopping spikes, and over-range lock-up.

The two amplifiers are the main amplifier and the nulling amplifier. Both amplifiers have offset-null capability. The main amplifier is connected continuously from the input to the output. The nulling amplifier, under control of the chopping oscillator and clock circuit, alternately nulls itself and the main amplifier.

The nulling connections, which are MOSFET gates, have inherently high impedance. The two external capacitors ($C_{EXTA}$ and $C_{EXTB}$), provide the required storage of the nulling potentials and the necessary nulling-loop time constant. The nulling circuits operate over the full common-mode and power-supply ranges and are independent of the output level. This provides high CMRR (140 dB), PSRR (140 dB), and AVOL (150 dB).

## 4.4   Intermodulation Problems

Some chopper-stabilized amplifiers suffer from intermodulation effects between the chopper frequency and input signals. These problems occur because the finite ac gain of the amplifier requires a small ac signal at the input. This is "seen" by the zeroing circuit as an error signal, which is chopped and fed back, thus injecting sum and difference signals. This can cause disturbances to the gain and phase-versus-frequency characteristics near the chopping frequency.

In troubleshooting circuits with this IC (or a similar IC), notice that intermodulation effects are substantially reduced in the IC by feeding the nulling circuit with a dynamic current that corresponds to the compensating-capacitor current. The correction current is applied to cancel that portion of the input signal that arises from the ac gain. Because this is the major error contribution to the IC, the intermodulation and gain/phase disturbances are held to very low values and can generally be ignored. This is not true in other circuits, where intermodulation can be a problem. The most common method of combatting intermodulation effects is to use a clock frequency that is substantially different from the signal frequencies involved (including any harmonics).

## 4.5   Null/Storage Capacitor Problems

As shown in Fig. 4-5, the null/storage capacitors are connected to the $C_{EXTA}$ and $C_{EXTB}$ pins, with a common connection to the $C_{RETN}$ pin. In either testing or troubleshooting, test connections must be made directly by either a separate wire or PC trace. This will avoid injecting load-current $IR$ drops into the capacitor circuits. The outside foil of each capacitor (where available) should be connected to the $C_{RETN}$ pin.

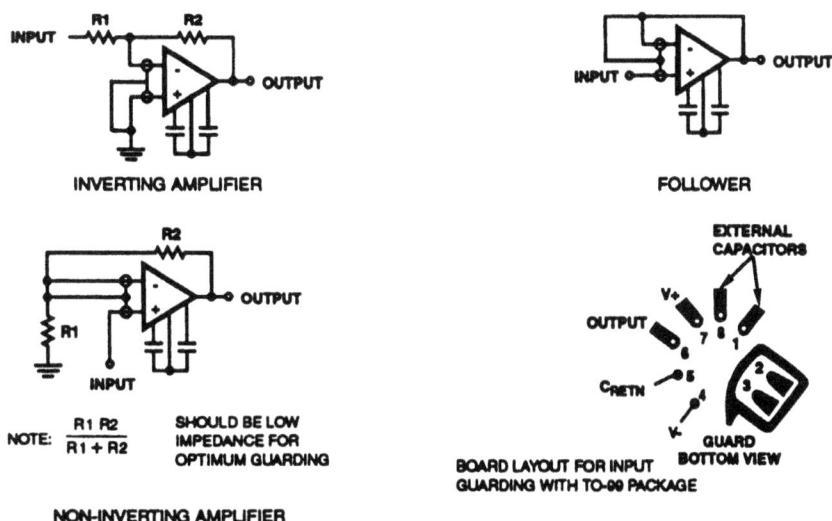

NOTE: $\dfrac{R1\ R2}{R1 + R2}$   SHOULD BE LOW IMPEDANCE FOR OPTIMUM GUARDING

**Figure 4.5.**   Connections for chopper-stabilized amplifiers (Harris Semiconductor, *Linear & Telecom ICs,* 1994, p. 2-703).

## 4.6    Overload Recovery-Time Problems

The overload or overdrive recovery problem covered for CFAs in Section 3.14.5 is sometimes a problem with chopper-stabilized amplifiers. In this IC, the problem can be minimized by the OUTPUT CLAMP circuit. When the OUTPUT CLAMP pin is tied to the inverting (−) input pin, or summing junction, a current path between this point and the output pin occurs, just before the IC output saturates. Because of this current path, uncontrolled input differentials are avoided, together with the consequent charge build-up on the correction-storage capacitors. As a result, the output swing is slightly reduced.

## 4.7    Chopper Clock Problems

The IC has an internal oscillator with a typical chopping frequency of 250 Hz. (This internal clock is available, for external use, at the CLOCK OUT pin on the 14-pin device.) The IC can also be operated by an external clock. The INT/EXT pin has an internal pullup and can be left open for normal operation. To use an external clock, the INT/EXT pin must be tied to $V-$, thus disabling the internal clock. An external clock signal can then be applied at the EXT CLOCK IN pin.

In either testing or troubleshooting, note that an internal divide by two provides the desired 50% input-switching duty cycle. Because the capacitors are charged only when EXT CLOCK IN is high, a 50% to 80% duty cycle is recommended for the external clock signal. The external clock can swing between $V+$ and $V-$. The logic threshold will be at or about 2.5 V below $V+$. Notice that a signal of about 400 Hz, with a 70% duty cycle, will be present at the EXT CLOCK IN pin with INT/EXT high or open. This is the internal clock signal before being fed to the divider.

### 4.7.1    Using a strobe signal to avoid capacitance misbalance

In applications where a strobe signal is available, it is possible to avoid capacitor misbalance during overload. This is done by connecting the strobe signal to the EXT CLOCK IN so that the strobe is low during the time that the overload is being applied to the amplifier. Under these conditions, neither capacitor will be charged. Because leakage at the capacitor pins is quite low at room temperature, the typical amplifier circuit will drift less than 10 μV/s and relatively long measurements can be made with little change in offset.

## 4.8    Null/Storage Capacitor Values

External capacitors $C_{\text{EXTA}}$ and $C_{\text{EXTB}}$ have optimum values for this IC (and for any similar IC) that depend on the clock or chopping frequency. The following is a summary of the factors to be considered when selecting external capacitors for testing or troubleshooting.

For the preset internal clock of about 250 Hz (in this IC), the correct value is 0.1 μF. This value should be scaled approximately in proportion if an external clock is used. This will maintain the same relationship between the chopping frequency and the nulling time constant. A high-quality film-type capacitor (such as mylar) is recommended, although a ceramic or other lower-grade capacitor might be good for some applications. For the fastest settling time after turn-on, low-dielectric-absorption capacitors (such as polypropylene) should be used. With ceramic capacitors, several seconds might be required to settle to 1 μV.

## 4.9    Static Protection Problems

All pins of the IC are static protected by input diodes (typical for most present-day IC devices). However, strong static fields and discharges should be avoided. Such fields and discharges might result in increased input-leakage currents.

## 4.10    Latchup Problems

Under certain circumstances, the CMOS junction at the IC pins might be triggered into a low-impedance state, resulting in excessive supply current. In testing or troubleshooting, be certain that no voltage greater than 0.3 V beyond the supply is applied to any pin. In general, the amplifier supplies must be established either at the same time or before any input signals are applied. If this is not possible, the circuits driving the inputs must limit input-current flow to less than 1 mA to avoid latchup—even under fault conditions.

## 4.11    Output Load Problems

The output circuit of the IC is a high-impedance type (about 18 kΩ). When the output load is less than 18 kΩ, the IC acts somewhat like a transconductance amplifier, where open-loop gain is proportional to load resistance (as described for CFAs in Section 2.4). For example, the open-loop gain will be 17 dB lower with a 1-kΩ load than with a 10-kΩ load.

In troubleshooting, if the IC is used strictly for dc, the lower-gain figure is of little importance. This is because the dc gain is typically greater than 120 dB—even with a 1-kΩ load. However, for wideband applications, the best frequency response will be with a load resistor of 10 kΩ or higher. This results in a smooth 6-dB/octave response from 0.1 Hz to 2 MHz, with phase shifts of less than 10° in the transition region, where the main amplifier takes over from the null amplifier.

## 4.12    Thermoelectric Problems

The ultimate limitations to ultra-high precision dc amplifiers are the thermoelectric or Peltier effects that occur in thermocouple junctions of dissimilar metals, alloys, silicon, etc. Unless all junctions are at the same temperature, thermoelectric voltage (typically about 0.1 μV/°C, but as much as tens of mV/°C for some materials) will be generated. This condition requires special precautions to take full advantage of the low offset voltages provided by a chopper-stabilized amplifier. The following is a summary of the precautions that should be considered in troubleshooting (especially in experimental circuits).

All components should be enclosed to eliminate air movement—especially that caused by power-dissipating elements in the system. Low thermoelectric-efficient connections should be used where possible. Power-supply voltages and power dissipation should be kept to a minimum (to minimize heat). High-impedance loads are preferable; good separation from surrounding heat-dissipating elements is advisable.

## 4.13    Guarding (Chopper Stabilized)

Extra care must be taken when assembling PC boards to take full advantage of the low input currents found in chopper-stabilized amplifiers. Boards must be thoroughly cleaned with TCE or alcohol and blown dry with compressed air. After cleaning, the boards should be coated with epoxy or silicone rubber to prevent contamination.

In troubleshooting, leakage currents can still cause trouble—even with properly cleaned and coated boards. This is especially a problem where input pins are adjacent to pins at supply potentials. The leakage can be significantly reduced by guarding techniques (to lower the voltage difference between the inputs and adjacent metal runs).

Figure 4-5 shows input guarding of the 8-pin TO-99 package. With this configuration, the holes adjacent to the inputs are empty when the IC is inserted in the board. The guard, which is a conductive ring surrounding the inputs, is connected to a low-impedance point. (The low-impedance point is at approximately the same voltage as the inputs.) Leakage currents from high-voltage pins are then absorbed by the guard.

## 4.14    Pin Compatibility

The basic pinout of the 8-pin IC corresponds, where possible, to that of the industry-standard 8-pin op amps (LM741, LM101, etc.). The external capacitors are connected to pins 1 and 8, usually used for offset-null or compensation capacitors (or simply not connected for many 8-pin op amps). In the case of the OP-05 and OP-07 op amps, the replacement of the offset-null pot (connected between pins 1/8 and $V+$) can be done by connecting the external capacitors to pins 1 and 8, with the junction or return lead of the capacitors connected to pin 5. This same configuration can be used for replacement of the compensation capacitor used between pins 1 and 8 of an LM108. When the IC is used to replace op amps, such as the LM101 or $\mu$A748, the junction between the external capacitors need not be connected to pin 5.

The 14-pin ICL7605S pinout corresponds most closely to that of an LM108 op amp (because of the NC pins used for guarding between the inputs and other pins). The LM108 does not use any of the extra pins and has no provision for offset nulling, but it does require a compensating capacitor. As a result, some changes in layout are required to use an ICL7650S in place of an LM108.

## 4.15   Basic Chopper-Stabilized Circuit Troubleshooting

Figures 4-6 and 4-7 show the basic connections for noninverting and inverting amplifier circuits, respectively, using the ICL7650S. Notice that both circuits show a dotted line for the clamp circuits, indicating that the clamp connection is optional.

In troubleshooting either circuit, the values of $R_1$, $R_2$, and $R_3$ are not crucial, except that the parallel combination of $R_1$ and $R_2$, plus $R_3$, must be greater than 100 kΩ for the noninverting circuit shown in Fig. 4-6. The parallel combination of $R_1$ and $R_2$ must be greater than 100 kΩ for the inverting circuit of Fig. 4-7. The first trial values for the external capacitors should be 0.1 μF, as shown.

The most likely causes of trouble are the external capacitors. If the capacitors values are not substantially the same, there will be an unbalance applied to the null amplifier. The same is true if either capacitor is leaking, even slightly. The resistors can also be a problem if the values are not correct. As in the case of an op amp, it is the ratio of input and feedback resistors that sets the circuit gain.

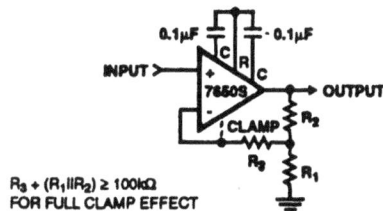

**Figure 4.6.**   Noninverting chopper-stabilized amplifier (Harris Semiconductor, *Linear & Telecom ICs,* 1994, p. 2-704).

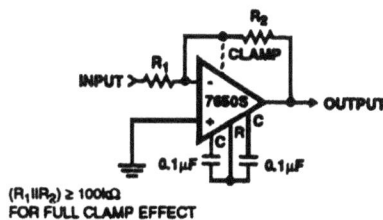

**Figure 4.7.**   Inverting chopper-stabilized amplifier (Harris Semiconductor, *Linear & Telecom ICs,* 1994, p. 2-704).

## 4.16   Chopper-Stabilized Booster Circuit Troubleshooting

The circuits shown in Figs. 4-6 and 4-7 can be used as replacements for most op-amp circuits, with two limitations. The output-drive capability is limited to a 10-kΩ load to get full output swing. The supply voltage is limited to ±8 V maximum. Both of these limitations can be overcome with a booster circuit, such as that shown in Fig. 4-8. This circuit combines the full output capabilities of the 741 (or other standard op amp) with the low input characteristics of the ICL7605S. However, in troubleshooting, remember that the two ICs shown form a composite device. As a result, watch the loop-gain stability when a feedback network is added. Again, the external capacitors are the most likely cause of trouble, with circuit gain being set by the input/feedback resistors connected to the 741.

## 4.17   Comparator Circuit Troubleshooting

Figure 4-9 shows how the clamp function of an ICL765OS can be used to form a zero-offset comparator. The usual problems in using a chopper-stabilized amplifier as a comparator are avoided (because of the clamp function). The clamp circuit forces the inverting input to follow the input signal. Notice that the threshold input must be capable of tolerating the output-clamp current (approximately equal to $V_{IN}/R$) without disturbing other portions of the system.

In troubleshooting, the external capacitors are the most likely cause of trouble. However, if the problem is that the circuit does not follow the input, try a different value for $R$.

## 4.18   Chopper/Bipolar Op-Amp Circuit Troubleshooting

Figure 4-10 shows a circuit that combines the superior dc performance of a dual chopper-stabilized op amp (the Linear Technology LTC1051) and the ultra low-noise voltage of a precision bipolar op amp (the Linear Technology LT1007). Again, as described for the circuit shown in Fig. 4-8, the output characteristics (low noise) are set by the bipolar LT1007, with the input characteristics set by the LTC1051 (low $V_{OS}$, drift, input-bias current, etc.). Although the circuit of Fig. 4-10 can be used whenever stability and low noise are required, the circuit is well suited to such applications as a strain-gauge amplifier. The circuit should be used with source resistances less than 1 kΩ to maintain good noise performance.

In troubleshooting, capacitors C1 and C2 are the most likely causes of trouble (leakage is a particular problem). Because of the unique nature of the circuit, certain points must be considered in troubleshooting. The following is a summary.

Any input-offset voltage in the LT1007 is offset by a dc-correction voltage applied at pin 8 through divider R2/R3. This function is performed by one half of the LTC1051 (which integrates the LT1007 input-offset voltage before application

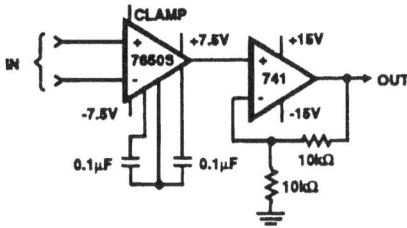

**Figure 4.8.**  Booster for chopper-stabilized amplifier (Harris Semiconductor, *Linear & Telecom ICs,* 1994, p. 2-704).

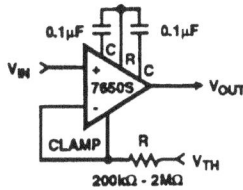

**Figure 4.9.**  Low-offset comparator with chopper-stabilized amplifier (Harris Semiconductor, *Linear & Telecom ICs,* 1994, p. 2-704).

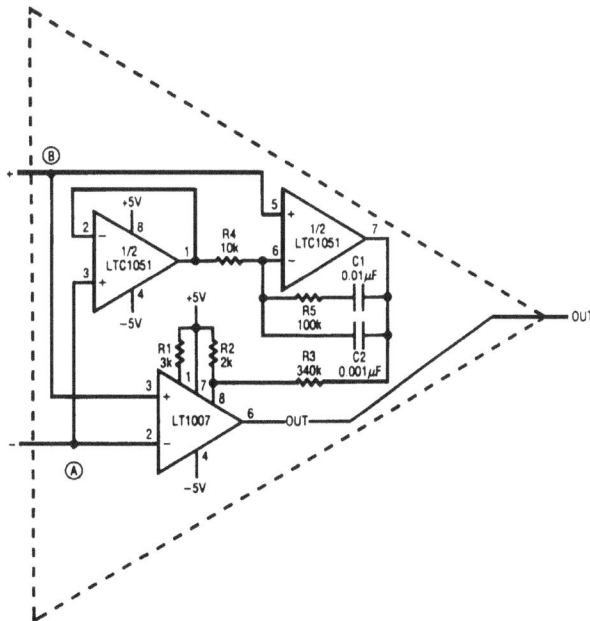

**Figure 4.10.**  Combined chopper/bipolar op amp (Linear Technology, Design Note 36, p. 1).

at pin 8). The other half of the LTC1051 buffers the $V_{OS}$-nulling circuits to eliminate loading at pin A. In troubleshooting any input-offset problems, suspect C1 and C2 first. Then check R1, R2, and R3. These resistors allow the integrator full output swing, ensuring proper offset correction to the LT1007.

If there is a problem of chopper noise getting into the LT1007 through pin 8, it is possible that the ratio of R3 to R2 is not adequate to limit the noise. The total measured input offset of the combined circuit is 2 µV with a 10 nV/°C drift.

Figures 4-11 and 4-12 show the recorded peak-to-peak noise of the Fig. 4-10 circuit for a 10-second interval and a 10-minute interval, respectively. These illustrations are included to provide a comparison for troubleshooting. During the 10-second interval (Fig. 4-11), the peak-to-peak noise is about 100 nV for both the DC-1Hz and DC-10Hz bandwidths. During the 10-minute interval (Fig. 4-12), the peak-to-peak noise is about 0.2 µV for both the DC-1Hz and DC-10Hz bandwidths. This represents a seven to nine times improvement over the DC-10Hz noise when the LTC1051 is operated alone and tested under the same conditions. The 0.2-µV also represents a 2.5 times improvement over the DC-1Hz noise of the LTC1051.

## 4.19   Troubleshooting Chopper-Stabilized Op Amps

Although troubleshooting for chopper-stabilized amplifiers is essentially the same as for bipolar op amps (Chapter 1), there are differences in performance. Consider these differences when testing or troubleshooting. (Do not expect the two types of ICs to perform in an identical manner—even in identical circuits.) The following is a summary of the comparison between bipolar and chopper-stabilized op-amp circuits. Figures 4-13 and 4-14 list a number of chopper and bipolar op amps, respectively. Figure 4-15 compares the advantages of the two types of op amps.

In wideband applications, bipolar op amps are superior—even though some chopper slew rates are 4 V/µs with bandwidths of 2.5 MHz. However, choppers have clock-frequency spikes, chopping-frequency spikes, aliasing errors, millisecond overload recovery, and high wideband noise.

In all parameters (except noise) the chopper-stabilized op amps are superior. (A 5-µV maximum offset voltage $V_{OS}$ and a 0.05-µV/°C maximum drift ($T_{CVOS}$) are guaranteed for all but the micropower LTC1049.) Changes with time and temperature cycling are near zero. These parameters cannot be measured accurately, but can be guaranteed by design, assuming that the auto-zero chopper loop (Fig. 4-10) is working properly. The best tightly specified bipolar op amps can only approach this performance, at the cost of much testing versus yield expense.

Where maximum output swing is required, the bipolar op amp generally has the edge. Typical chopper-stabilized amplifiers are usually limited to about ±9 V for the power supplies. This in turn limits the output swing to something less than 9 V. An exception is the LTC1150, which operates with standard

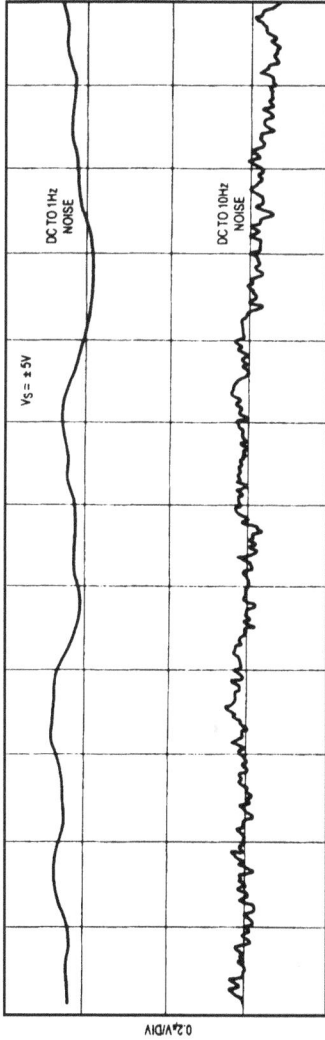

**Figure 4.11.** Noise recorded for a 10-s period (Linear Technology, Design Note 36, p. 2).

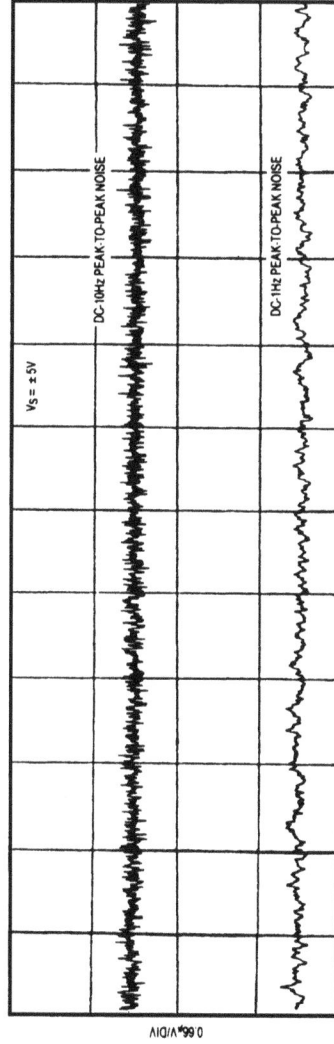

**Figure 4.12.** Noise recorded for a 10-minute period (Linear Technology, Design Note 36, p. 2).

| PART NUMBER | DESCRIPTION | MAX $V_{OS}$ (25°C) | MAX $TCV_{OS}$ | TYPICAL 0.1Hz TO 10Hz NOISE | EXTERNAL CAPS REQUIRED | MAXIMUM SUPPLY VOLTAGE |
|---|---|---|---|---|---|---|
| LTC1049 | Single, Micropower | 10µV | 0.10µV/°C | 3.0µVp-p | No | ± 9V |
| LTC1050 | Single, Low Power | 5µV | 0.05µV/°C | 1.6µVp-p | No | ± 9V |
| LTC1051 | Dual, Low Power | 5µV | 0.05µV/°C | 1.5µVp-p | No | ± 9V |
| LTC1052 | Single, 7652 Upgrade | 5µV | 0.05µV/°C | 1.5µVp-p | Yes | ± 9V |
| LTC1053 | Quad, Low Power | 5µV | 0.05µV/°C | 1.5µVp-p | No | ± 9V |
| LTC1150 | Single, ± 15V Operation | 5µV | 0.05µV/°C | 1.8µVp-p | No | ± 18V |

**Figure 4.13.** Typical chopper-stabilized op amps (Linear Technology, Design Note 42, p. 2).

| DESCRIPTION | SINGLE | DUAL | QUAD |
|---|---|---|---|
| Low Cost, Optimum Performance | LT1001 LT1012 LT1097 | LT1013 LT1078 | LT1014 LT1079 |
| Low Noise, Wideband | LT1007 LT1028 LT1037 | | |
| Low Noise, Audio | LT1115 | | |
| Single Supply, Low Power | LT1006 | LT1013 | LT1014 |
| Single Supply, Micropower | LT1077 | LT1078 LT1178 | LT1079 LT1179 |

**Figure 4.14.** Precision bipolar op amps (Linear Technology, Design Note 42, p. 2).

| PARAMETER | ADVANTAGE | | COMMENTS |
|---|---|---|---|
| | CHOPPER | BIPOLAR | |
| Offset Voltage | ✓ | | ⎫ |
| Offset Drift | ✓ | | ⎬ No Contest |
| All Other DC Specs | ✓ | | ⎭ |
| Wideband, 20Hz to 1MHz | | ✓ | See Details in Text |
| Noise | | ✓ | See Details in Text |
| Output: Light Load    Heavy Load | ✓ | ✓ | Rail to Rail Swing 2mA Limit on Choppers |
| Single Supply Application | ✓ | | Inherent to Choppers Needs Special Design Bipolars |
| ±15V Supply Voltage | | ✓ | Except LTC1150 |
| Prejudice/Tradition | | ✓ | Still a Chopper Problem |
| Cost | | ✓ | Unless DC Performance Needed |

**Figure 4.15.** Comparison of chopper-stabilized and precision bipolar op amps (Linear Technology, Design Note 42, p. 1).

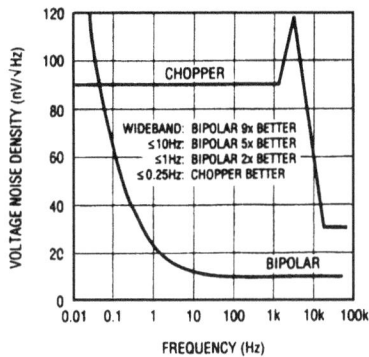

**Figure 4.16.** Noise comparison of bipolar and chopper-stabilized amplifiers (Linear Technology, Design Note 42, p. 1).

±15-V supplies (typical for most bipolar op amps). The LTC1150 still guarantees 5-µV offset and the 0.05-µV/°C drift, as shown in Fig. 4-13.

Where noise performance is crucial, bipolar op amps are superior. As shown in Fig. 4-16, bipolar noise is nine times better from 10 Hz to 1 kHz. This comparison is for the industry-standard LT1001 and the OP-07. Bipolar designs optimized for low noise (such as the LT1007, LT1028, LT1037, or LT1115) have 36 to 100 times lower noise than choppers. However, chopper-amplifier noise is flat, whereas bipolar noise depends on frequency (noise increases with decreases in frequency). Therefore, if the bandwidth is limited, chopper noise gets comparatively better.

# Wideband Transconductance Amplifier Circuit Troubleshooting

This chapter is devoted to troubleshooting for wideband transconductance amplifiers (WTAs). It starts with a comparison of WTAs to OTAs and CFAs to show the differences in troubleshooting. With this out of the way, a detailed description of a typical WTA is provided, emphasizing how circuit functions are related to troubleshooting. Then the chapter describes why WTA circuits fail in either the experimental or final form.

## 5.1 Differences in WTA Troubleshooting

WTAs are similar to OTAs and CFAs in that internal characteristics (such as gain) can be controlled by external components. However, unlike OTAs and CFAs, the WTA does not require any feedback components to produce accurate gain. In troubleshooting, this means that there is no closed-loop phase shift (because there is no external negative feedback circuit). As mentioned, the phase shift from output back to input is the primary cause of oscillation in conventional amplifiers (op amp, OTA, CFA, etc.), especially at higher frequencies.

Another factor that affects the troubleshooting of WTA circuits is amplifier gain. The output of a WTA is a current that is proportional to the applied differential voltage. This provides inherent circuit protection for the outputs. Of greater importance, circuit gain is set by the ratio of two impedances, and an internally set current-gain factor ($K$). This makes WTAs substantially different from other amplifiers in both testing and troubleshooting.

## 5.2 The Basic WTA Circuit

Figures 5-1 and 5-2 show the pinouts and typical operating circuits, respectively, for two WTAs (the Maxim MAX435 and MAX436). As shown, the

MAX435 has a differential output, whereas the MAX436 is single-ended. However, both ICs have a differential input. Because the output is a current, and because there is no negative feedback that might shift in phase, bandwidths and slew rates are considerably higher than for op amps, OTAs, and CFAs.

For example, the MAX435 has a bandwidth of 275 MHz, an 800 V/μs slew rate, and a 1% settling time of 18 ns to a 0.5-V step or pulse input. The MAX436 has the same settling time, a bandwidth of 200 MHz, and an 850 V/μs slew rate. Both ICs have a CMRR of 53 dB at 10 MHz and an input-offset voltage of 300 μV. Unlike CFAs, both WTAs have fully symmetrical, high-impedance (about 750 kΩ) inputs that tolerate wide differential-input voltage without destructive failure or amplifier saturation.

### 5.2.1   WTA input/output relationship

Figure 5-3 shows operation of the two WTAs from an input/output standpoint. In effect, WTAs are essentially voltage-controlled current sources. Signal gain is set by the ratio of two impedances: the user-selected transconductance element or network ($Z_t$) and the output-load impedance ($Z_L$). The WTA output is a current that is proportional to the differential input voltage and inversely proportional to the impedance of the user-selected $Z_t$.

A differential input voltage ($V_{IN}$) applied between the input terminals causes current to flow in the transconductance element ($Z_t$). This current is equal to $V_{IN}/Z_t$. The current in the transconductance element is multiplied by the preset current gain ($K$) of the WTA and it appears at the output terminal(s) as a

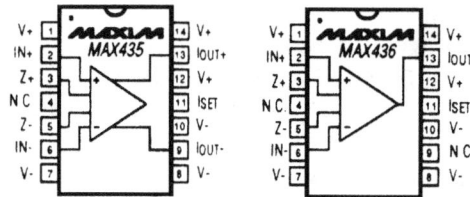

**Figure 5.1.**   Pin configuration of typical WTA (*Maxim New Releases Databook,* 1994, p. 8-5).

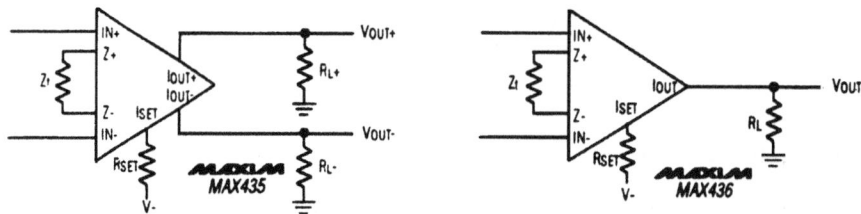

**Figure 5.2.**   Typical WTA operating circuits (*Maxim New Releases Databook,* 1994, p. 8-5).

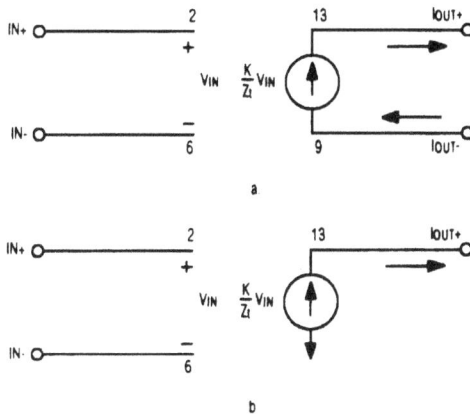

Figure 5.3. Operation of WTAs from an input/output standpoint (*Maxim New Releases Databook*, 1994, p. 8-13).

current equal to $K$ times $(V_{IN}/Z_t)$. The current flows through the load impedance to produce an output voltage according to the following equation: $V_{OUT} = K(Z_L/Z_t)V_{IN}$, where $K$ is the WTA current-gain ratio, $Z_L$ is the output-load impedance, $Z_t$ is the transconductance element impedance, and $V_{IN}$ is the differential input voltage.

With proper selection of component values, it is possible to implement an amplifier circuit that accepts differential input voltages covering the entire input-voltage range ($-2.5$ V to $+2.5$ V, in the case of these WTAs) without overloading the output stage. This characteristic makes WTAs ideal for use in wideband instrumentation amplifiers, differential amplifiers/receivers, and settling-time measurement circuits. Because the input is symmetrical in both versions, opposite signal polarity can be obtained by interchanging the input terminals.

## 5.3  WTA Gain Problems

WTAs produce an output current by multiplying a differential input voltage $(V_{IN})$ by the transconductance $(K/Z_t)$. The voltage gain $(A_V)$ is set by the impedance of the transconductance network $(Z_t)$ and the output-load impedance $(Z_L)$, according to the equation: $A_V = K(Z_L/Z_t)$.

In troubleshooting, remember that the factor $K$ in the gain equation refers to the current gain of the WTA IC. $K$ is factory trimmed to produce a low-drift, stable current gain. ($K$ is trimmed to 4.0% ±2.5% for the MAX435 and to 8.0% ±2.5% for the MAX436.) Factor $Z_t$ is the impedance of a user-selected, two-terminal transconductance element or network connected across the WTA $Z+$ and $Z-$ terminals (Fig. 5-2). Circuit gain and frequency shaping are set by selection of the transconductance network and the output impedance, so any gain problems can usually be traced to the transconductance network and/or the output impedance.

In an experimental circuit, remember that there are limits to changes in $Z_t$ or output impedance (to change circuit gain). For example, in these WTAs, the transconductance network should be selected so that the current flowing in the network (equal to $V_{IN}/Z_t$) does not exceed 2.5 mA under worst-case conditions of maximum input voltage and minimum transconductance-element impedance $(Z_t)$. Also, output current should not exceed ±10 mA per output for the MAX435 or ±20 mA for the MAX436.

## 5.4   Effect of WTA Internal Current

The supply current, power dissipation, and output-current drive capability of a WTA are set by the internal current. In turn, internal current is controlled by an external resistor $(R_{SET})$ connected from the $I_{SET}$ pin to the $V-$ supply, as shown in Fig. 5-2. A typical value for these WTAs is an $R_{SET}$ of 5.9 kΩ. This provides an output drive capability of ±10 mA per output (MAX435) and ±20 mA for the single-ended MAX436.

In troubleshooting an experimental circuit, the data sheet provides a number of graphs for typical values of load current and power dissipation for different values of $R_{SET}$. In the absence of such information, use the recommended 5.9-kΩ value for these WTAs. Also, as a practical matter, the current source should be decoupled with a 0.22-µF ceramic capacitor connected from the $I_{SET}$ pin to $V-$.

In troubleshooting any WTA circuit, remember that a larger-value $R_{SET}$ reduces WTA supply current, power dissipation, and output-current drive capability. A smaller $R_{SET}$ value increases the output-current drive capability. However, you must take care that the power-dissipation rating for the IC package type is not exceeded for the specific circuit operating conditions. It is especially important to consider the maximum load current and ambient operating temperature of the circuit.

## 5.5   Shutdown During Troubleshooting

When troubleshooting an experimental circuit, the power dissipation of the WTA can be reduced significantly by disconnecting the $R_{SET}$ resistor from the $I_{SET}$ pin. This will reduce the supply current of the MAX435 to about 450 µA, and the MAX436 to about 850 µA. Of course, the output load will not be driven. Simply reconnect the $R_{SET}$ resistor to the $I_{SET}$ pin to reactivate the WTA.

## 5.6   DC Accuracy Problems

The dc accuracy of WTA circuits is affected by several factors, including input-offset voltage, output-offset voltage, output-offset current, accuracy of current-gain factor $K$, and the tolerance of external transconductance and load impedances. The following is a summary of how these factors affect troubleshooting.

### 5.6.1   Effect of input-offset voltage (WTA)

Any input-offset voltage $(V_{OS})$ causes a current to flow in the transconductance element $(Z_t)$ when no differential voltage (signal) is applied at the input terminals. The current is then multiplied by $K$ to produce an error current at the output terminal(s), and results in a continuous dc output with no input. $V_{OS}$ is caused by a mismatch of base-emitter voltages between the transistors in the WTA input circuit. (These are the same factors that produce input-offset voltage in bipolar op amps, Chapter 1.)

You can measure $V_{OS}$ as the voltage between the transconductance terminals (from $Z+$ to $Z-$), with each side of the input ($IN+$ and $IN-$) grounded, and no transconductance element in the circuit ($Z+$ and $Z-$ are open). However, you cannot change or eliminate $V_{OS}$ because the factor is built into the WTA IC.

### 5.6.2  Effect of input-offset current (WTA)

The output-offset current ($I_{OS}$) is a second source of dc errors in WTA circuits. $I_{OS}$ is measured with no transconductance element in the circuit ($Z+$ and $Z-$ are open) and the input terminals ($IN+$ and $IN-$) are grounded (the same as for $V_{OS}$). $I_{OS}$ is the current that flows from the output terminal of the WTA under these conditions, and is the result of imperfect matching of devices in the output-current mirror circuits of the WTA. Notice that output current caused by the input-offset voltage ($V_{OS}$) is not included in the $I_{OS}$ specification because that component of the total output current varies with the value of the transconductance element.

### 5.6.3  Total dc output-voltage error (WTA)

The total dc output-voltage error (sometimes listed as the $V_{ERR}$) depends on $I_{OS}$, $V_{OS}$, $K$, output-load resistance $R_L$, and transconductance-element resistance $R_t$. In troubleshooting, only $R_L$ and $R_t$ can be altered or corrected because the remaining factors are part of the WTA IC.

### 5.6.4  Effect of current-gain factor $K$ (WTA)

In addition to affecting the dc accuracy of WTAs, the $K$ factor is also a source of gain error in WTA circuits. Again, you cannot change the $K$ factor. However, the $K$ for each of the WTAs is factory trimmed to an initial tolerance of $\pm 2.5\%$. The variation of $K$ with operating temperature is listed in the data sheet.

### 5.6.5  Effect of finite output impedance (WTA)

Figure 5-4 shows how output impedance for WTAs is calculated. In addition to affecting the dc accuracy, the finite output impedance also affects WTA gain. The output(s) of the WTAs are voltage-controlled current sources. An ideal current source would have an infinite output impedance so that the output current would be independent of the load impedance. In the real world, assume that the MAX435/436 output impedance is about 3.5 k$\Omega$ for troubleshooting purposes.

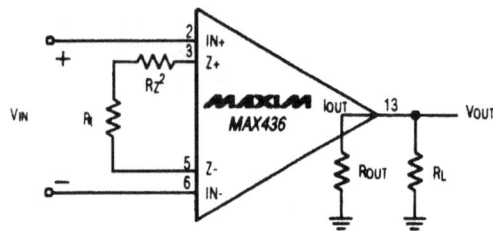

$$R_{LEQ} = R_{OUT} \parallel R_L = \left[ \frac{(R_{OUT})(R_L)}{R_{OUT} + R_L} \right]$$

$$V_{OUT} = (V_{IN}) \left( \frac{K}{R_t + R_Z} \cdot \right) \left[ \frac{(R_{OUT})(R_L)}{R_{OUT} + R_L} \right]$$

$R_Z = 0.15\Omega$
$R_{OUT} = 3.5k\Omega$
* $R_Z$ IS $Z$- TERMINAL INPUT IMPEDANCE  SINCE $R_Z$ IS TYPICALLY 0.15$\Omega$, IT CAN USUALLY BE IGNORED

**Figure 5.4.** Finite output-impedance of WTAs (*Maxim New Releases Databook,* 1994, p. 8-15).

### 5.6.6    Calculating WTA circuit gain and voltage-gain error

The theoretical circuit gain ($A_V$) and voltage-gain error ($dA_V$) of a WTA circuit can be calculated using the equations of Fig. 5-4. In troubleshooting, these factors can then be compared against actual circuit performance. Notice that the WTA output impedance ($R_{OUT}$) is parallel with the circuit load-impedance ($R_L$), reducing the equivalent load impedance. When troubleshooting any WTA circuit, remember that you cannot change $R_{OUT}$, but you can change $R_L$ (in most circumstances).

After accounting for the finite WTA output impedance ($R_{OUT}$), the actual circuit gain is calculated as:

$$A_V = V_{OUT}/V_{IN} = (K R_{OUT} R_L)/(R_t) (R_{OUT} + R_L)$$

The voltage-gain error ($dA_V$), with respect to the theoretical gain $A_V = K$ times ($R_L/R_t$) is equal to:

$$dA_V/A_V = R_L/(R_L + R_{OUT})$$

### 5.7    Board Layout and Power-Supply Bypassing

As in the case of any amplifier, good PC-board layout and power-supply bypassing are important to WTA performance. These factors become crucial as the frequency increases. Although a WTA eliminates closed-loop phase shift as a course of circuit oscillation (a major problem in all other IC amplifiers), careful high-frequency techniques should be used to optimize the performance of WTA circuits. The following is a summary of the major points to be considered, especially when troubleshooting experimental circuits.

It is recommended that a ground plane be used. The ground plane should include the entire portion of the PC board that is not dedicated to a specific signal trace.

Sockets are not recommended with WTAs because the additional pin-to-pin capacitance introduced by sockets degrades the wideband performance.

Keep the length of traces connected to the WTA input terminals as short as possible to minimize signal reflections and/or inductive coupling of high-frequency signals to the WTAs. If the signal input signals must travel more than a few inches, use controlled-impedance lines or coaxial cables. All signals should be properly terminated. Minimize the PC-board pad area for input connections to prevent capacitive coupling of stray high-frequency signals.

Passive components used in WTA circuits should (preferably) be surface mounted to minimize stray inductance. If surface-mount components are not used, keep component lead lengths to an absolute minimum.

Bypass each power supply directly to the ground plane with a 0.22-$\mu$F ceramic capacitor, placed as close to the supply pins as possible. Bypass the $I_{SET}$ pin with a 0.22-$\mu$F ceramic to the $V+$ pin. Keep capacitor lead lengths as short as possible to minimize series inductance. Surface-mount (chip) capacitors are ideal for this purpose.

## 5.8   Capacitive Load Problems (WTA)

Because the WTA requires no feedback, phase shift (because of capacitive loading of the output) does not degrade the output. However, capacitive loading does reduce slew rate and bandwidth. This is because these factors are limited by the rate at which the WTA output current can charge the capacitive load. In troubleshooting, especially with experimental circuits, avoid capacitive coupling from the WTA output terminals to the input or transconductance terminals. Such connections introduce high-frequency feedback and could result in oscillation.

## 5.9   WTA Circuit Troubleshooting Examples

The remainder of this chapter is devoted to troubleshooting a variety of WTA circuits. All of the troubleshooting techniques covered thus far should be used. For simplicity, the bypass capacitors and the $R_{SET}$ resistor are not shown on the following circuit schematics.

Except for the last circuit, the value of $R_{SET}$ is the recommended 5.9 k$\Omega$. Each power supply ($V+$ and $V-$) is bypassed to ground with a 0.22-$\mu$F ceramic capacitor. The $I_{SET}$ pin is also bypassed with a 0.22-$\mu$F ceramic.

## 5.10   Coaxial-Cable Driver (WTA) Circuit Troubleshooting

Figures 5-5 and 5-6 show the MAX435 and MAX436, respectively, connected as coaxial-cable drivers. Figures 5-7 and 5-8 show the pulse response for the MAX435 and MAX436 circuits, respectively.

To maximize power transfer and minimize distortion of high-speed signals, transmission lines must be terminated at both the source and receiving ends

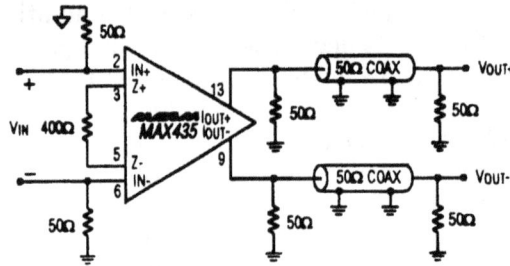

**Figure 5.5.** MAX435 coaxial-cable driving circuit (*Maxim New Releases Databook*, 1994, p. 8-15).

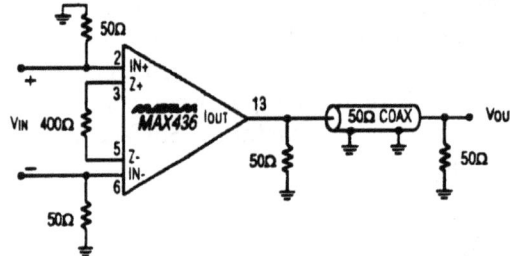

**Figure 5.6.** MAX436 coaxial-cable driving circuit (*Maxim New Releases Databook*, 1994, p. 8-16).

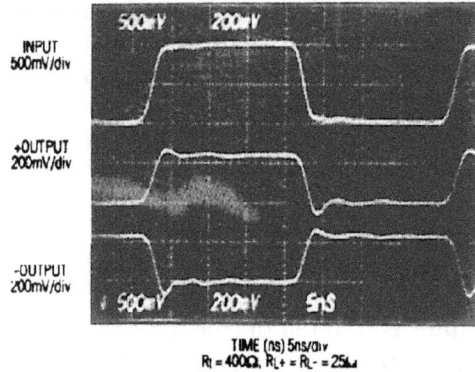

**Figure 5.7.** MAX435 pulse response (*Maxim New Releases Databook*, 1994, p. 8-16).

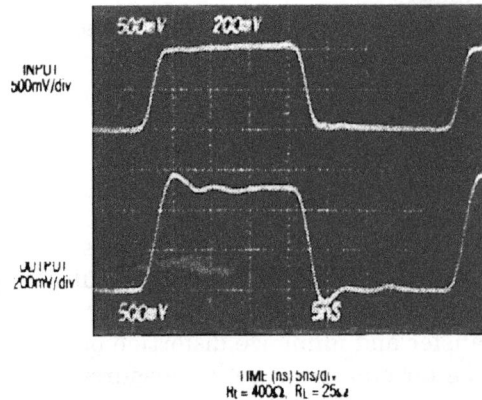

**Figure 5.8.** MAX436 pulse response (*Maxim New Releases Databook*, 1994, p. 8-16).

with the characteristic impedance of the line. (A 50-$\Omega$ coaxial cable requires an impedance of 50 $\Omega$ at each end for optimum performance.)

In troubleshooting these circuits in experimental form, the only resistance value that could be changed is the resistance between pins 3 and 5. All of the remaining resistance values are set by the characteristic impedance of the cables.

There is a basic difference in troubleshooting these WTA circuits from similar circuits using op amps and CFAs. With voltage-mode amplifiers, the transmitting end of a coaxial cable is typically back-terminated with a resistor in series with the amplifier output (as shown for the CFA cable driver in Fig. 3-30). This is because the output of a voltage-mode amplifier (typically) has a low impedance. Because the output of a WTA is a current source, the output impedance is relatively high (about 3.5 k$\Omega$ in this case). As a result, when driving coaxial cables with WTAs, the back-termination resistor should be in parallel with the output impedance of the WTA, as shown in Fig. 5-5 and 5-6. Notice that back-terminating cables in this manner reduces the effective voltage gain of the circuit by a factor of 2.

## 5.11    Summing Amplifier Circuit Troubleshooting

Figure 5-9 shows how two or more signals can be summed together using WTAs. The output voltage $V_{OUT}$ is the sum of the two output voltages, as shown by the equations. In troubleshooting, two factors could produce an incorrect $V_{OUT}$. If the internal $K$ factors of the two WTAs are not matched, there will be an unbalance, and $V_{OUT}$ will not be a true sum. The same is true for resistances $R_{t1}$ and $R_{t2}$. If it is crucial that $V_{OUT}$ is the true sum of the two input signals, it is possible to adjust the value of either $R_{t1}$ or $R_{t2}$ to offset any unbalance in $K$ factors (which cannot be changed).

## 5.12    Lowpass Amplifier Circuit Troubleshooting

Figure 5-10 shows a single-ended MAX436 connected as a lowpass amplifier. Figure 5-11 shows the gain/frequency response. (Notice that Fig. 8 in Fig. 5-11 refers to Fig. 5-10.)

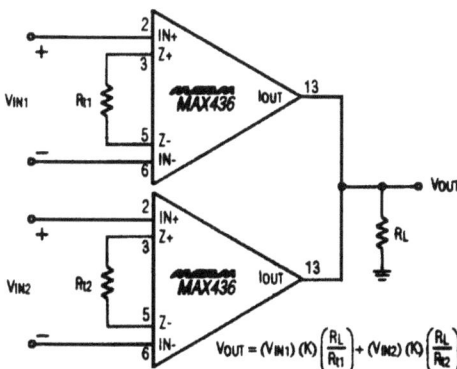

$$V_{OUT} = (V_{IN1})(K)\left(\frac{R_L}{R_{t1}}\right) + (V_{IN2})(K)\left(\frac{R_L}{R_{t2}}\right)$$

Figure 5.9. · WTA summing amplifier (*Maxim New Releases Databook,* 1994, p. 8-16).

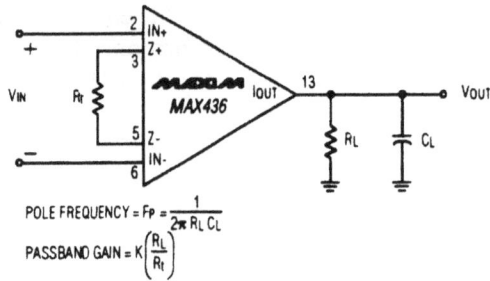

POLE FREQUENCY = FP = $\frac{1}{2\pi\,R_L\,C_L}$

PASSBAND GAIN = $K\left(\frac{R_L}{R_t}\right)$

**Figure 5.10.** WTA lowpass amplifier (*Maxim New Releases Databook,* 1994, p. 8-17).

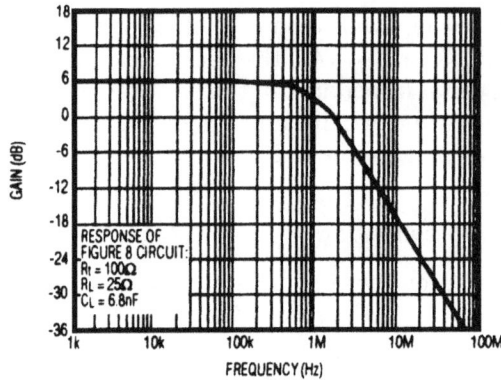

**Figure 5.11.** WTA lowpass-amplifier characteristics (*Maxim New Releases Databook,* 1994, p. 8-17).

In troubleshooting, if the pole (or corner) frequency is incorrect, but the gain is good, suspect $C_L$. If the gain is incorrect, but the pole frequency is good, suspect $R_t$ (or the WTA $K$ factor). If both gain and frequency are incorrect, $R_L$ is the most likely cause of trouble. However, a leaking $C_L$ could cause problems in both gain and frequency.

## 5.13  Highpass Amplifier Circuit Troubleshooting

Figure 5-12 shows a single-ended MAX436 connected as a highpass amplifier. Figure 5-13 shows the gain/frequency response. (Note that Fig. 10 in Fig. 5-13 refers to Fig. 5-12.)

In troubleshooting, if the pole (or corner) frequency is incorrect, but the gain is good, suspect a problem at $C_t$. If the gain is incorrect, but the pole frequency is good, suspect $R_L$ (or the WTA $K$ factor). If both gain and frequency are incorrect, $R_t$ is suspect. However, a leaking $C_t$ could cause problems in both gain and frequency.

## 5.14  Bandpass Amplifier Circuit Troubleshooting

Figure 5-14 shows a single-ended MAX436 connected as a bandpass amplifier. Figure 5-15 shows the gain/frequency response. (Notice that Fig. 12 in Fig. 5-15 refers to Fig. 5-14.)

$$\text{CORNER FREQUENCY} = F_C = \frac{1}{(2\pi)(R_t)(C_t)}$$

$$\text{PASSBAND GAIN} = K\left(\frac{R_L}{R_t}\right)$$

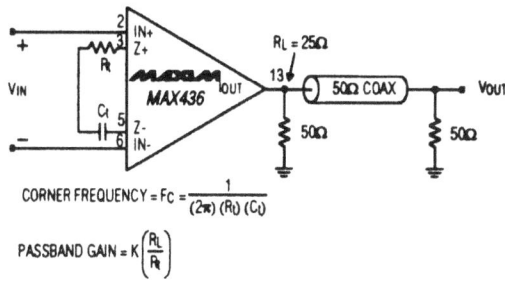

**Figure 5.12.** WTA highpass amplifier (*Maxim New Releases Databook,* 1994, p. 8-17).

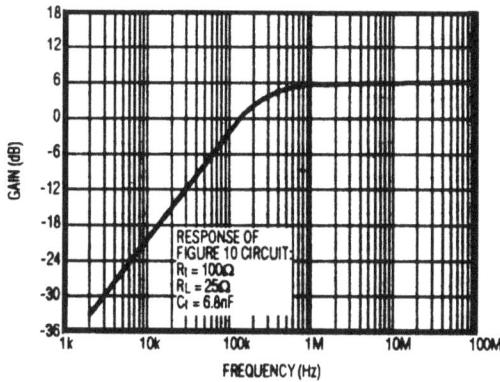

**Figure 5.13.** WTA highpass-amplifier characteristics (*Maxim New Releases Databook,* 1994, p. 8-17).

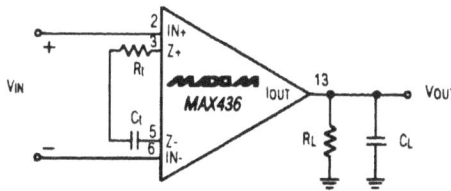

$$\text{LOW CORNER FREQUENCY} = F_L = \frac{1}{(2\pi)(R_t)(C_t)}$$

$$\text{POLE FREQUENCY} = F_P = \frac{1}{(2\pi)(R_L)(C_L)}$$

$$\text{PASSBAND GAIN} = K\left(\frac{R_L}{R_t}\right)$$

**Figure 5.14.** WTA bandpass amplifier (*Maxim New Releases Databook,* 1994, p. 8-18).

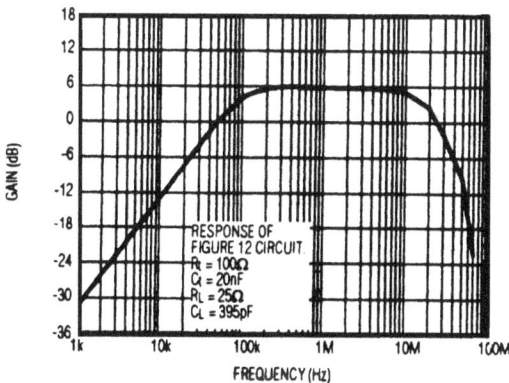

**Figure 5.15.** WTA bandpass-amplifier characteristics (*Maxim New Releases Databook,* 1994, p. 8-18).

In troubleshooting, notice that the low corner frequency is set by $R_t/C_t$, and the pole frequency is set by $R_L/C_L$. The passband gain is set by $K$, $R_L$, and $R_t$. Because there is no interaction between the transconductance-network impedance and the output-load impedance, poles and zeros in the WTA transfer function can be independently set by the two impedance networks. For example, if only the low corner frequency is incorrect, suspect $C_t$. If only the pole frequency is incorrect, suspect $C_L$. If the frequency response is good, but the gain is incorrect, suspect the WTA $K$ factor (which cannot be changed).

## 5.15    Tuned Amplifier Circuit Troubleshooting

Figure 5-16 shows a single-ended MAX436 connected as a tuned amplifier. Figure 5-17 shows the gain/frequency response. (Notice that Fig. 14 in Fig. 5-17 refers to Fig. 5-16.)

In troubleshooting, notice that the high corner frequency is set by the transconductance network of $L_t/C_t$. The resonant frequency of the transconductance network is determined by the equation shown. The impedance of the transconductance network is minimum at the resonant frequency. This provides maximum amplifier gain at that frequency. The $Q$ (or tuning sharpness) of the amplifier is a function of the parasitic components associated with the $L_t/C_t$ net-

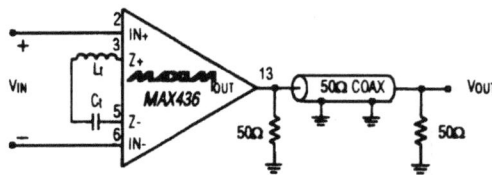

$$\text{HIGH CORNER FREQUENCY} = F_H = \frac{1}{2\pi \sqrt{(L_t)(C_t)}}$$

Q IS A FUNCTION OF PARASITICS OF $L_t$ AND $C_t$

**Figure 5.16.** WTA tuned amplifier (*Maxim New Releases Databook,* 1994, p. 8-18).

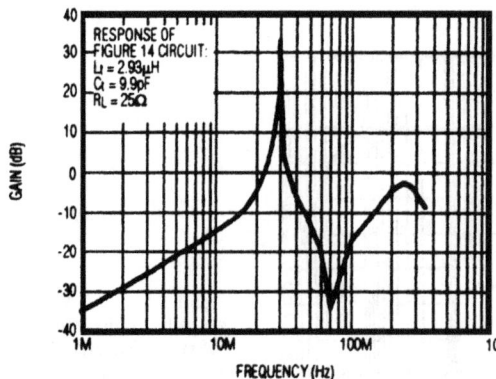

**Figure 5.17.** WTA tuned-amplifier characteristics (*Maxim New Releases Databook,* 1994, p. 8-18).

work. If the $L_t/C_t$ network contains much resistance, the tuning will be broad (low $Q$). A minimum of resistance in the $L_t/C_t$ network produces sharp tuning (high $Q$).

## 5.16    Crystal-Tuned Amplifier Circuit Troubleshooting

Figure 5-18 shows a single-ended MAX436 connected as a crystal-tuned amplifier. Figure 5-19 shows the gain/frequency response. (Notice that Fig. 16 in Fig. 5-19 refers to Fig. 5-18.)

In troubleshooting, notice that the crystal replaces the $L_t/C_t$ network shown in Fig. 5-16. A crystal provides a much higher $Q$ (sharp tuning) and more accurate control of the tuned-amplifier frequency. The crystal impedance is minimum at the resonant frequency (25 MHz), resulting in maximum gain for the amplifier at that frequency. The accuracy of the selected frequency for this circuit depends entirely on the accuracy of the crystal. Gain for the circuit is set primarily by the WTA $K$ factor and to a lesser extent by the output load resistance.

## 5.17    Driver/Receiver Circuit Troubleshooting

Figure 5-20 shows the MAX435 and MAX436 connected as a driver/receiver circuit for transmission of video signals over a twisted-pair line (instead of a coaxial cable). One differential input of the MAX435 drives the balanced twisted

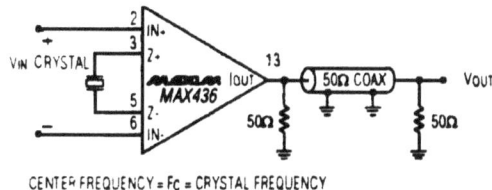

**Figure 5.18.**   WTA crystal-tuned amplifier
(*Maxim New Releases Databook,* 1994, p. 8-19).

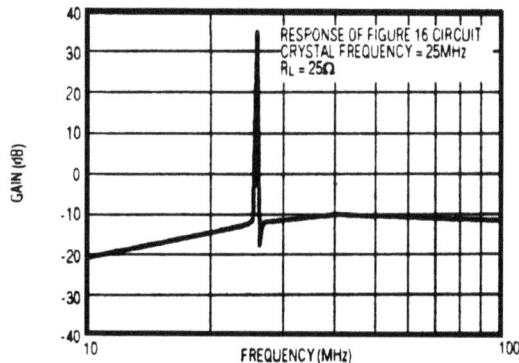

**Figure 5.19.**   WTA crystal-tuned amplifier characteristics (*Maxim New Releases Databook,* 1994, p. 8-19).

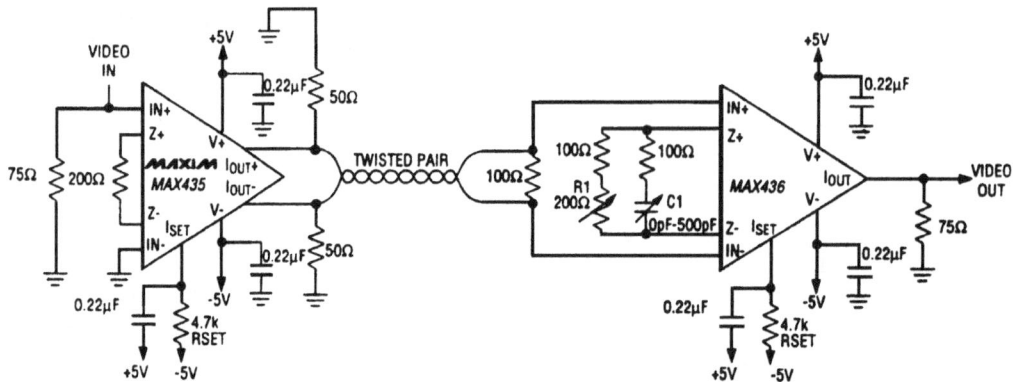

**Figure 5.20.** WTA video twisted-pair driver/receiver (*Maxim New Releases Databook,* 1994, p. 8-19).

pair from a ground-referred input signal, eliminating the need for a balun transformer or for two single-ended output drivers. The circuit is good for distances up to 5000 feet, where a single channel of baseband (composite) video is to be transmitted. The twisted-pair line is far less expensive than a corresponding length of coaxial cable.

In troubleshooting, two special problems must be considered, in addition to the usual factors (proper bypassing, correct values for transconductance network, etc.). First, the transmission line must be balanced (or differential) to minimize common-mode noise. The differential output of the MAX435 and differential input of the MAX436 make this possible. Second, the twisted-pair line must be properly terminated to minimize reflections. The circuit shown and the MAX435/436 characteristics (CMRR of 53 dB and high input/output impedances) meet both of these requirements.

The receiver portion of the circuit uses the single-ended output of the MAX436 for balanced-to-single-ended line conversion. The 100-$\Omega$ resistor from IN+ to IN− provides proper line termination. The transconductance network (from $Z+$ to $Z-$) provides for gain adjustment (set for +6 dB), as well as line equalization. (Line equalization is sometimes required to boost the high-frequency gain of the receiver to make up for the limited bandwidth of the twisted pair.)

In troubleshooting, if the circuit does not perform properly with all of the components known to be good, the problem could be improper adjustment. The circuit requires line-equalization adjustment to get proper brightness and color for the video signals. R1 is adjusted for proper brightness (to boost overall gain to compensate for resistance losses in the twisted pair). C1 is adjusted for best color (to extend bandwidth). Both adjustments are performed while viewing the screen (TV, monitor, etc.) at the receiver end.

# 6

# Audio-Amplifier Circuit Troubleshooting

This chapter is devoted to troubleshooting audio amplifiers. It starts with a review of audio-amplifier basics (bias circuits, operating point, distribution, frequency limitations, coupling methods, etc.) from a troubleshooting standpoint. This is followed by an example of audio-amplifier troubleshooting that covers both discrete and IC circuits. It then covers typical IC audio amplifiers, emphasizing how circuit functions are related to troubleshooting.

## 6.1  Audio-Amplifier Bias Circuits

Figure 6-1 shows a typical audio-amplifier bias network. (Similar bias circuits are also used in amplifiers operating at frequencies higher than the audio range.) The bias networks serve more than one purpose. Typically, the bias-circuit resistors (1) set the operating point, (2) stabilize the circuit at the operating point, and (3) set the approximate input-output impedances of the circuit. The following is a summary of how these functions must be considered during troubleshooting.

### 6.1.1  Amplifier operating point

The basic purpose of the bias network is to establish collector-base-emitter voltage and current relationships at the operating point of the amplifier circuit. The operating point is also known as the *quiescent point, Q point, no-signal point, idle point,* or *static point,* and is covered in Section 6.2.

In troubleshooting experimental circuits, it might be necessary to change the values of the bias-circuit resistors. Because transistors rarely operate at the $Q$ point, the basic bias networks are generally used as a reference or starting point for design. The actual circuit configuration (and especially the bias network values) are generally selected on the basis of dynamic circuit conditions (desired output voltage swing, expected input signal level, etc.). If any of these characteristics are incorrect, suspect the bias network.

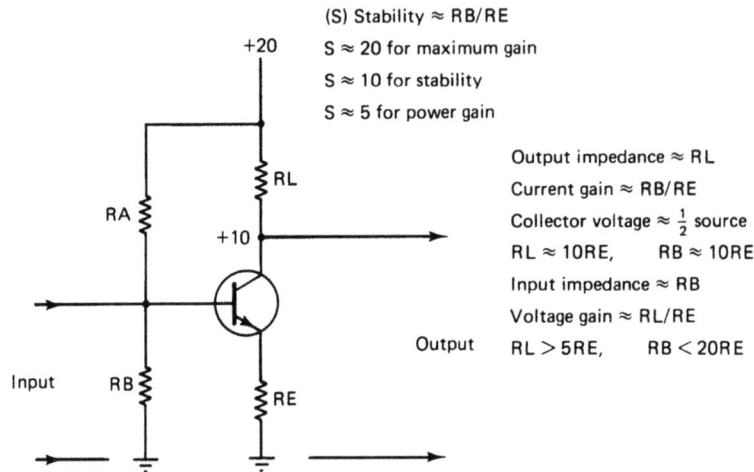

**Figure 6.1.** Typical audio-amplifier bias network.

### 6.1.2  Bias stabilization

The next function of the bias network is to stabilize the amplifier circuit at the desired operating point. Although there are many bias-network configurations, each with advantages and disadvantages, one major factor must be considered for any network. The network must maintain the desired current relationships in the presence of temperature and power-supply changes (and possible transistor replacement). In some cases, frequency changes (and changes caused by component aging) must also be offset by the bias network.

### 6.1.3  Thermal runaway

The most undesirable effect of poor bias stabilization is thermal runaway. When current passes through a transistor junction, heat is generated. In turn, this causes more current to flow through the junction—even though the voltage, circuit values, etc., remain the same. The additional current causes even more heat, with a corresponding increase in current flow. If the heat is not dissipated by some means (IC or transistor case characteristics, heatsink, etc.), the transistor burns out.

In troubleshooting a circuit with a repeated burnout, first suspect the bias network. A possible exception is where the transistor or IC should have a heatsink. Adequate bias stabilization prevents any drastic change in junction currents, despite changes in temperature, voltage, etc. (within practical limits).

### 6.1.4  Input/output impedances

The resistors used in bias networks also have the function of setting the input and output impedances of the amplifier circuit. In theory, the input/output impedances of an amplifier circuit are set by many factors (transistor beta, transistor input/output capacitance, etc.). In practical troubleshooting, the input/output impedances of a resistance-coupled amplifier (at frequencies up to about 100 kHz) are set by the bias network resistors. For example, the output impedance of the amplifier shown in Fig. 6-1 is set by the value of resistance $R_L$.

### 6.1.5  Basic bias-stabilization techniques

Bias stabilization involves negative feedback or inverse feedback. The most common form of such feedback in audio circuits is *emitter feedback* (also known as *inverse-current feedback,* such as shown in Fig. 6-1. Base current (and, consequently, collector current) depends on the differential in voltage between base and emitter. If the differential voltage is lowered, less base current (and, consequently, less collector current) flows. The opposite is true when the differential is increased.

In troubleshooting this basic circuit, remember that all current flowing through the collector (ignoring collector-base leakage) also flows through the emitter resistor. The voltage across the emitter resistor therefore depends (in part) on the collector current. Should the collector current increase for any reason (an increase in junction temperature, for example), emitter current and the voltage drop across the emitter resistor also increase. This tends to decrease the differential between base and emitter, thus lowering the base current. In turn, the lower base current tends to decrease the collector current and offset the initial collector-current increase.

### 6.2  Amplifier Classifications Based on Operating Point

Amplifiers are often classified as to *operating point* (the amount of current flow under no-signal conditions). The following is a brief summary of the four basic operating-point classifications.

In troubleshooting any amplifier circuit, remember that the base-collector junction is always reverse biased at the operating point. No base-collector current flows (with the possible exception of reverse-leakage current). On the other hand, the base-emitter junction is biased so that base-emitter current flows under certain conditions and possibly under all conditions. When base-emitter current flows, emitter-collector current also flows.

### 6.2.1  Class-A amplifier

As shown in Fig. 6-2, a class-A amplifier operates only over the linear portion of the transistor characteristic curve. (The curve represents the relationship between base voltage, or input, and collector current, or output.) At no point of the input signal cycle does the base become so positive or negative that the transistor operates on the nonlinear portion of the curve. The transistor collector current is never cut off nor does the transistor ever reach saturation.

The main advantage of the class-A amplifier is the relative lack of distortion. The output waveform follows that of the input waveform, except in amplified

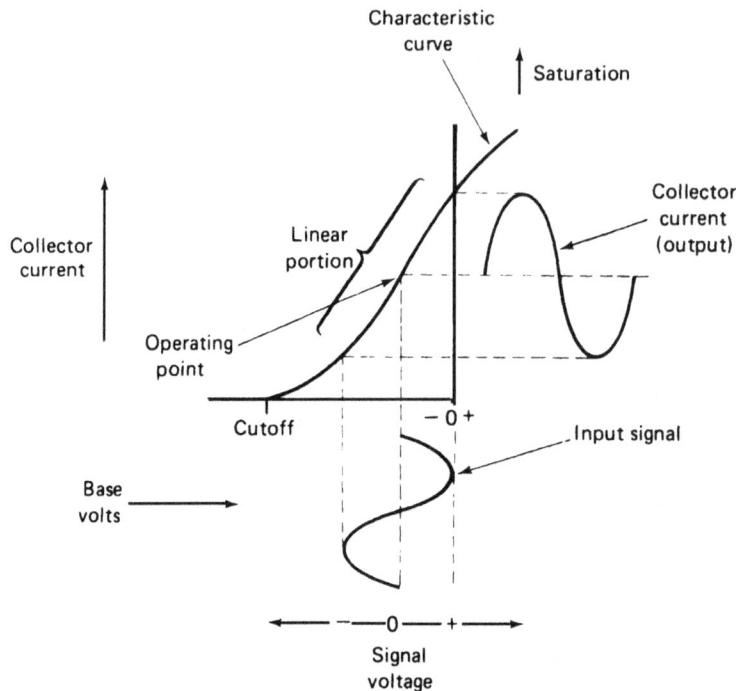

**Figure 6.2.**  Typical class-A amplifier characteristic curve.

form. However, any class of amplifier has some distortion. The main disadvantages of class-A amplifiers are relative inefficiency (lower power output for a high power input) and the inability to handle large signals. Rarely is a class-A amplifier more than about 35% efficient. If the power input to a class-A amplifier is 1 W (generally, the maximum power-dissipation capability of a single transistor), the output is less than 0.3 W.

The peak-to-peak output signal-voltage swing of a class-A amplifier is limited to something less than the total supply voltage. Because the output voltage must swing both positive and negative, the peak output is less than one half of the supply voltage.

For example, assume that the supply is 15 V and the amplifier is biased so that the $Q$-point collector voltage is one-half the supply, 7.5 V. (Such a $Q$ point is generally typical for a class-A amplifier.) Under these conditions, the output voltage cannot exceed ±7.5 V. If distortion is to be at a minimum (the usual reason for using class-A amplifiers), the output is usually about ±5 V (with a 15-V supply). This keeps the transistor on the linear portion of the curve.

Typically, the curve becomes nonlinear near the cutoff and saturation points. However, this can be determined only from an actual test of the amplifier circuit. In any case, the input voltage swing of a class-A amplifier is limited by the output voltage-swing capability and the voltage-amplification factor. For example, if the output is limited to ±7.5 V and the voltage amplification factor is 100, the input is limited to ±0.075 V (75 mV).

In troubleshooting any audio-amplifier circuit, remember that class A is generally used as a voltage amplifier because of the limitations described. Typically, a class-A amplifier stage is used ahead of a power-amplifier stage in audio circuits.

### 6.2.2 Class-B amplifier

As shown in Fig. 6-3, a class-B amplifier operates only on one half of the input signal. Class-B operation is produced when the base-emitter bias is set so that the operating point coincides with the transistor cutoff point. For an NPN transistor, this means making the base more negative than for class-A operation. (For PNP transistors, class B is produced when the base is more positive than for class A.) Either way, the base-emitter reverse bias is increased for class-B operation.

As shown in Fig. 6-3, when the input signal is zero, the collector current does not flow. During the positive half cycle of the signal voltage (Fig. 6-3 is for an NPN transistor), the collector current rises to the peak, then falls back to zero in step with the variations of that half cycle. During the negative half cycle of the signal voltage, there is no collector current because the base-emitter reverse bias is always greater than the transistor cutoff voltage. The collector current flows only during half of the input signal cycle.

Considerable distortion occurs if a single transistor is operated as class B. This is because the waveform of the resulting collector current resembles that of the positive half cycle of the input signal and, consequently, does not resemble the complete waveform of the input. Class B is generally used when two transistors are connected in push-pull. This makes it possible to reconstruct an output waveform that resembles the full waveform of the input.

Class-B amplifiers are generally used as power amplifiers, rather than voltage amplifiers. In a typical audio-amplifier circuit using discrete components (they still exist!), two push-pull transistors are operated in class B, preceded by a single class-A amplifier stage. The class-A stage provides voltage amplification, whereas the class-B stage produces the necessary power amplification.

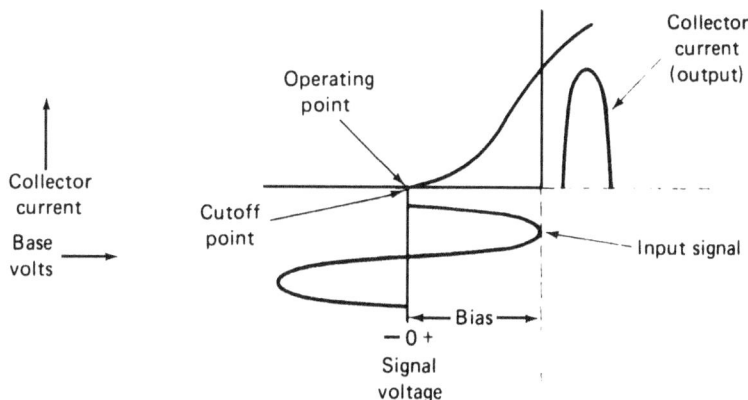

**Figure 6.3.**  Typical class-B amplifier characteristic curve.

### 6.2.3  Class-AB amplifier

Class B is the most efficient operating mode for audio amplifiers because the least amount of current is drawn. That is, the transistors are cut off at the $Q$ point and draw collector current only in the presence of an input signal. However, true class-B operation often results in crossover distortion.

Figure 6-4 shows the effects of crossover distortion. Compare the input and output waveforms. In true class-B operation, the transistor remains cut off at very low inputs (because transistors have a low current gain at cutoff) and turns on abruptly with a large signal. There is no current flow when the base-emitter voltage is below about 0.65 V (for silicon transistors). During the instantaneous pause, when one transistor stops conducting and the other starts conducting, the output waveform is distorted.

In troubleshooting, signal distortion is not the only bad effect of crossover distortion. The instantaneous cutoff of collector current can set up large volt-

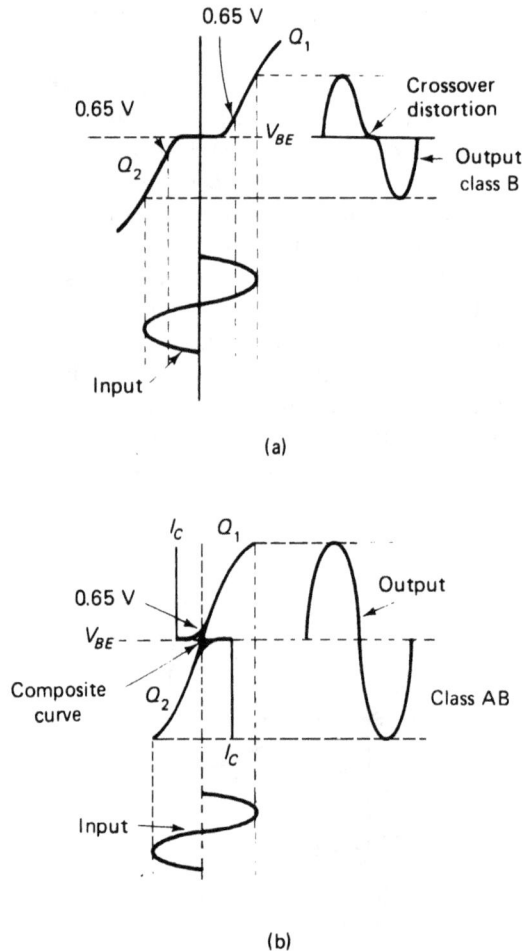

(a)

(b)

**Figure 6.4.**  Effects of crossover distortion.

age transients equal to several times the size of the supply voltage. This can cause the transistor to break down.

### 6.2.4   Minimizing crossover distortion

Although crossover distortion cannot be completely eliminated, it is possible to minimize the effects by operating the output stage as class AB (or somewhere between B and AB). As shown in Fig. 6-4B, the transistors are forward biased just enough for a small amount of collector current to flow at the $Q$ point. Some collector current flows at the lowest signal levels and there is no abrupt change in current gain. The combined collector currents produce a composite curve that is essentially linear at the crossover point, resulting in a faithful reproduction of the input (at least as far as the crossover point is concerned). Of course, class AB is less efficient than class B because more current must be used.

In troubleshooting audio power circuits, you could find an alternative method to minimize crossover distortion. This technique involves putting diodes in series with the collector or emitter leads of the push-pull transistors. Because the voltage must reach a certain value (typically 0.65 V for silicon diodes) before the diode conducts, the collector-current curve is rounded (not sharp) at the crossover point.

### 6.2.5   Class-C amplifier

Figure 6-5 shows the characteristic curve of a typical class-C amplifier. Notice that the transistor is reverse biased considerably below the cutoff point. During the positive half cycle of the input signal, the signal voltage starts from zero, rises to the positive peak value, and falls back to zero. (Figure 6-5 is for an NPN transistor.) Notice that a portion of the input signals causes the base-emitter junction to be forward biased. There is a flow of collector current for a portion of half the input cycle. The negative half cycle of the input signal lies well below the cutoff point of the transmitter.

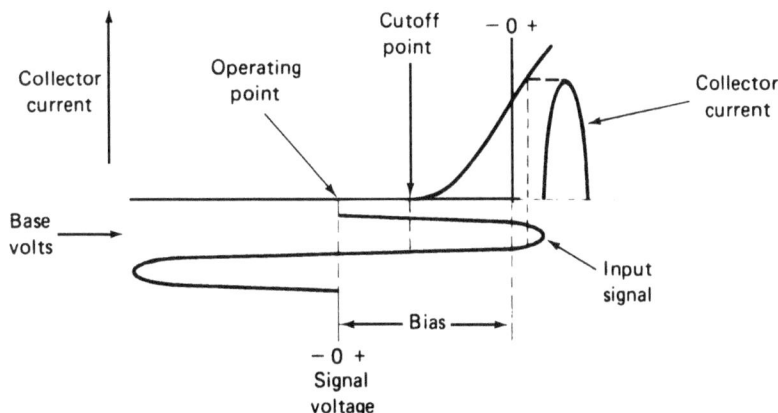

**Figure 6.5.** Typical class-C amplifier characteristic curve.

Collector current flows only during that portion of the positive half cycle of the input signal between the cutoff point and the peak. The resulting collector current is a pulse, the duration of which is considerably less than a half cycle of the input signal. Obviously, the waveform of the output signal cannot resemble that of the input signal. Nor can this resemblance be restored by the push-pull method (as with class B or AB). Class C is limited to those applications where distortion is of no concern. Generally, class C is limited to use in RF amplifiers.

## 6.3  Amplifier-Circuit Distortion Problems

For these purposes, *distortion* is defined as that condition when the output signal of an amplifier is not identical to the waveform of the input signal. A small amount of distortion is generally present in all amplifiers. However, amplifier circuits are usually designed to keep such distortion within acceptable limits. In some cases, amplifiers have circuits that introduce a form of distortion. This is generally to offset or compensate for distortion already present in the signal.

In addition to crossover distortion (Section 6.2.4), intermodulation distortion (Section 1.5.15), and harmonic distortion (Section 1.5.14), other specific types and causes of distortion occur in amplifier circuits. These include amplitude distortion, frequency distortion, and phase distortion. Any of these, either separately or in combination, might be present in an amplifier of any type. The following is a summary of distortion problems from a troubleshooting standpoint.

### 6.3.1  Amplitude-distortion problems

*Amplitude distortion* occurs in the transistor and is the result of operating the transistor over the nonlinear portion of the characteristic curve.

In troubleshooting an experimental circuit, the usual remedy for amplitude distortion is to use a base-emitter bias that places the operating point well within the linear portion of the curve, preferably at the center of the linear portion. Also, the amplitude of the input signal must be small enough so that the positive and negative half cycles do not drive the transistor beyond the linear portion. Generally, a low input signal and proper operating point (at the center of the linear portion) cause gain to be sacrificed. In any event, an overdriven amplifier (used to get maximum gain) almost always results in some amplitude distortion.

### 6.3.2  Frequency-distortion problems

*Frequency distortion* occurs because the input signal rarely, if ever, is at a single frequency. Instead, the input signal usually contains components of several frequencies, making the signal waveform somewhat complex. In addition to a transistor, an amplifier circuit is composed of resistors, capacitors, and possibly inductances (coils and transformers). Capacitors and inductances have reactance. Because reactance is a function of frequency, the different frequencies of the input signal encounter different reactances. The high and low frequencies of the signal can be impeded in different degrees or amounts. This distorts the signal waveform from the original.

For example, assume that the input signal is a complex waveform, composed of three frequencies: 10, 100, and 1000 Hz, all of the same amplitude. The reactance of coupling capacitors between stages is different for each of the three frequencies. Capacitive reactance increases with a decrease in frequency. The 10-Hz signal is attenuated more than the 100-Hz signal and much more than the 1000-Hz signal. Even though the transistor amplifies all three frequencies equally, the signals are no longer of equal amplitude and the output waveform is different (distorted) from the input.

To minimize the effect of frequency distortion, amplifier circuits are usually designed to eliminate unwanted capacitance and inductance. In troubleshooting an amplifier circuit with poor frequency response (assuming that the components, such as transistors and capacitors, are good and of the correct value), the problem is almost always the result of poor PC-board layout or component placement.

### 6.3.3   Phase-distortion problems

The fact that the input signal contains components of different frequencies is also responsible for *phase distortion* in an amplifier circuit. When a signal flows through a capacitor or an inductor, the signal encounters a shift in phase. The degree of this phase shift is a function of frequency. The high- and low-frequency components of the signal are phase shifted by different amounts. These different phase shifts cause a distortion of the signal waveform.

As is the case with frequency distortion, phase distortion in amplifier circuits is generally the result of poor PC-board layout or other component-placement problems.

## 6.4   Frequency Limitations of Circuit Components

Were it not for reactance, a transistor (by itself) should be capable of operating at any frequency from zero (direct current) on up. The top frequency limit should be set only by the transit time of electrons across the transistor junctions. However, limitations are placed on the operating frequency of any transistor by the transistor characteristics. The other components (capacitors, resistors, inductors, etc.) used in amplifier circuits also limit the operating frequency.

Every electronic component has some impedance and is thus frequency sensitive. The component does not attenuate (or pass) signals of all frequencies equally. Even a simple PC trace or component lead has impedance. Metal, being a conductor, has some resistance. If alternating current is passed through the trace or lead, there is some inductive reactance. If the trace is near another conductor (or other metal object) there is some capacitance between the two conductors, and thus some capacitive reactance. The reactance and resistance combine to produce impedance, which, in turn, varies with frequency.

Many of the impedances presented by components are of little practical concern in troubleshooting. However, certain impedances have a very pronounced effect on amplifier-circuit operation, particularly as the frequency increases. Four major components are used in amplifier circuits: transistors, resistors, capacitors, and inductances (coils and transformers). The following is a summary of how these components affect troubleshooting.

### 6.4.1    Transistor frequency limitations

As shown in Fig. 6-6, all transistors have some capacitance between the junctions (emitter-base and collector-base). If any of the elements is common or ground, the remaining elements have some capacitance to ground. For example, in a common-emitter amplifier, there is some capacitance from base to ground (CBE across the input) and collector to ground (CCE across the output) and capacitance from the collector to base (which forms a feedback path from output to input).

Capacitive reactance decreases with an increase in frequency and vice versa. A capacitance in series with a conductor presents less attenuation to the signal as the frequency increases. A capacitance across a conductor (for example, in parallel from the conductor to ground) acts as a short to signals of increasing frequency.

All transistors have some inductance in their leads. This produces inductive reactance in series with the transistor elements. Inductive reactance increases with frequency, thus attenuating signals at higher frequencies.

From a troubleshooting standpoint, the input and output capacitances of transistors have little effect at audio frequencies. Typically, all signals up to about 20 kHz (or higher) are amplified by the same amount. However, amplification begins to drop at some frequency. The top frequency limit of a circuit is set by the transistor with the lowest frequency capability.

### 6.4.2    Resistor frequency limitations

Resistors offer relatively few problems in audio circuits because resistors attenuate signals equally. Only at higher frequencies, where the resistor leads and body could produce some kind of reactance or impedance, is there any particular concern about the frequency limits imposed by resistors. However, resistors do produce voltage drops that can be a problem in interstage coupling (Section 6.5) and when used with coupling capacitors.

### 6.4.2    Bypass-capacitor frequency limitations

As shown in Fig. 6-7, bypass capacitors are used to provide a signal path around high resistances. For example, if the power supply of an audio amplifier does not have a filter capacitor, or if a battery is used, the collector-emitter

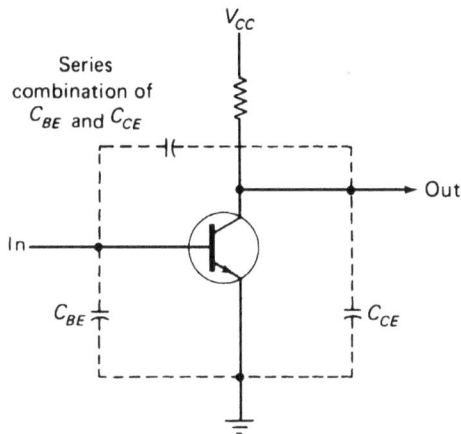

**Figure 6.6.** Capacitances associated with transistor elements.

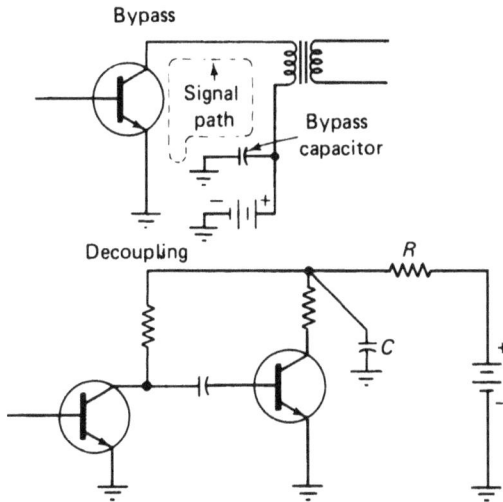

**Figure 6.7.** Examples of bypass and decoupling capacitors.

current must pass through a high resistance. This can impede the ac component of the signal. A bypass capacitor provides the signal path.

When several stages of amplification are connected, the stages all join at one point (the common power supply). In multistage amplifiers, there is the possibility of one stage feeding back through the power supply to a previous stage, thus causing interference with the signal. To avoid this feedback, one or more of these stages can be decoupled from the power supply.

Figure 6-7 shows a typical decoupling network. Resistor R is placed in series between the load resistors of the stages and the power supply. This produces a high-resistance path for the ac signal to the power supply. Capacitor C offers a low-resistance path for the signal, and thus decouples (or bypasses) the signal to ground.

In troubleshooting, the functions of bypass and decoupling capacitors are the same and the terms are interchanged. In either case, the main concern is that the reactance must be low at the lowest frequency involved. This requires a capacitor of a given value, which increases as the frequency decreases. For example, assume that the lowest frequency involved is 100 Hz and the minimum required reactance is 100 $\Omega$. This requires a capacitance value of about 16 $\mu$F. If the frequency is decreased to 10 Hz, the capacitance must be raised to 160 $\mu$F to keep the reactance below 100 $\Omega$.

### 6.4.3  Coupling-capacitor frequency limitations

Figure 6-8 shows some typical coupling-capacitor circuits. As covered in Section 6.5, coupling capacitors are often used at the input and output of amplifier stages to block direct current. The values of coupling capacitors

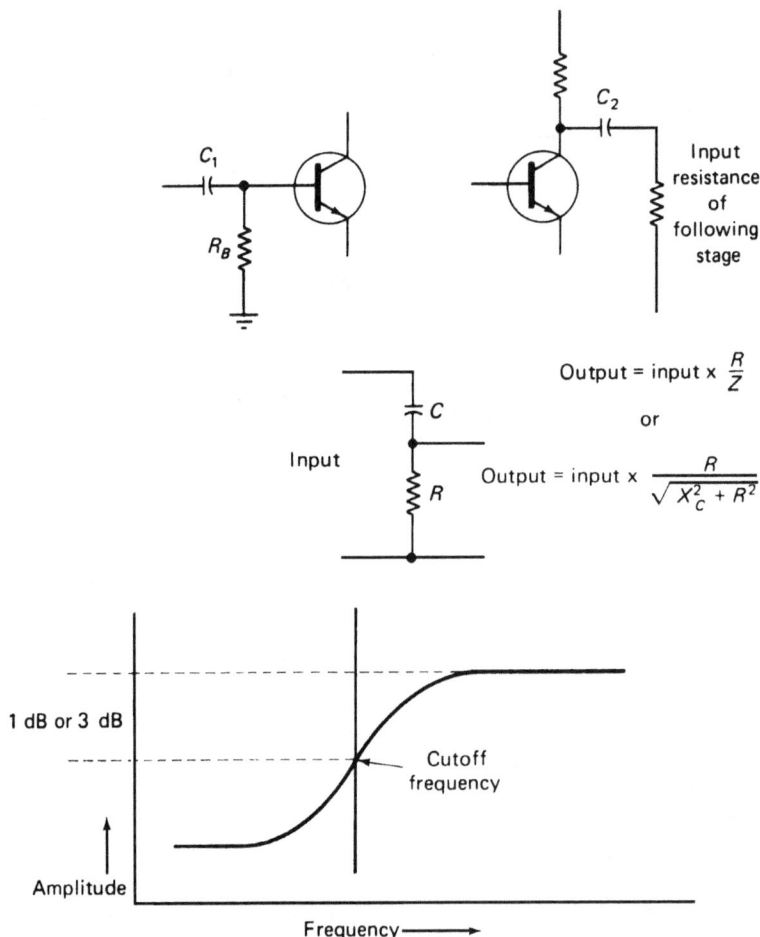

$$\text{Output} = \text{input} \times \frac{R}{Z}$$

or

$$\text{Output} = \text{input} \times \frac{R}{\sqrt{X_C^2 + R^2}}$$

**Figure 6.8.**  Formation of high-pass (low-cut) RC filter by coupling capacitors and related resistances.

depend on the low-frequency limit at which the amplifier operates and on the resistances with which the capacitors operate.

As the frequency increases, capacitive reactance decreases, and the coupling capacitors become (in effect) a short to the signal. As a result, the high-frequency limit need not be considered in audio circuits. In Fig. 6-8, C1 forms a highpass RC filter with $R_B$. Capacitor C2 forms another highpass filter with the input resistance of the following stage (or the load).

Input to the filter is applied across the capacitor and resistor in series. The filter output voltage is taken across the resistance. The relation of input voltage to output voltage is:

$$\text{Output voltage} = \text{input voltage time (R / Z)}$$

where $R$ is the dc resistance and $Z$ is the impedance (obtained by the combination of series capacitive reactance and dc resistance).

In troubleshooting, the following guideline can be used to determine the approximate effect of coupling capacitors on frequency. When the reactance drops to about one half of the resistance, the output drops to about 90% of the input (about a 1-dB loss). Using the 1-dB loss as the low-frequency cutoff point, the value of $C_1$ or $C_2$ can be found by: *capacitance* $= 1 / (3.2\ FR)$ where *capacitance* is in farads, $F$ is the low-frequency limit in Hz, and $R$ is resistance in ohms.

If a 3-dB loss at the low-frequency cutoff point can be tolerated, the value of $C_1$ or $C_2$ can be found by: *capacitance* $= 1 / (6.2\ FR)$.

### 6.4.4  Inductance frequency limitations

As covered in Section 6.5, inductances (coils and transformers) are sometimes used in audio circuits. The coils are used in place of the collector resistor as a load. This permits the collector to be operated at a higher voltage. Transformers are used for coupling between stages (mostly in older audio circuits) or to match the final output stage of an audio amplifier to a loudspeaker.

In troubleshooting, remember that the inductive reactance of coils and transformers increases with frequency. This increase in reactance also increases impedance and lowers gain.

### 6.4.5  Stray impedance frequency limitations

As covered, any conductor (PC traces, component leads, wiring, terminals, etc.) can have resistance, reactance, and impedance. You must be careful when placing conductors in the circuit to minimize the effects of this (stray) impedance. The effects of stray impedances can alter the characteristics of components. A classic example of this is stray capacitance, which is added to the input and output capacitances of transistors (Fig. 6-6).

From a troubleshooting standpoint, the effects of stray impedances are usually not crucial at audio frequencies, but they do become important as the frequency increases. That is why proper PC-board layout is stressed throughout this book.

## 6.5    Effect of Coupling in Audio Circuits

Figure 6-9 shows some classic coupling methods in audio circuits. All amplifiers require some form of coupling. Even a single-stage audio amplifier must be coupled to the input and output devices. If more than one stage is involved, it must have interstage coupling. Notice that the four basic coupling methods shown (capacitor, inductive, direct, and transformer) also require resistors and could be called *resistance coupling*. However, the term *resistance-coupled* is generally used to indicate that the circuit does not have inductances or transformers between stages, and that the input and/or output impedance is formed by a resistance. Capacitor coupling is often called *resistance-capacitance* or *RC coupling*.

This section covers how different coupling methods affect the operation of audio circuits from a testing or troubleshooting standpoint.

### 6.5.1    Direct coupling

With the direct coupling found in Fig. 6-9A, the collector of one transistor is connected to the base of the following amplifier. The outstanding characteristic

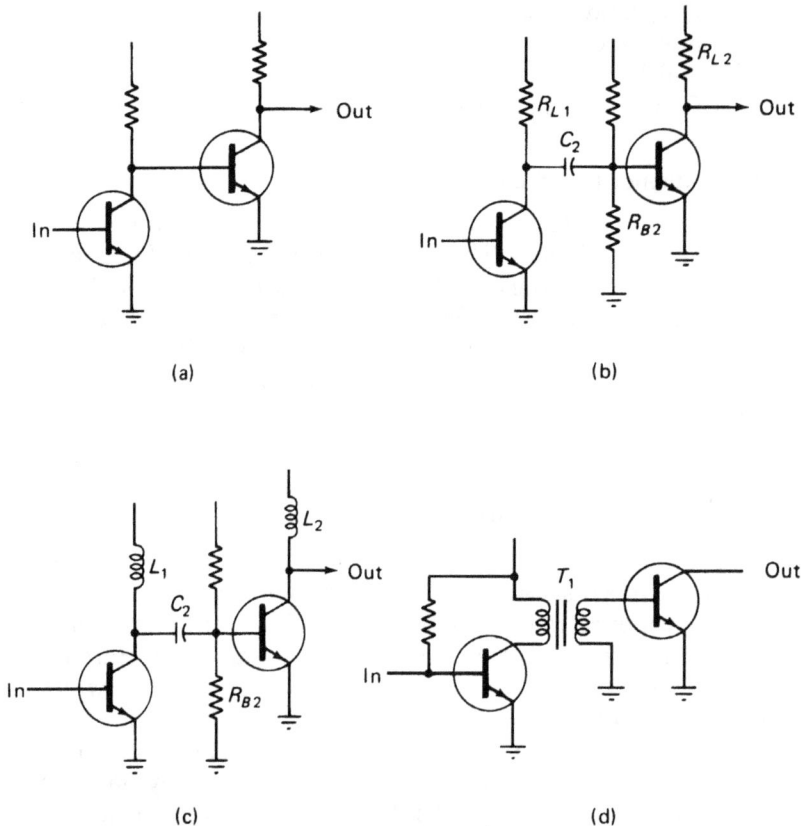

(a)                                    (b)

(c)                                    (d)

**Figure 6.9.**  Some classic coupling methods in audio circuits.

of a direct-coupled amplifier is the ability to amplify direct current and low-frequency signals. As covered, in troubleshooting, one of the major problems with direct coupling is that the circuit cannot tell the difference between a change in supply voltage and a change in input signal. This is particularly a problem in feedback circuits, as covered throughout the remainder of this chapter.

### 6.5.2    Capacitive or RC coupling

As shown in Fig. 6-9B, capacitor or RC coupling is formed by load resistor RL1 of stage 1, base resistor RB2 of stage 2, and coupling capacitor C2. The input signal is acted upon by stage 1 and appears in amplified form as the voltage drop across RL1. The dc component of the amplified signal is blocked by C2, which passes the ac component to the input section of stage 2 for further amplification.

In troubleshooting circuits with RC coupling, notice that amplification is generally uniform over nearly the entire audio range because resistor values are independent of frequency changes. However, as covered in Section 6.4.3, RC-coupled amplifiers have a low-frequency limit imposed by the reactance of the capacitor (which increases as the frequency decreases). One disadvantage of RC coupling is that the supply voltage is dropped (usually one half) by the load resistance, so the collectors operate at a reduced voltage.

### 6.5.3    Inductive or impedance coupling

As shown in Fig. 6-9C, the load resistors are replaced by inductors L1 and L2. The advantage of impedance coupling over resistance coupling is that the resistance of the load inductor is less than that of the load resistor. For a power supply of a given voltage, the collector voltage is higher (thus, the possible output voltage is higher).

The main disadvantage of inductive coupling is frequency discrimination. At very low frequencies, the gain is low because of the coupling capacitor reactance (as in the RC-coupled circuit). The gain increases with frequency, leveling off at the middle frequencies of the audio range. (However, the frequency spread of this level portion is not as great as for the RC circuit.) At very high frequencies, the gain drops off because of the increased reactance. Impedance coupling is rarely, if ever, used at frequencies above the audio range. In present-day equipment, inductive coupling is used only in applications with a special need for the inductor or coil.

### 6.5.4    Transformer coupling

Compared to the RC-coupled circuit, the transformer-coupled amplifier of Fig. 6-9D has essentially the same advantages and disadvantages as the impedance-coupled amplifier. The transistor collectors can be operated at higher voltages. The impedances are set by the transformer primary and secondary windings. However, transformers are frequency sensitive (impedances can change with frequency), so the frequency range of transformer-coupled amplifiers is limited.

Transformer coupling is most effective when the final amplifier output must be fed to a low-impedance load. For example, the impedance of a typical loudspeaker is about 4 to 16 $\Omega$, whereas the output impedance of a transistor stage is several hundred (or thousand) $\Omega$. A transformer at the output of an audio amplifier can offset the obviously undesired effects of such a mismatch.

## 6.6    Basic Amplifier-Circuit Troubleshooting

The basic troubleshooting procedure for amplifier circuits is the same as for most other multistage circuits. The input and output waveforms of each stage can be monitored with a scope. Any stage showing an abnormal waveform (in amplitude, waveshape, etc.) or the absence of an output waveform, with a known-good input signal, points to a defect in that stage. Voltage measurements on all transistor (or IC) elements then pinpoints the problem.

The problem of troubleshooting amplifier circuits with feedback is not quite that easy. Measuring stage gain is especially difficult. For example, if you try opening the feedback loop to make a gain measurement, you usually find so much gain that the amplifier saturates, and the measurements are meaningless. On the other hand, if you make waveform measurements on a closed-loop circuit, you often find the input and output signals are normal (or near normal) even though the waveforms inside the loop are distorted. For this reason, feedback loops (especially internal-stage feedback loops) require special attention.

### 6.6.1    Direct-coupled amplifier with feedback

Figure 6-10 shows a direct-coupled amplifier with interstage feedback. (The feedback is provided by the 47-k$\Omega$ resistor.) The waveforms indicate that the input and output signals are sine waves. Approximately, a 15% distortion is inside the feedback loop (between Q1 and Q2), but only a 0.5% distortion is at the output. This is only slightly greater distortion than at the input (0.3%). Open-loop gain for this circuit is about 4300; closed-loop gain is about 1000. The gain ratio (open loop to closed loop) of 4 to 1 is typical for feedback amplifiers.

Transistor Q1 has a varying signal on both the emitter and base. In a non-feedback amplifier, the signal usually varies at one element, either the emitter or base. Because most feedback systems use negative feedback, the signals at both the base and emitter are in phase. The resultant gain is much less than when one of these elements is fixed (no feedback, open loop). This accounts for the large gain increase when the loop is opened. Either the base or the emitter of the transistor stops varying and the base-emitter control elements "see" a much larger effective input signal.

To understand circuit operation for troubleshooting, assume that clipping is introduced in Q1 (as shown in Fig. 6-11), and that Q2 is perfect. (Section 6.6.2 describes how clipping might occur.) The signals applied to the base and emitter of Q1 are not identical. The resultant signal at the control point of Q1 is quite distorted. In effect, the distortion is a mirror image of the distortion introduced by Q1.

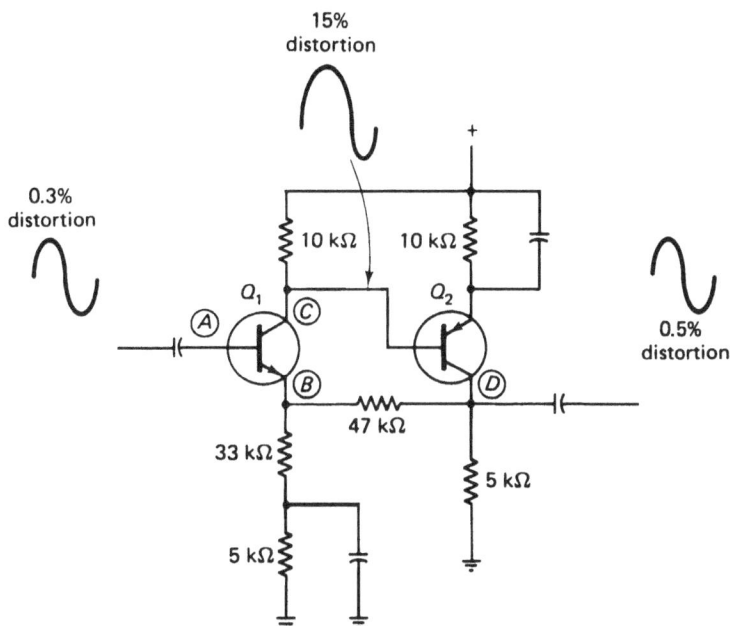

**Figure 6.10.** Direct-coupled amplifier with interstage feedback.

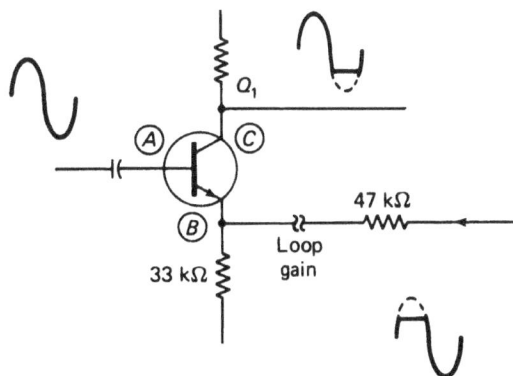

**Figure 6.11.** Amplifier-induced distortion (clipping).

Transistor Q1 amplifies the distortion and adds its own counterdistortion. After many trips around the loop, distortion is still inside the loop, but the distortion is counterbalanced by the feedback. The final output from Q2 is undistorted or is relatively free of amplifier-induced distortion. The higher the amplification (and the greater the feedback), the more effective the cancellation becomes, and the lower the output distortion becomes.

## 6.6.2   Some classic distortion problems

The three most common causes of distortion in amplifier circuits are: overdriving, operating the transistor at the wrong bias point, and the inherent nonlinearity of any solid-state device. The following section summarizes these problems.

*Overdriving* can be caused by many factors (too much input signal, too much gain in the previous stage, etc.). However, the net result is that the output signal is clipped in both peaks. It is clipped in one peak because the transistor is driven into saturation and on the other peak because the transistor is driven below cutoff.

In some cases, the transistor operates at the wrong bias point, but of only one peak. For example, if the input signal is 1 V and the transistor is biased at 1 V, the input swings from 0.5 to 1.5 V. Assume that the transistor saturates at any point above 1.6 V and is cut off at any point below 0.4 V. No problems occur with the correct bias (1 V).

Now assume that the bias point is shifted (because of component aging, transistor leakage, etc.) to 1.3 V. The input now swings from 0.8 to 1.6 V, and the transistor saturates when one peak goes from 1.6 to 1.8 V. If, on the other hand, the bias point is shifted down to 0.7 V, the input swings from 0.2 to 1.2 V, and the opposite peak is clipped as the transistor goes into cutoff.

Even if the transistor is not overdriven, it is still possible to operate a transistor on a nonlinear portion of the curve because of wrong bias. All transistors have some portion of the input/output curve that is more linear than other portions. That is, the output increases (or decreases) directly in proportion to input. An increase of 10% at the input produces an increase of 10% at the output. Ideally, transistors are operated at the center (Fig. 6-2). If the bias point is changed, the transistor can operate on a portion of the curve that is less linear than the desired point.

The inherent nonlinearity of any solid-state device (diode, transistor, IC, etc.) can produce distortion even if a stage is not overdriven and is properly biased. That is, the output never increases (or decreases) directly in proportion to the input. For example, an increase of 10% at the input can produce an increase of 13% (or any other percent) at the output. This is one of the main reasons why feedback is used in amplifiers where low distortion is required.

A negative-feedback loop operates to minimize distortion, in addition to stabilizing gain. The feedback takeoff point has the minimum distortion of any point within the loop. From a practical troubleshooting standpoint, if the final output (last stage or IC output) distortion and the overall gain are within limits, all stages or ICs within the loop can be considered as operating properly. Even if one or more stages have some abnormal gain, the overall feedback system has compensated for the problem. Of course, if the overall gain and/or distortion are not within limits, the individual stages must be checked.

### 6.6.3    Troubleshooting tips for feedback-amplifier circuits

The following areas should be given special attention when troubleshooting any feedback-amplifier circuit.

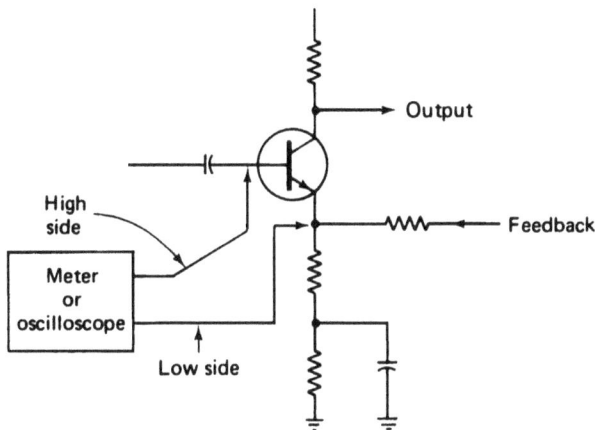

**Figure 6.12.**  Measuring input signal voltages or waveforms.

Take special care when measuring the gain of amplifier stages in a feedback-amplifier circuit. For example, in the circuit of Fig. 6-10, if you measure the signal at the base of Q1, the base-to-ground voltage is not the same as the input voltage. To get the correct value, connect the low side of the measuring device (meter or scope) to the emitter and the other lead to the base (as shown in Fig. 6-12). Measure the signal across the base-emitter junction to get the effect of feedback.

Avoid opening the feedback loop. In some cases, this can cause circuit damage. Even if there is no damage, the technique is rarely effective. Open-loop gain is usually so high that some stage blocks or distorts badly. If the loop must be opened (possibly in op-amp or OTA circuits), distortion is increased. A normally closed-loop amplifier can show considerable distortion when operated as an open loop—even though the amplifier circuit is good.

Suspect the transistors when the transistor voltages are good, but gain is low and/or distortion is present in a feedback-amplifier circuit that once worked properly. If closed-loop gain is low, this usually means that the open-loop gain has fallen far enough that the resistor ratios no longer set the gain. For example, if the base of Q2 in Fig. 6-10 is lowered, the open-loop gain is lowered. Also, the lower beta lowers the input impedance of Q2, which, in turn, reduces the effective value of the load resistor for Q1. This also has the effect of lowering the overall gain.

In troubleshooting a feedback-amplifier circuit where the waveforms show low gain, but element voltages are normal, try replacing the transistors. However, never overlook the possibility of open or badly leaking emitter-bypass capacitors. If the capacitors are open or leaking (acting as a resistance in parallel with the emitter resistor), there is considerable negative feedback and little gain. Of course, a completely shorted emitter-bypass capacitor produces an abnormal dc voltage indication at the transistor emitter.

### 6.6.4  Distortion caused by transistor leakage

One problem sometimes overlooked in a feedback-amplifier circuit with a distortion failure pattern is overdriving because of transistor leakage. (The problem of transistor leakage is covered further in Section 6.7.)

It is often assumed that collector-base leakage reduces gain because the leakage is in opposition to the signal-current flow. Although this is true in the case of a single stage, it might not be true where more than one feedback stage is involved. Whenever this is collector-base leakage, the base assumes a voltage nearly equal to that of the collector (nearer than is the case without leakage). An increase in the transistor current causes lower input resistances, which, in turn, causes the stage gain to increase. At the same time, a reduction in input resistance causes a reduction in common-emitter input resistance. This might or might not cause a gain reduction (depending on where the transistor is located in the amplifier).

The effects of feedback are increased if the circuit is direct coupled because the operating point (bias) of the following stage is changed, possibly resulting in distortion. For example, the collector of Q1 is connected directly to the base of Q2 (Fig. 6-10). If Q1 starts to leak (or if the collector-base leakage increases with age), the base of Q2 (as well as the collector of Q1) shifts the $Q$ point (no-signal voltage level).

## 6.7   The Subtle Effects of Leakage in Amplifier Circuits

If an amplifier circuit has considerable leakage, the gain is reduced to zero and/or the signal waveform is drastically distorted. These indications make the problem easy or relatively easy to locate. The real difficulty occurs when there is just enough leakage to reduce amplifier gain, but not enough leakage to seriously distort the waveform or produce transistor voltages that are way off.

A classic example is collector-base leakage (such as shown in Fig. 6-13). This condition has the same effect as a resistance between the collector and base. The base assumes the same polarity as the collector (although at a lower value) and the transistor is forward biased. If leakage is sufficient, the forward bias can be enough to drive the transistor into or near saturation. When a transistor is operated at or near the saturation point, the gain is reduced for a single stage (as shown in Fig. 6-14).

Excessive transistor leakage can be spotted easily if all transistor voltages are abnormal. For example, as shown in Fig. 6-13, the base and emitter are high and the collector is low when measured in reference to ground. However, if the normal operating voltages are not known, the transistor can appear to be good because all voltage relationships are normal. That is, the collector-base junction is reverse biased (collector more positive than the base for an NPN), and the emitter-base junction is forward biased (emitter less positive than the base for NPN).

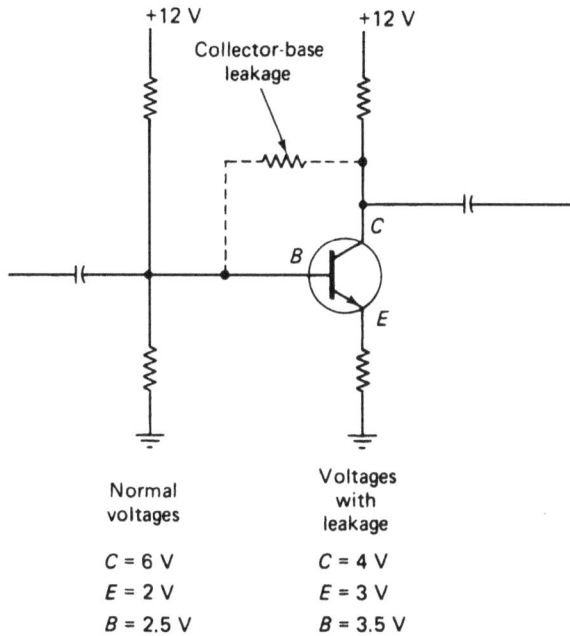

Normal voltages

$C = 6$ V
$E = 2$ V
$B = 2.5$ V

Voltages with leakage

$C = 4$ V
$E = 3$ V
$B = 3.5$ V

**Figure 6.13.** Effect of collector-base leakage on transistor element voltages.

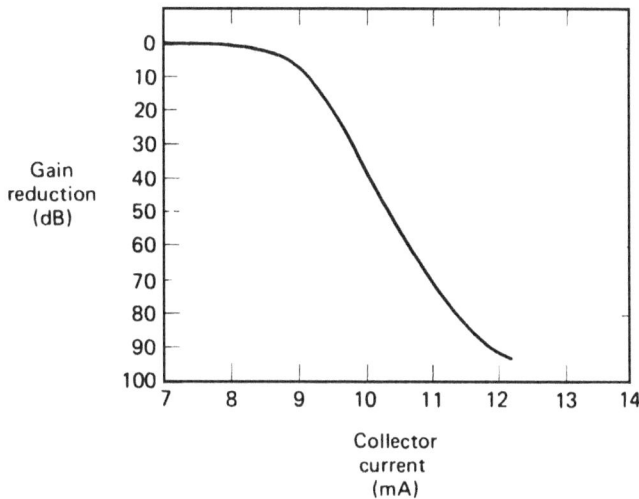

**Figure 6.14.** Relative gain at various average collector-current levels.

### 6.7.1  A simple check for transistor leakage

Figure 6-15 shows a simple way to check transistor leakage. Measure the collector voltage to ground. Then short the base to the emitter and remeasure the collector voltage. If the transistor is not leaking, the base-emitter short turns the transistor off, and the collector voltage rises to the same value as

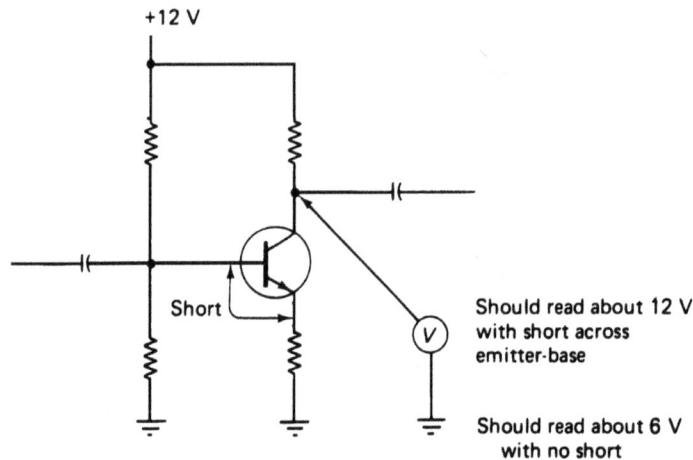

**Figure 6.15.** Checking for transistor leakage in amplifier circuits.

the supply. If there is any leakage, a current path remains (through emitter-base short, collector-base leakage path, and collector resistor). Some voltage drop is across the collector resistor and the collector's voltage is at some value lower than the supply.

As a troubleshooting example, assume that in the circuit of Fig. 6-15, the supply is 12 V, the collector resistance is 2 kΩ, and the collector measures 4 V with respect to ground. This means that an 8-V drop is across the collector resistor and it has a collector current of 4 mA (8 / 2000 = 4 mA). Normally, the collector is operated at about one half the supply voltage of 6 V. However, simply because the collector is at 4 V (instead of 6 V), it does not make the circuit faulty. Some circuits are designed that way, so the transistor must be checked for leakage.

Further assume that the collector voltage rises to 10 V when the base and emitter are shorted. This shows that the transistor is cutting off, but some current is still flowing through the collector resistor, about 1 mA (2 / 2000 = 1 mA). A 1-mA current flow is high for a meter. However, to confirm that a transistor is leaky, connect the meter through a 2-kΩ resistor to the 12-V supply, preferably at the same point where the collector resistor connects to the supply.

Finally, assume that the indication is 11.7 V through the external resistor. This shows that there is some transistor leakage. The amount of leakage is not particularly important from a practical troubleshooting standpoint. The presence of any current flow with a transistor supposedly cut off is sufficient to replace the transistor.

## 6.8  The Subtle Effects of Capacitors in Amplifier Circuits

This section covers the effects of capacitors in amplifier circuits from a troubleshooting standpoint.

### 6.8.1   Emitter-bypass capacitors

The emitter resistor shown as R4 in Fig. 6-16 is used to stabilize the transistor gain and prevent thermal runaway. With an emitter resistor in the circuit, any increase in collector current produces a greater drop in voltage across the resistor. With all other factors the same, the change in emitter voltage reduces the base-emitter forward-bias differential, thus tending to reduce collector current flow.

The emitter resistor is not bypassed when circuit stability is more important than gain. When signal gain must be high, the emitter resistance is bypassed to permit the passage of the signal. If the emitter-bypass capacitor (such as C2 in Fig. 6-16) is open, stage gain is reduced drastically, although the transistor dc voltages remain substantially the same.

In troubleshooting, a low-gain symptom is in any amplifier with an emitter bypass and the voltages appear normal, check the bypass capacitor. This can be done by shunting the bypass with a known-good capacitor of the same value. As a precaution, shut off the power before connecting the shunt capacitor; then reapply power. This prevents damage to the transistor (because of large current surges). If an emitter bypass is shorted or leaking, the emitter voltage is not correct, thus localizing the problem. In this case, shunting a good capacitor across the suspected capacitor has little effect on the gain. The suspected capacitor must be tested by substitution.

### 6.8.2   Coupling capacitors

The coupling capacitor shown as C1 in Fig. 6-16 is used to pass signals from the previous stage to the base of Q1. If C1 is shorted or leaking badly, the voltage from the previous stage is applied to the base of Q1. This forward biases Q1, causing heavy current flow and possible burnout of the transistor. In any event, Q1 is driven into saturation and the stage gain is reduced. If C1 is open, there is little or no change in the voltages at Q1, but the signal from the previous stage does not appear at the base of Q1.

In troubleshooting, a shorted or leaking C1 shows up as abnormal voltages (and probable distortion of the signal waveform). If you suspect C1 of being shorted or leaking, replace C1. An open C1 shows up as a lack of signal at the base of Q1, with a normal signal at the previous stage. If an open C1 is suspected, replace C1 or try shunting C1 with a known-good capacitor, whichever is convenient.

### 6.8.3   Decoupling/bypass capacitors

The *decoupling capacitor* (also known as a *bypass capacitor*), shown as C3 in Fig. 6-16, is used to pass signals to ground (to provide a return path) and to prevent signals from entering the power-supply line or other circuits connected to the line. In effect, C3 and R5 form a lowpass filter that passes dc and very low-frequency signals (well below the frequency of the circuits) through the power-supply line. Higher-frequency signals are passed to ground and do not enter the supply line.

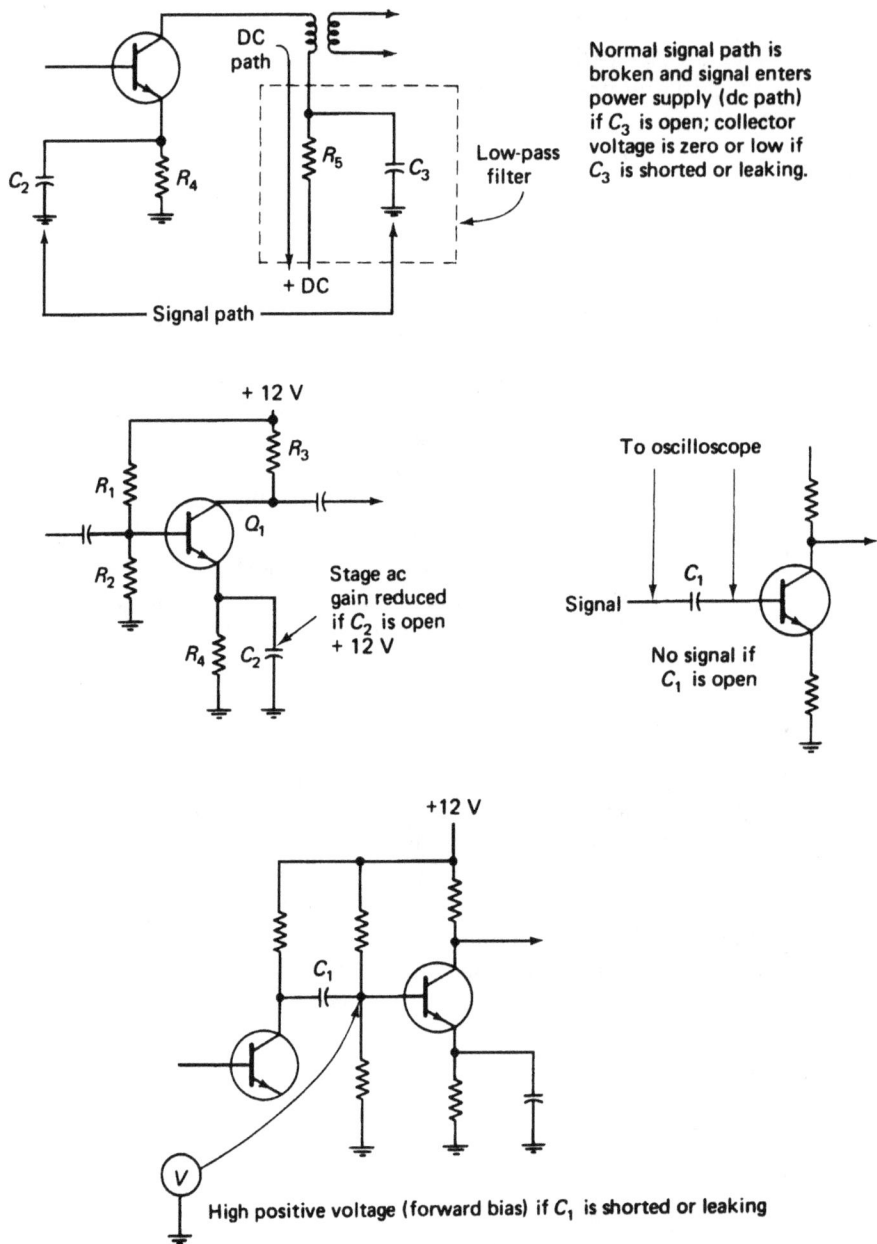

DC path

Normal signal path is broken and signal enters power supply (dc path) if $C_3$ is open; collector voltage is zero or low if $C_3$ is shorted or leaking.

$C_2$

$R_4$

$R_5$

$C_3$

Low-pass filter

+ DC

Signal path

+ 12 V

$R_3$

$R_1$

$Q_1$

$R_2$

Stage ac gain reduced if $C_2$ is open

$R_4$   $C_2$

+ 12 V

To oscilloscope

$C_1$

Signal

No signal if $C_1$ is open

+12 V

$C_1$

V

High positive voltage (forward bias) if $C_1$ is shorted or leaking

**Figure 6.16.** Effect of capacitor failure in amplifier circuits.

In troubleshooting, if C3 is shorted or leaking badly, the supply voltage is shorted to ground or greatly reduced. This reduction of collector voltage makes the stage totally inoperative or reduces the output, depending on the amount of leakage in C3. If C3 is open, there is little or no change in the voltages at Q1. However, the signals appear in the supply lines. Also, signal gain is reduced and the signal waveform is distorted. In some cases, at higher signal frequencies, the signals simply cannot pass through the supply circuits. Because there is no path through an open C3, the signal does not appear on the collector circuit in any form. In practical circuits, the result of an open C3 depends on the values of $R_5$ (and other supply components), as well as on the signal frequency involved.

## 6.9   IC Amplifier Troubleshooting

Many amplifier circuits are manufactured in IC form. From a practical troubleshooting standpoint, components external to the IC package are the only ones that can be tested or replaced. Fortunately, major disasters are relatively rare in a well-protected IC because input overloads never drive the circuit into saturation. Also, when such major failures occur, the failures are relatively easy to troubleshoot. The problems are easy to spot by normal signal tracing with waveforms or by voltage measurements at the IC terminals. For example, a major failure shows up as a normal input, but with no output at a particular stage (or at the input and output of the IC). However, IC amplifiers are often plagued with such problems as hum, drift, and noise. This section describes the most likely causes for such problems, with practical approaches for locating the faults.

### 6.9.1   Hum and ripple problems

In IC amplifiers, any hum or ripple almost always comes from the dc power supplies feeding the amplifier. A possible exception is when hum is picked up because of poor shielding or badly grounded leads.

The first step in troubleshooting a hum or ripple problem is to short the input terminals and monitor the output with a scope. If the hum or ripple is removed when the input terminals are shorted, the hum is probably being picked up by the leads or at the terminal. Look for loose shields, loose ground terminals, and cold-solder points where lead shielding is attached to feed-through terminals.

If the hum or ripple is not removed when the input terminals are shorted, the hum is probably coming from the power supply. Monitor the supply voltages at the point where the voltages enter the IC. If the supply shows an abnormal amount of ripple, the problem is in the power supply. However, because the amplifier has considerable gain, the ripple (as monitored at the amplifier output) might be much greater than at the power supply.

If both the supply and the IC appear to be good, but there is still excessive hum or ripple, suspect the decoupling/bypass capacitors that are connected to the IC at the supply pins. Hum or ripple can occur if these capacitors are open or leaking. If hum is in an experimental circuit, it is possible that the capacitors are good, but that the capacitance is insufficient.

### 6.9.2  Drift problems

Look in several places when trying to track down the causes of drift. However, the best places to start are with the power supplies and PC boards.

Well-designed IC amplifiers do not usually drift (change output characteristics) over time, but they are extremely sensitive to power-supply stability. For example, with IC power amplifiers, the typical dual-supply voltages (required for differential-amplifier circuits) are ±12 or ±15 V. For satisfactory operation, the drift should be less than 1 mV/minute.

In troubleshooting, because of the low voltages involved, power-supply stability measurements are best made with five- or six-place digital meters. Such meters can be connected to the monitoring point and checked at least once every minute over at least a 5-minute interval. If the drift is less than 1 mV/minute over this time interval, the power supply is probably satisfactory for most applications.

Although the IC package is sealed, all points in the amplifier circuit are not sealed. With contamination on the PC board at the points, drift is possible. (The feedback or summing point is a classic example.) When you remember that the input current at the feedback point of some IC amplifiers is in the order of $10^{-11}$ A, it is easy to see why any contamination from fingerprints (providing leakage paths into the junction) can cause annoying output instability.

In troubleshooting, try to handle PC boards as little as possible, and then only when wearing cotton gloves. Try to avoid touching any IC pins with your bare hands. Boards suspected of being contaminated should be washed carefully with a clean degreasing solvent and dried with warm, dry air. (Never blow boards dry with an air hose because air lines invariably contain oil and water.)

### 6.10  Audio-Amplifier Circuit Controls

The most common operating controls for audio circuits used with music- or voice-reproduction equipment (hi-fi, stereo, public address, etc.) are the volume (or loudness) control and the tone (treble and bass) controls. The other most common audio control is the gain control. Volume and gain controls are often confused because both controls affect the amplifier-circuit output.

A true gain control sets the gain of one stage in the amplifier. A true volume control sets the level of the signal passing through the amplifier, without affecting the gain of any or all stages. A gain control is usually part of a stage, whereas a volume control is usually located between stages or at the input of the first stage.

Most stereo amplifiers have some form of balance control so that both channels can be balanced. Also, most hi-fi stereo amplifiers have some form of playback equalization for tape and phono playback. Equalization, tone, and balance circuits are covered in Section 6.11.

### 6.10.1  Volume control

As shown in Fig. 6-17, the basic volume control is a variable resistor or potentiometer (pot) connected as a voltage divider. The voltage output (or signal level) depends on the pot setting. If the audio circuit is to be used with voice or music, the pot is usually of the audio-taper type, where the voltage output is not linear throughout the setting range. This produces a nonlinear voltage output to compensate for the human ear's nonlinear response to sound intensity. (The human ear has difficulty in hearing low-frequency sounds at low levels and responds mainly to the high-frequency components.) When the audio circuit is not used with voice or music (such as when audio circuits are used to control operation of motors), the volume control is usually of the linear type (unless there is some special circuit requirement). With such a control, the actual voltage or signal is directly proportional to the control setting.

In most circuits, the volume control is isolated from the circuit with coupling capacitors (as shown). If a volume control is part of the circuit (such as a collector or base resistance) any change in volume setting can change impedance, gain, or bias. Unfortunately, the capacitors create a low-frequency response problem. As in the case of coupling capacitors (Section 6.8.2), capacitor C1 forms a high-pass RC filter with volume control R1, and C2 forms another high-pass filter with the input resistance of the following stage (as shown by the equations). Keep this in mind when troubleshooting a circuit where the frequency response appears to shift drastically with a change in volume-control setting.

In most audio circuits, the volume control is found at the amplifier input stage or between the first and second stages. In troubleshooting an experi-

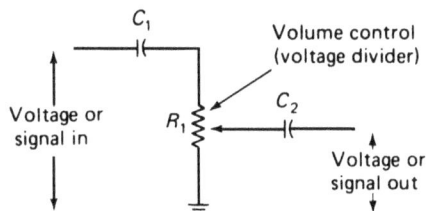

$$C = \frac{1}{3.2FR} \quad \text{for 1 dB}$$

$$C = \frac{1}{6.2FR} \quad \text{for 3 dB}$$

$C$ in farads, $F$ in hertz, $R$ in ohms

**Figure 6.17.**  Basic volume-control circuit.

mental circuit, remember that the input impedance of the circuit is set by the volume control, when the control is located at the input (thus the impedance changes when the control setting is changed). When the control is between stages, the resistance value of the pot is selected to match the output impedance of the previous stage.

### 6.10.2    Gain control

As shown in Fig. 6-18, the basic gain control is a variable resistance or pot, serving as one resistance element in the amplifier circuit. The emitter resistance (or source resistance, in the case of a FET amplifier) is the logical choice for a gain control. A variable emitter has minimum effect on the input or output impedance of the stage, but it directly affects both current and voltage gain. (Variable base or collector resistances would affect stage input and output impedance, respectively.)

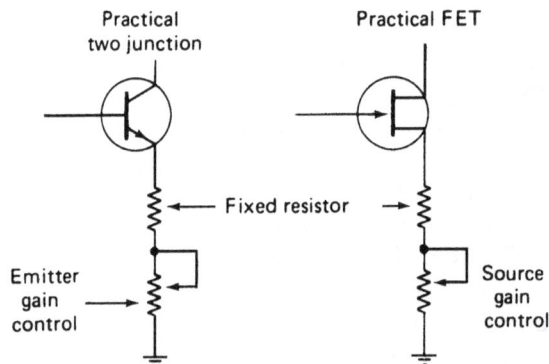

Figure 6.18.    Basic gain-control circuits.

When the emitter resistance is decreased, both current and voltage gain is increased, and vice versa. Usually, the emitter resistance is in series with a fixed resistance. If the gain control is set to minimum resistance (zero), some resistance still provides gain stabilization and prevents thermal runaway. The gain pot is generally not of an audio-taper type, unless the circuit has some special requirement.

### 6.10.3    Electronic volume control

When audio circuits are used in present-day consumer-electronic equipment, the amplifier is usually provided with electronic volume controls. Such controls are in IC form and are (in turn) under control of a microprocessor. Figure 6-19 shows the electronic volume-control circuits (left channel only) for a typical hi-fi system. The volume up, down, muting, and loudness functions are controlled by IC604 (sometimes called a *volume control,* other times called an *attenuator*). The volume is adjusted in 40 steps, including full muting. IC604 is under control of microprocessor IC901.

IC901 generates a volume-control code (Fig. 6-20) in response to commands from front-panel switches (or system control from a remote). The code includes clock, data, and strobe information at pins 10, 100, and 12 of IC604, respectively. The code is applied to decoder circuits within IC604 and causes IC604 switches and attenuators to be selected. For example, assume that the volume is already at −10 dB, without the loudness function, and you want to turn the loudness function on, with a volume of −8 dB. The loudness button is pressed (once), and the volume-up button is held until the desired −8 dB is obtained. Under these conditions, bits 1, 2, 3, 8, 9, and 20 are selected (data line high).

When the loudness button is pressed, the loudness switch in IC604 closes, connecting a loudness network between pins 2, 3, and 5 of IC604. The loudness circuit attenuates the midrange audio frequencies and passes the bass and treble frequencies. This has the effect of supplying (or reinforcing) positive feedback to the audio at pin 5 of IC604 (at high and low frequencies). When the mute button is pressed, the mute switch in IC604 is opened to interrupt the audio. However, this does not affect the attenuators. When the mute button is pressed again, the audio is restored at the same level of attenuation (unless the attenuation is changed during mute).

### 6.10.4  Electronic volume control troubleshooting

The following is an example of troubleshooting for an electronic volume-control circuit, such as shown in Figs. 6-19 and 6-20.

Start by checking for audio at pin 5 of IC604. If there is no audio, check the circuits ahead of IC604. Typically, the electronic volume control is in the audio path between the amplifier input and output.

If pin 5 of IC604 does have audio, check for audio at pin 4 of IC604 while operating the volume-up and volume-down buttons. The audio volume should increase in 10-dB steps at pin 4 each time you press the corresponding button. (If you hold the volume buttons, the audio should increase or decrease steadily in 10-dB steps.) If no audio is at pin 4 with audio at pin 5, suspect that IC604 has failed. If audio is at pin 4, but there is no change when the volume buttons are pressed, monitor strobe (pin 12), data (pin 13), and clock (pin 14) outputs from IC901 while holding the volume buttons. Although you probably cannot decode the information on the control lines, the presence of pulse activity on the lines usually indicates that IC901 is good.

If any one of the control lines shows no activity with the volume buttons pressed, suspect that IC901 has failed. If pulse activity is on all three lines, but the volume at pin 4 of IC604 does not change (with the volume buttons held), suspect that IC604 has failed. If audio is at pin 4 of IC604 and the volume changes, check for audio at pin 7 of IC704. If it is absent, suspect that IC605 has failed. If audio is at pin 4, also check for audio at pin 8, and be sure that the audio changes in 2-dB steps when the volume buttons are operated.

If there is no audio at pin 8, suspect IC604. If there is audio but there is no change with the volume buttons operated, suspect either IC901 or IC604. (IC604 is the most likely suspect if the audio is good at pin 7, but it is possible

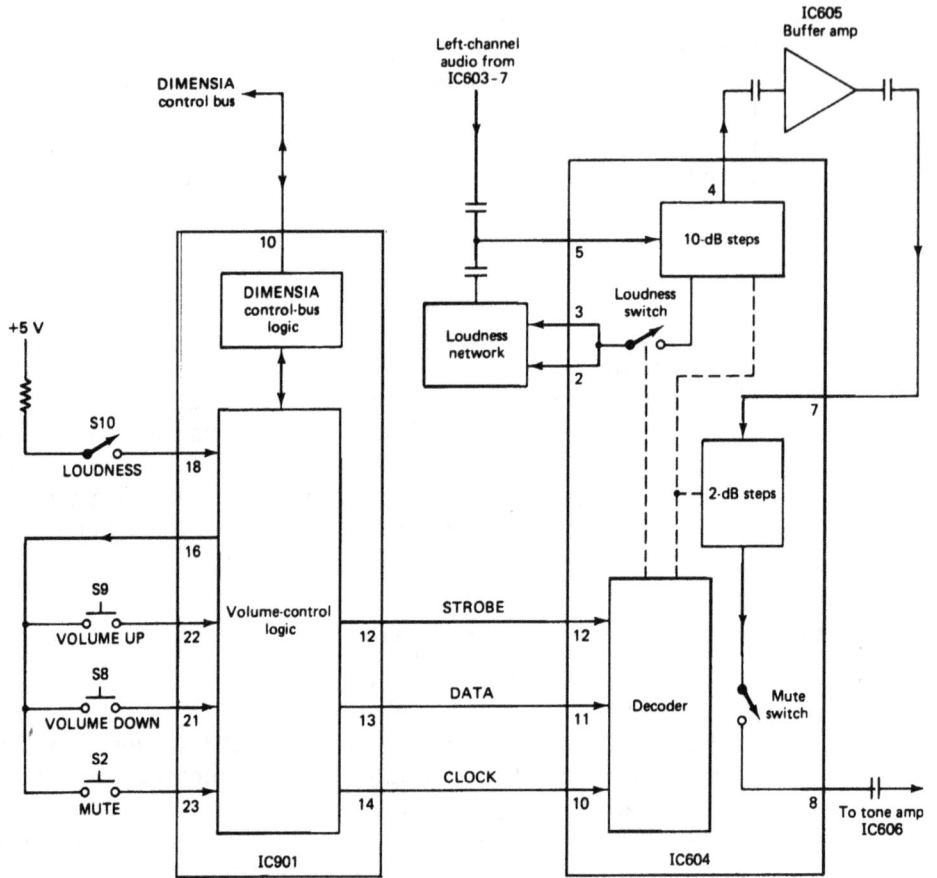

**Figure 6.19.**   Typical electronic volume-control circuits.

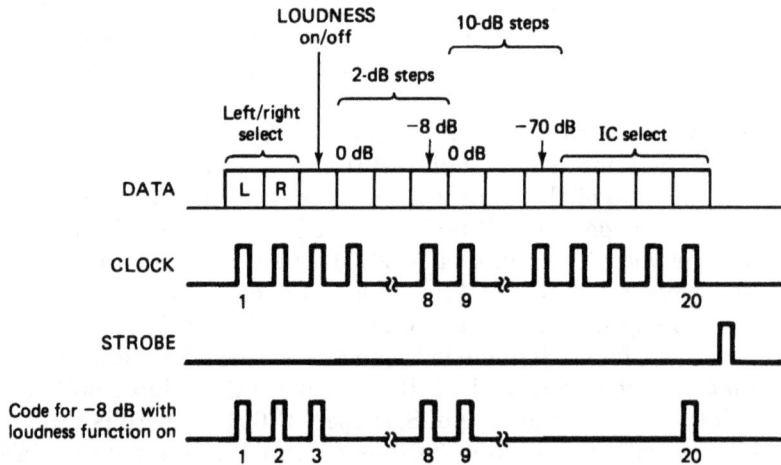

**Figure 6.20.**   Typical electronic volume-control code.

that IC901 is not generating the correct code to produce 2-dB changes in volume.) Again, check for pulse activity on the control lines.

If audio is at pin 8 and the audio changes in 2-dB steps when the volume buttons are operated, press the mute button and check that the audio is cut off at pin 8. If not, suspect a problem in the mute switch circuit, IC901 or IC604. You can also check for pulse activity on the control lines. Press the mute button again and check that audio is restored at pin 8 and is at the same level (if you have not pressed the volume buttons during the mute condition). If audio is not restored, suspect a problem in the mute circuit, IC901 or IC604.

### 6.10.5  Loudness circuits

The loudness circuit or network shown in Fig. 6-19 is a Fletcher-Munson network that attenuates the midrange audio frequencies and passes the bass/treble frequencies. In this particular audio-amplifier circuit, the loudness network is part of the electronic volume control (although there are external components). In some cases, the loudness network is separate from the volume control. No matter what the configuration, a failure symptom for the loudness functions in any audio circuit is usually very difficult to define. This is because the loudness function attenuates the midrange signals so that the ear hears what appears to be the same level across the audio range. (In rare cases, the loudness function boosts the treble and bass.) Unfortunately, all ears are not the same and not all loudness circuits define midrange at the same frequencies. Usually, the customer complains that "there is no difference when I play the tape or recording with the loudness function on or off."

To troubleshoot such a symptom, apply an audio signal (for example, between 7 and 10 kHz) with loudness off. Then press the loudness button and check for a drop of about 20 dB in level at pin 4 of IC604. Repeat the test at 50 Hz and 20 kHz. There should be substantially no change in level (at pin 4 of IC604) at the low and high ends of the audio range (unless a boost circuit is used) even though a change occurs at the midrange. No matter what type of loudness circuit is used, if there is no change in audio level at any frequency when the loudness function is switched in and out, a problem is in the loudness circuit. Start by checking the loudness switch circuit, IC901, IC604, and the network connected at pins 2 and 3 of IC604.

## 6.11  Audio-Amplifier Output Circuits

The tone, balance, and equalization controls for most present-day audio amplifiers are associated with the output circuits (or between the preamp and output). The following is a summary of these circuits from a troubleshooting standpoint.

### 6.11.1  Tone controls

Figure 6-21 shows one channel of a tone-control network using an IC amplifier. This circuit has both treble and bass controls. The treble control provides for adjusting the high-frequency response of the amplifier. Such adjustments

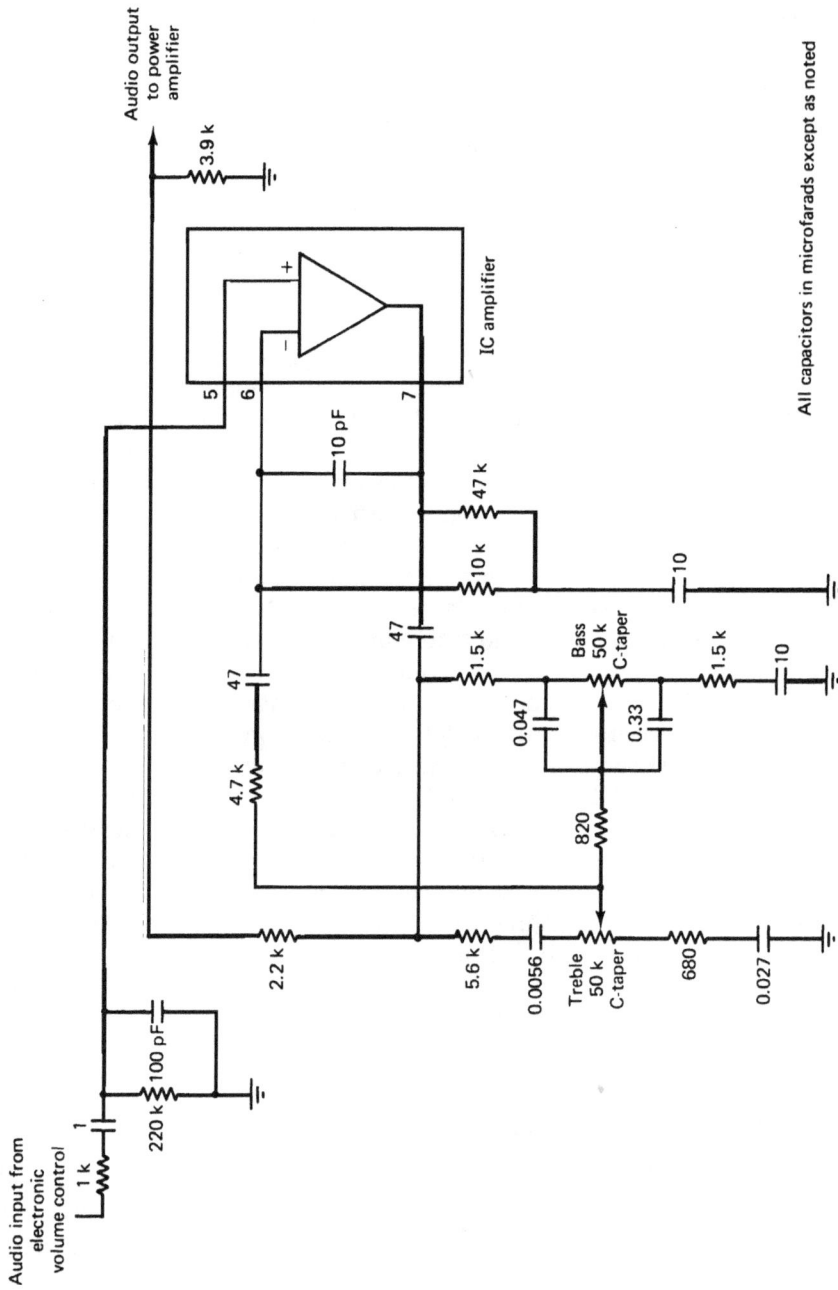

**Figure 6.21.** One channel of a tone-control network with IC amplifier.

might be necessary because of variation in response of the human ear or to correct the frequency response of a particular recording. The bass control provides for adjusting the low-frequency response. Such adjustments might be necessary because of the variation in response of the human ear, which does not respond as well to low-frequency signals (at low levels) as to high-frequency sounds at the same level. Also, coupling capacitors present high reactance to low-frequency signals.

These conditions require that the low-frequency signals be boosted (in relation to high-frequency signals), which is the primary job of the tone-control circuits. There are many circuit arrangements for tone controls. Some use adjustable feedback (mainly in the treble controls). Other circuits involve bypassing the coupling capacitors with adjustable reactances (mainly in bass controls). However, the most common tone controls are RC filters using audio-taper pots as the adjustable R portion of the filter.

The circuit of Fig. 6-21 presents no insertion loss because of the IC amplifier (a preamp, in this case). The circuit does not provide for volume control because an electronic volume control (Fig. 6-19) is used. The tone-control network of Fig. 6-21 is connected between the electronic volume control and the output or power amplifier.

### 6.11.2  Playback-equalization circuits

Playback-equalization circuits (for both turntables and tape players) usually have resistors and capacitors that form a feedback network. At any given frequency, the amount of feedback (and thus the frequency response) is set by selecting the appropriate RC combination. As the frequency increases, the capacitor reactance decreases, resulting in a change of feedback (and a corresponding change in frequency response).

Figure 6-22 shows the standard RIAA equalization curve for phonograph use (yes, turntables are still manufactured!). The recording curve is the inverse of the playback curve, so adding the two curves together produces a flat frequency-versus-amplitude response. In phonographic recording, the high frequencies are emphasized to reduce the effects of noise and the low inertia of the cutting stylus. The low frequencies are attenuated to prevent large excursions of the cutting stylus. It is the job of the frequency-selective feedback network to accomplish this addition of the recording and playback responses.

In troubleshooting any playback circuit, remember that it is impossible to have the playback network be the exact inverse of the recording because each recording system is slightly different. However, optional guidelines can be applied. A typical audio range is from 20 Hz to 20 kHz, so a rolloff is at both the low and high ends.

Figure 6-23 shows the standard NAB equalization curves for tape use. (Typical tape record/playback circuits are covered in Section 6.13). Again, the recording curve is the inverse of the playback curve, so adding the two produces a flat response. Likewise, the high frequencies are emphasized and the lows are attenuated. However, unlike the phono playback, tape playback tends to flatten out after about 3 to 4 kHz.

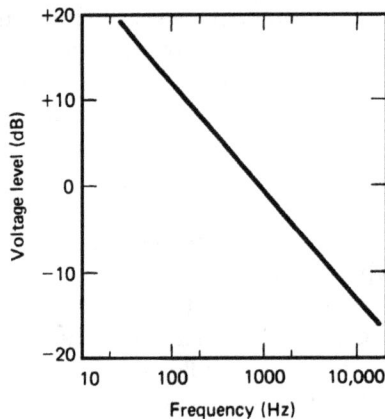

**Figure 6.22.** RIAA playback-equalization curve.

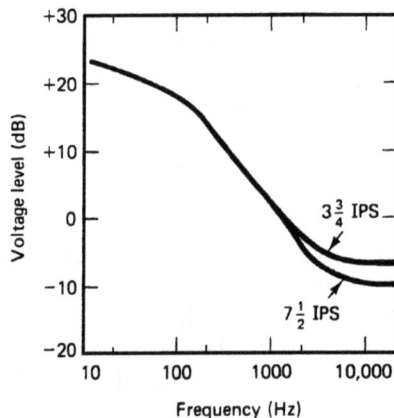

**Figure 6.23.** NAB playback-equalization curve.

Different tape speeds require a different response. Figure 6-23 shows the play-back-response curves for both $3^3/_4$ and $7^1/_2$ inches per second (ips). Up to about 1 kHz, the curves are almost identical. Because each tape speed has only one frequency breakpoint (where the curve must start to flatten), a simple RC compensation network is sufficient (instead of the multisection network for phono playback). The breakpoint for $3^3/_4$ ips occurs at about 1.85 kHz, whereas the breakpoint for $7^1/_2$ ips is at about 3.2 kHz.

In troubleshooting, notice that the accuracy of both the RIAA and NAB compensation is only as good as the components used. Typically, resistors and capacitors with a 1% (or better) tolerance are used in discrete-component playback networks. Also, it might be necessary to trim the value to get an exact (or near exact) performance curve. Remember this if you must replace any component in any playback-equalization network.

### 6.11.3   Balance and output circuits

Figure 6-24 shows the amplifier output circuits, including the balance and tone controls. The power amplifiers for both channels are contained in one

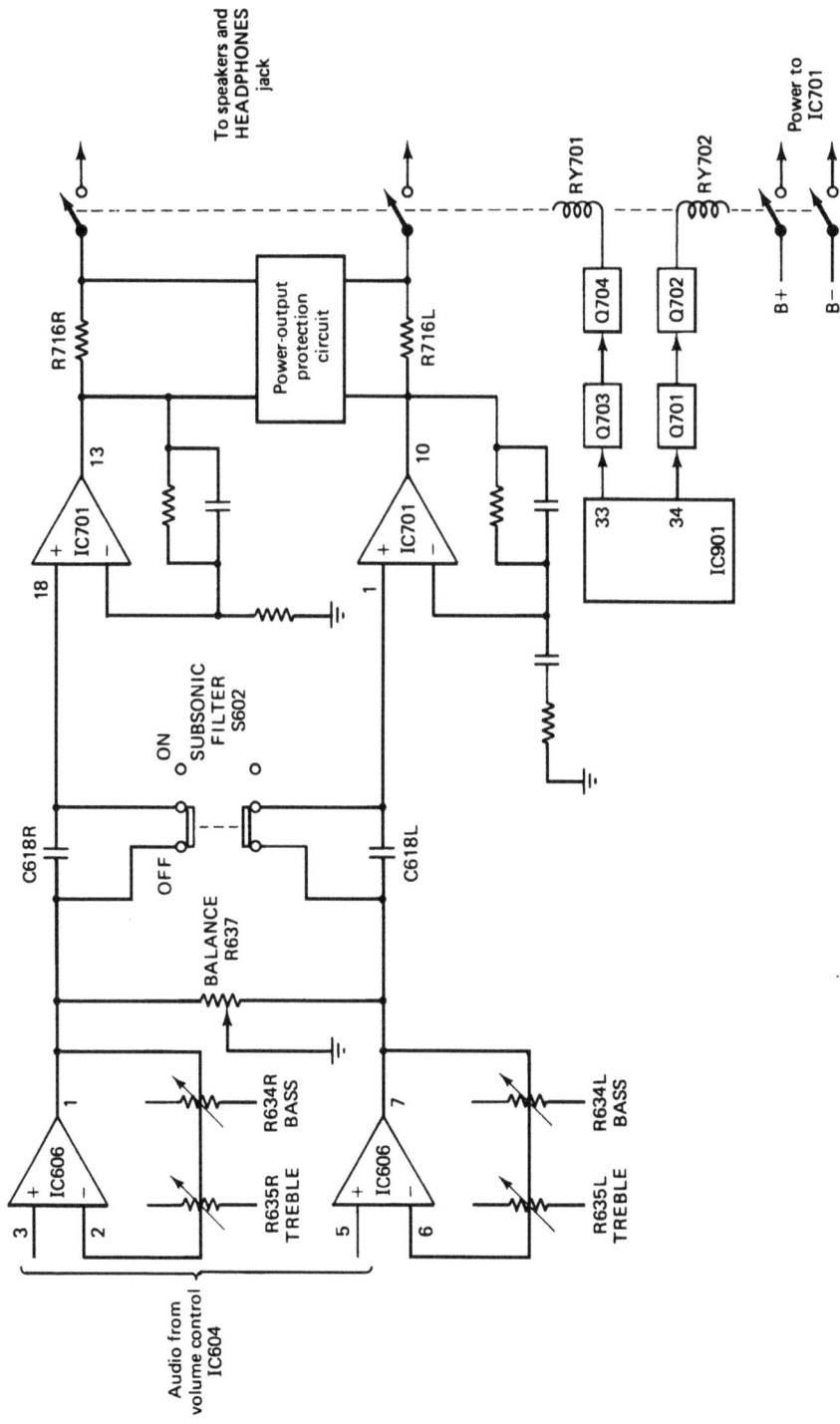

**Figure 6.24.** Typical audio-amplifier output circuits.

167

hybrid IC (IC701). Because of the power rating of 50 W, the IC is mounted on a heatsink. The tone amplifiers that provide for treble and bass adjustments are also contained within a single package (IC606).

Both the left and right channels use a subsonic filter consisting essentially of C618 and subsonic-filter switch S602. The subsonic filter attenuates frequencies below 20 Hz to reduce rumble caused by warped records or defective turntables. The audio output is monitored by protection circuits under control of IC901, as described in Section 6.12.

Audio from electronic volume control IC604 is applied to the noninverting inputs of IC606. Bass control R634 and treble control R635 are connected between the output and inverting input of IC606. Decreasing the bass or treble negative feedback has the effect of boosting the bass or treble, and vice versa. The output of IC606 is coupled to the balance control and subsonic filter. Balance control R637 is connected between the right- and left-channel audio, with the wiper connected to ground. Moving the wiper changes the impedance of both channels simultaneously. R637 is usually set to provide equal signal levels in both channels.

With subsonic filter switch S602 in the OFF position, C618 is out of the circuit, and low-frequency signals are not attenuated. With S602 in the ON position, C618 acts as a high-pass filter, attenuating all frequencies below 20 Hz. The output of the subsonic filter is applied to the noninverting input of IC701. An RC network is connected between the output of IC701 and the inverting input of IC701 to prevent oscillation. The audio output from IC701 is applied through R716, which acts as the sensing resistor for the protection circuits (Section 6.12). The audio from R716 is applied to the speaker terminals and the headphone jack.

### 6.11.4  Audio-circuit troubleshooting

Before troubleshooting the circuits of Fig. 6-24, check that audio is at pins 3 and 5 of IC606. If no audio is present, check the volume-control circuits (Fig. 6-19). Start by setting the bass, treble, and balance controls to midrange. This simple act of faith has been known to cure a "no audio or weak audio in one channel" symptom.

If a total loss of audio is in one or both channels, check the speaker switches. This also solves many "no audio" problems. This amplifier has two sets of outputs (A and B) for each channel. Of course, not all amplifiers have such a configuration, but the speaker switches should be checked first.

When you are sure that all switches and controls are properly set, check for audio at pins 1 and 7 of IC606. If it is absent, but audio is at pins 3 and 5, suspect that IC606 has failed. Also check the bass and treble controls because they are in the negative-feedback path of IC606. If audio is at pins 1 and 7 of IC606, check for audio (at about the same level) at pins 1 and 18 of IC701. If the audio level at IC701 is substantially different from the level at IC606, trace the audio path through the subsonic filter.

The subsonic filter switch S602 setting should have little or no effect on the audio level, except at very low frequencies (below about 20 Hz). If you notice a drastic change in audio level at different settings of S602 from about 1 kHz and up, look for problems in the subsonic filter circuit (such as leakage in C618).

If audio is at pins 1 and 18 of IC701, but not at pins 10 and 13, suspect that IC701 has failed. Before you pull IC701 (heatsink and all), be sure that the 45-V supply is applied to various IC701 terminals. The 45-V (both plus and minus) supply is applied through RY702. In turn, RY702 is turned on through Q701 and Q702 when pin 34 of IC901 goes low (when the amplifier power switch is pressed). Most of the other ICs receive operating power when the power cord is plugged in (whether or not the power button is pressed).

Because of the heavy current drain and high heat dissipation, power amplifier IC701 is turned on only during play. This is typical for most audio amplifiers, where the final power stage is a single IC in the 40- to 50-W range. If the 45-V supply is absent at IC701, suspect a problem with RY702, Q701, Q702, or IC901 (check pin 34 of IC901 for a low).

If audio is at pins 10 and 13 of IC701, but the audio does not reach the speakers or headphones, suspect a problem with RY701, Q703, or Q704. Also, check for a low at pin 33 of IC901. Pin 33 should go low at the same time as pin 34. Remember that the power-output protection circuits described in Section 6.12 are designed to cut off the audio in the event of an overload. Defective protection circuits can cut off the audio—even without an overload.

## 6.12    Output-Protection Circuits

Figure 6-25 shows the output-protection circuits. The overload-protection circuit prevents damage to IC701 when a low-impedance or shorted speaker is connected. The midpoint-protection circuit prevents damage to the speakers in case of a defective IC701 and is turned on when a dc potential (usually called *dc offset*) is present at the output of IC701. The thermal-protection circuit prevents damage to IC701 from excessive heat.

### 6.12.1    Thermal protection

Thermal-protection switch S703 is mounted on the IC701 heatsink and is normally closed. This keeps D707 reverse-biased and Q902 off. With Q902 not conducting, pin 3 of IC901 is high and pin 33 of IC901 remains low to keep the speakers connected to the IC701 output. If the temperature of the IC701 heatsink rises to 100°C, S703 opens, forward-biasing D707. This turns Q902 on and produces a low at pin 3 of IC901. Under these conditions, pin 33 of IC901 goes high to disconnect the speakers from IC701. Simultaneously, pins 36 through 40 of IC901 are pulsed to flash the function-display portion of FL901 (on the front panel).

### 6.12.2    Midpoint protection

The midpoint-protection circuit functions by monitoring the dc output from IC701. In theory, there should be no dc output from IC701 to the speakers. (Excessive direct current can damage the speaker coils.) However, there could be as much as ±1.7 V at the IC701 output without damage to the speakers. The

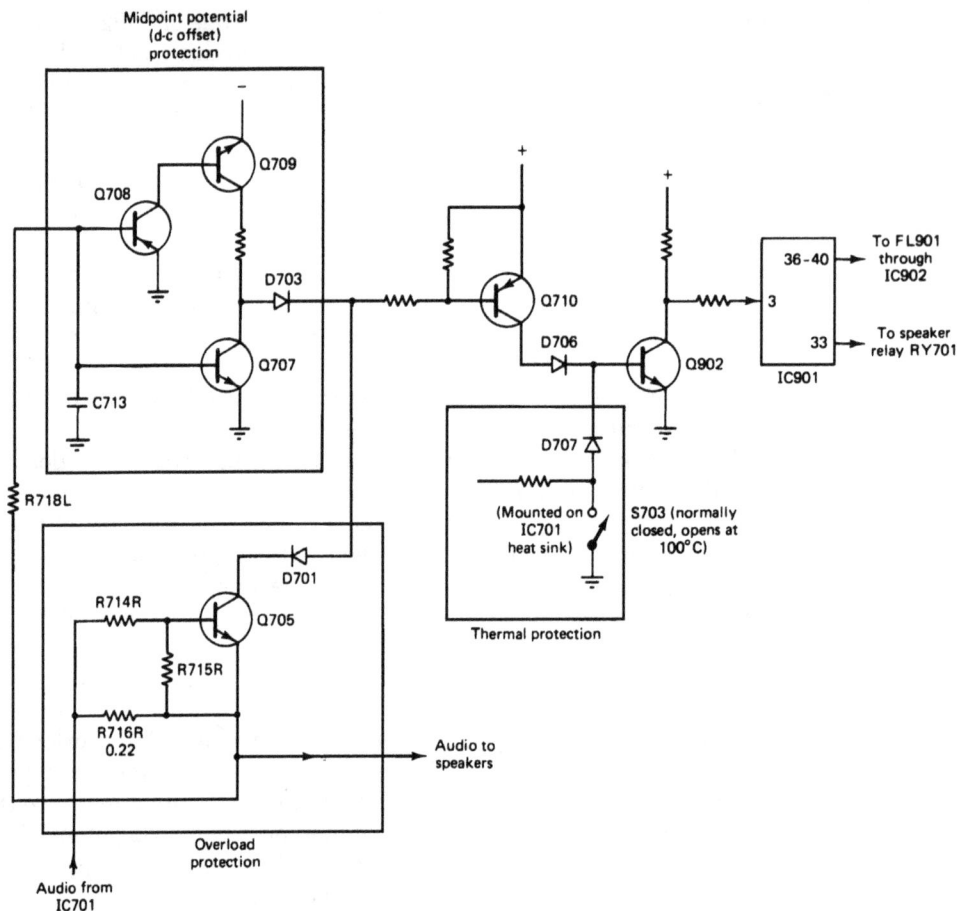

**Figure 6.25.**   Typical output-protection circuits.

midpoint-protection circuit (called the *dc-offset-protection circuit* in some amplifiers) is turned on if the 1.7-V value is exceeded.

If any dc output is from IC701 to the speakers, this potential causes C713 to charge through R718L and R.C713 charges to the average value of the speaker voltage. During normal operation, with the dc output from IC701 less than ±1.7 V, the midpoint-protection circuit is turned off. If the average charge across C713 increases above ±1.7 V, Q707 is turned on, forward-biasing D703. This applies a low to the base of Q710, turning Q710 on and forward-biasing D706. This applies a high to the base of Q902, turning Q902 on and causing pin 3 of IC901 to go low. IC901 then produces a high at pin 33 to disconnect the speakers (and to pulse the FL901 function display), as covered.

### 6.12.3   Overload protection

The overload-protection circuit is the same for both channels, so only the right channel is covered here. Audio output from IC701 to the speakers is applied through R716R, a 0.22-Ω resistor. This resistance is much smaller than the speaker load impedance (typically 6 to 16 Ω).

During normal operation, the voltage across R716R is very small. If a shorted or very low-impedance speaker is connected, excessive output current flows through R716R, and the voltage across R716R increases sharply. Resistors R714R and R715R are connected as a voltage divider across R716R to the base of Q705R. When the current through R716R increases (because of a short or a low-impedance load), the voltage applied to Q705R increases, turning Q705R on. This forward-biases D701, turns on Q710, forward biases D706, turns on Q902, and applies a low to pin 3 of IC901 to disconnect the speakers and pulse the FL901 display.

### 6.12.4  Output-protection troubleshooting

The front-panel function display (FL901) should flash on and off, and the speakers should be disconnected when any one of the following occurs: FL701 becomes overheated (the IC701 heatsink reaches 100°C), the constant (no audio) dc voltage applied to the speakers exceeds ±1.7 V, or the speaker output line is shorted (or is at any impedance less than that of the speakers).

Except for the low-impedance output, these conditions are difficult to simulate, making the circuits difficult to check. Also, if you do succeed in simulating any one of these conditions and the protection circuits are not functioning properly, you can damage the equipment (for example, burn out the speaker coil and/or overheat IC701).

If you must check the circuits, try shorting the speaker lines (either L or R, or both) to ground temporarily (very temporarily). Check that the function portion of FL901 flashes on/off and that the speakers are disconnected. If not, temporarily short pin 3 of IC901 to ground, and check for a flashing display with the speakers disconnected. If the display flashes and the speakers are cut with pin 3 of IC901 shorted, but not when the speaker lines are shorted, suspect a problem with Q705, Q707, Q708, Q709, Q710, and Q902. If the display does not flash and the speakers are not disconnected with pin 3 of IC901 shorted, suspect that IC901 has failed.

You can also check that the anode of D707 is at ground (unless the IC701 heatsink is at 100°C or higher). If it is not, suspect that S703 is open. You can also check the bases of Q707 and Q708. Both bases should be 0 V (ideally) but might be at some potential less than ±1.7 V, without triggering the protection circuits. If the bases are at some value in excess of ±1.7 V, pin 3 of IC901 should go low and the display should flash. If not, suspect a problem with Q707 through Q710 or Q902.

## 6.13  Audio Recording/Playback Circuits

This section describes troubleshooting for the circuits of a typical audio cassette player. Only those circuits associated with recording and playback are

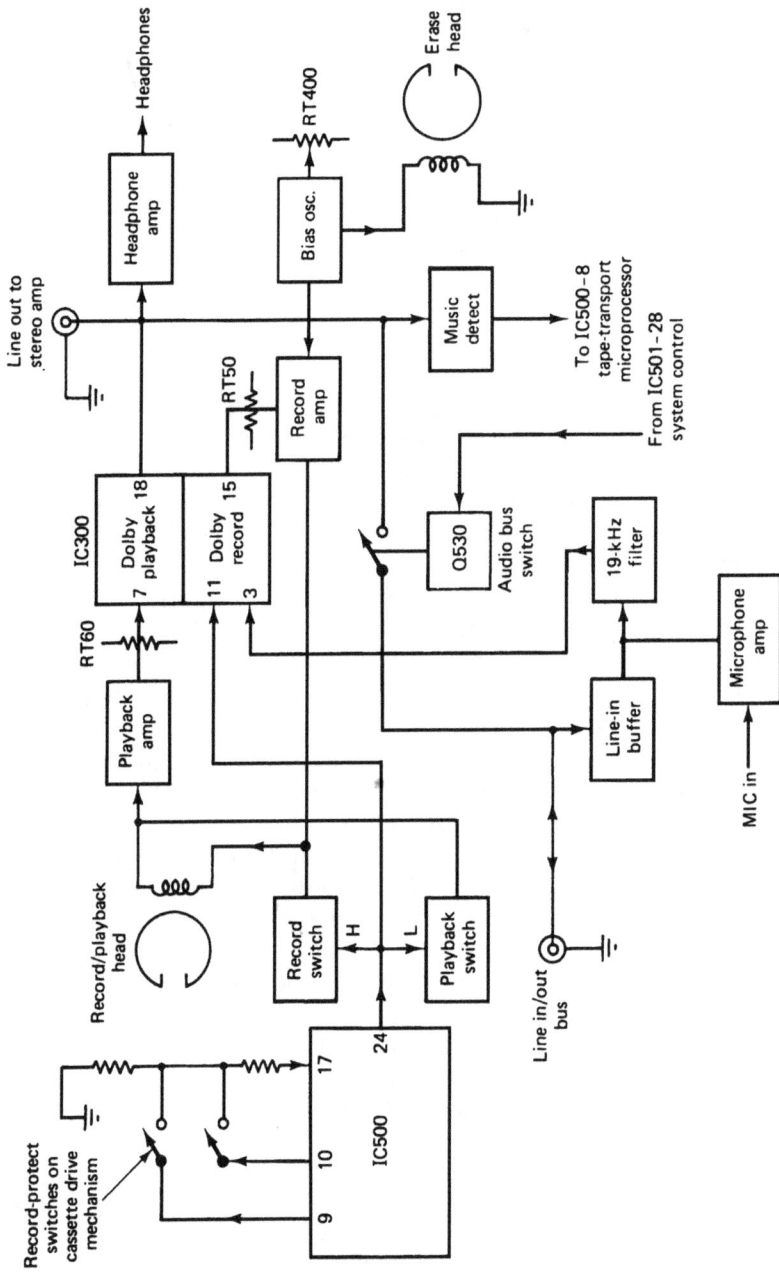

Figure 6.26.  Overall record/playback circuits.

covered. The audio cassette player is part of a modular home-entertainment system.

### 6.13.1 Overall recording/playback circuits

Figure 6-26 shows the overall record/playback circuits (for one channel) in block form. The record/playback heads are placed in the playback or record mode by the recording/playback switching circuit. This switching circuit is controlled by an output from pin 24 of IC500.

During the playback mode (IC500 pin 24 low), the head recovers the recorded audio signal from the tape and applies the signal to a playback amplifier. During the record mode (IC500 pin 24 high), the switching circuits place the recording/playback heads in the record mode so that both line and microphone audio can be recorded on tape.

### 6.13.2 Recording/playback switching during playback

Figure 6-27 shows the recording/playback switching, as well as the amplifier used during playback. With pin 24 of IC500 low, both Q62 and Q63 are turned on (grounding the recording input to the head), with Q61 turned off (permitting the recovered audio to pass to IC60).

As covered in Section 6.11.2, playback equalization circuits are used for both phonograph and tape players. The equalization network connected to IC60 is the standard equalization circuit for many audio cassette players. The RC time constants are: 3190 μs for R69 and C62, 120 μs for R64 and C62, and 70 μs for R61 and C61. The voltage gain of IC60 is about 49 dB at 400 Hz.

The output of IC60 is applied to the playback portion of the Dolby processing circuits in IC300 through playback-gain control RT60. After processing by the Dolby circuits, the audio is applied to the headphone amplifier and to the music-detect circuits.

### 6.13.3 Headphone and line-out

Figure 6-28 shows the headphone and line-out circuits. Notice that output level control RV40 controls audio to both the headphones and the line (stereo amplifier). The output of the headphone amplifier is typically 80 mV into an 8-Ω impedance when the circuit is tested with a Dolby calibration tape.

Audio-bus switch S530 permits the line in-out jack to be used as a bidirectional audio bus connection during system operation. Audio-bus switch IC01 is controlled by Q530, which, in turn, is controlled by the signal at pin 28 of IC501 (high for audio bus). During stand-alone operation (audio-bus switch IC01 open), the line in-out jack acts as a line-in function only. During system operation, the line in-out jack can be used as both an input and an output connection (in=switch open; out=switch closed) during playback.

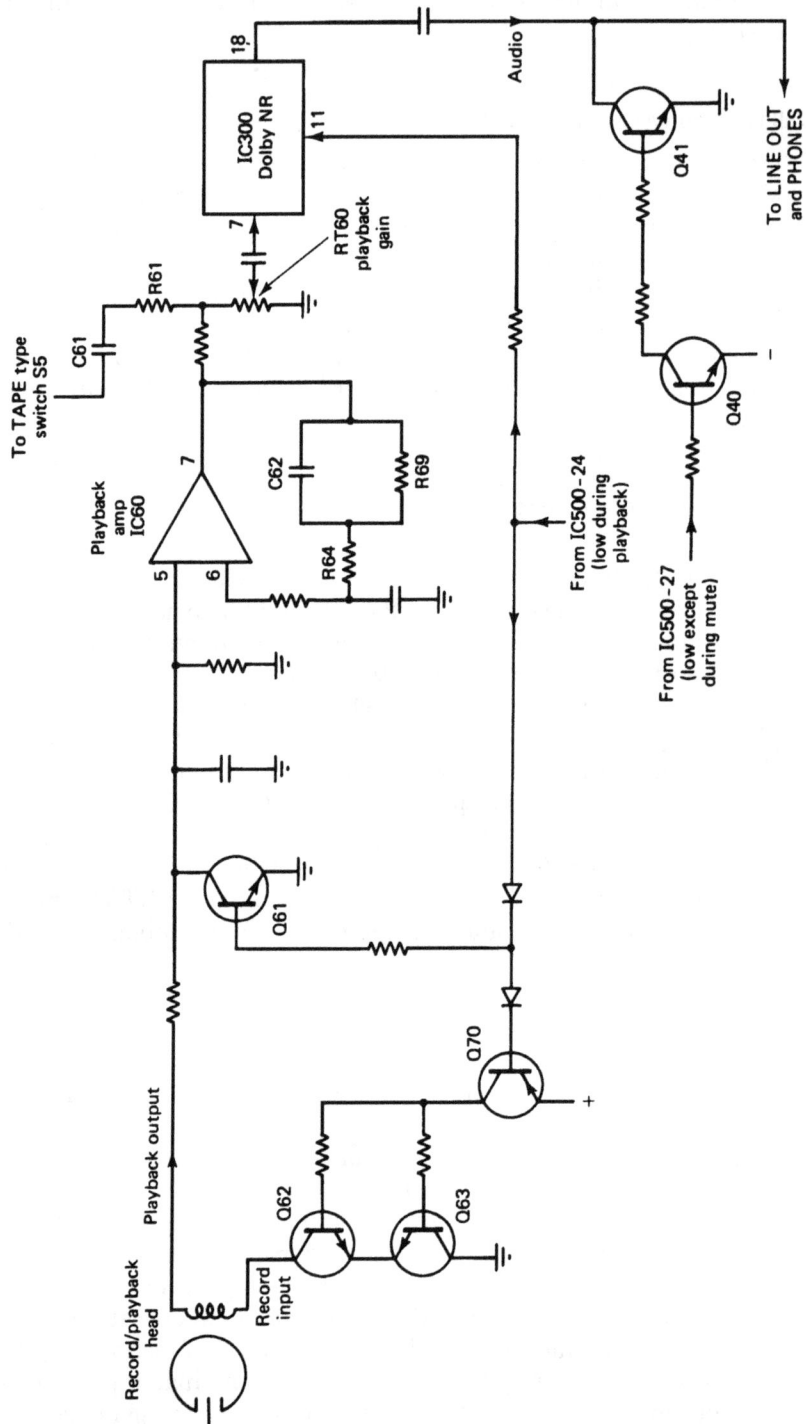

**Figure 6.27.** Record/playback switching and amplifier circuits.

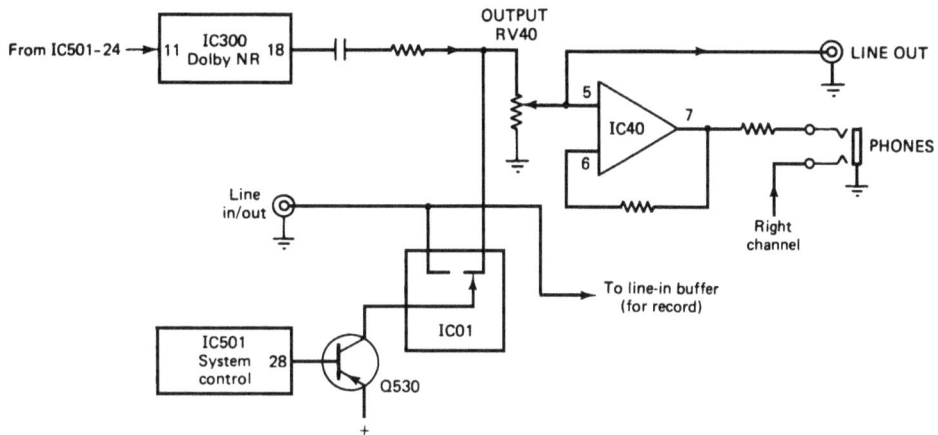

**Figure 6.28.**  Typical headphone and line-out circuits.

### 6.13.4  Recording/playback switching during recording

Figures 6-29 and 6-30 show the recording/playback switching, as well as the buffer, amplifier, and filter circuits used during recording. With pin 24 of IC500 high, Q70 is turned off, as are Q62 and Q63 (removing the ground from the recording input to the head). Q61 is turned on, completing the ground connection for the recording input of the head.

*Recording input*  As shown in Fig. 6-29, the audio at the line in-out jack (for system use) is applied to the line-in buffer circuits. The buffer keeps the line input at an impedance of about 200 Ω. The audio exits the buffer and is applied to a 19-kHz filter that removes 19-kHz pilot signals (from an FM multiplex broadcast) and/or bias signals (from the bias oscillator). Either of these signals can upset the response characteristics of the Dolby NR (noise-reduction package) IC300. Audio from the filter is applied to the recording portion of IC300. Audio from the microphone input can also be recorded.

*Recording output*  As shown in Fig. 6-30, audio from IC300 is applied to recording amplifier IC50 through record/playback-level adjustment RT50 together with signals from the bias oscillator. After processing by IC50, the audio is applied to the recording input of the recording/playback head and recorded on tape. Note that a fixed amount of bias current is applied to the erase head. However, the record head receives higher or lower bias current, depending on the position of the tape-type switch (S5). Bias current is also adjusted by RT400.

*Tape equalization*  The record amplifier consists of IC50, with the associated components to compensate for record-current requirements. As covered in Section 6.11.2, these components boost both the high and low ends of the frequency range. Tape-type switch S5 cuts in the components as necessary to provide the correct compensation for the three basic types of tape (normal, CR02, or chromium oxide, and metal).

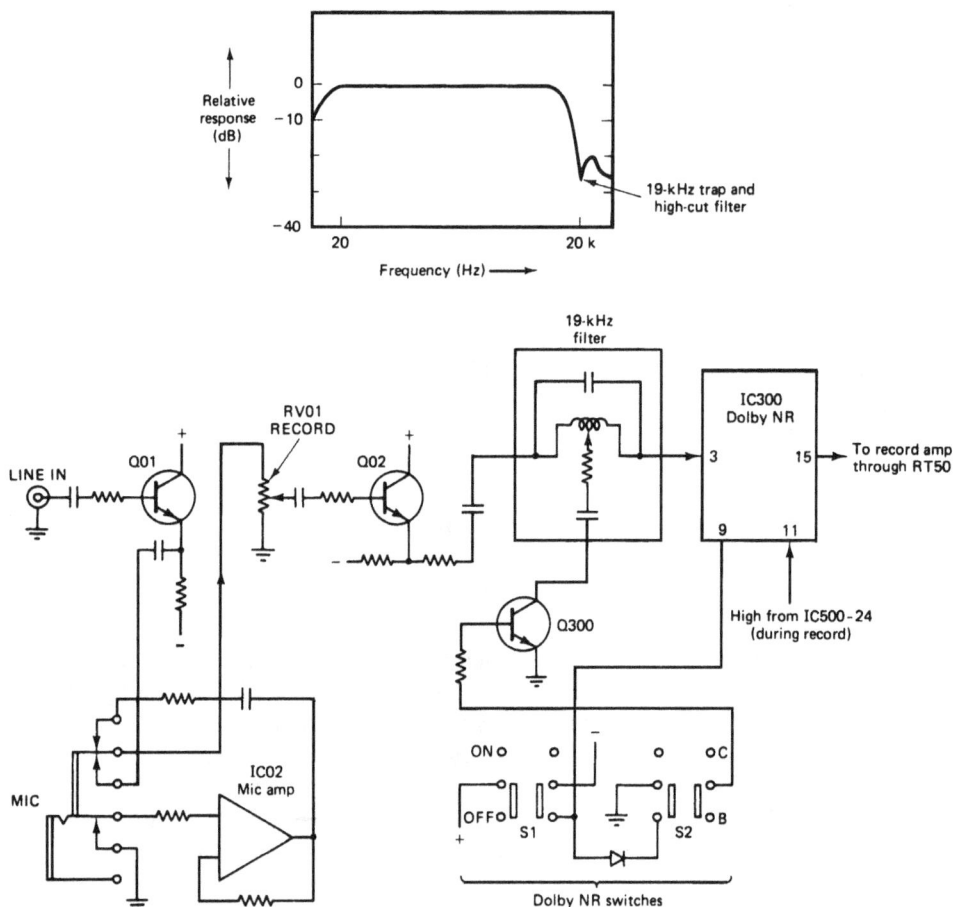

**Figure 6.29.** Record/playback switching and filter circuits.

*Record-protect functions*    As shown in Fig. 6-26, record-protect switches on the cassette-drive mechanism prevent IC500 from placing the deck in the record mode (prevent pin 24 from going high), no matter what commands are applied to IC500.

When a cassette with the tabs intact is installed, the record-protect switches are closed. Scan signals from pins 9 and 10 of IC500 are applied to pin 17, permitting normal control by IC500 (that is, IC500 can go high if so instructed by the front-panel controls or by a remote command).

When a cassette with the tabs removed is installed, the record-protect switches are open. This prevents scan signals from being applied to pin 17, and prevents record operation (pin 24 of IC500 cannot go high), no matter what commands are applied to IC500.

Note that there are separate record-protect switches for forward and reverse (on this particular deck). Remember this when troubleshooting a "no record in one direction" symptom. Check that both tabs are in place on the cassette. Try another cassette or put heavy tape over the missing tab area.

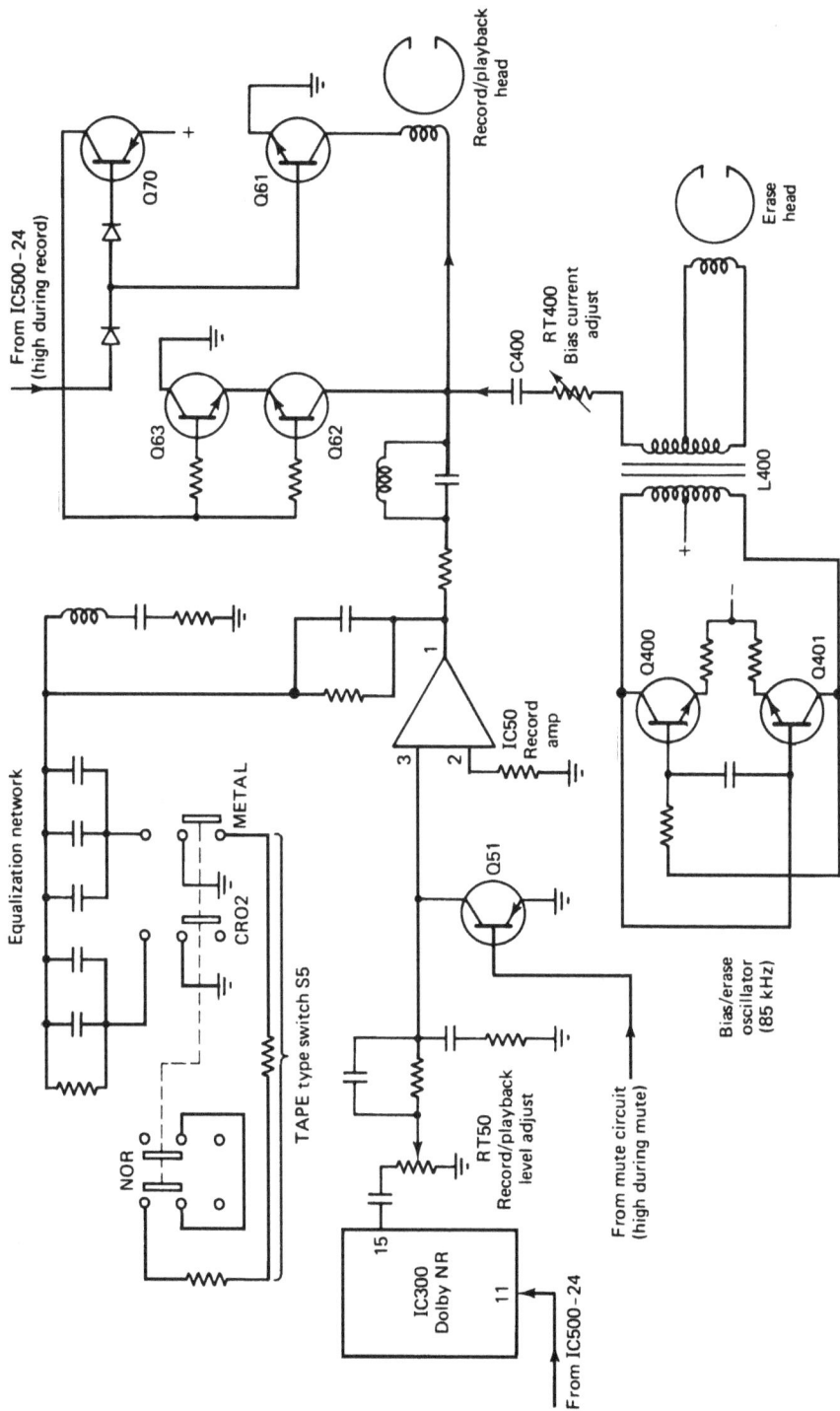

**Figure 6.30.** Record/playback amplifier and bias circuits.

## 6.13.5    Recording/playback troubleshooting

If audio is absent or abnormal during playback, but the tape transport appears to be normal (tape runs properly in both directions), first check the playback gain adjustment, as described in the service literature. Playback gain is adjusted by RT60 (as shown in Fig. 6-27). Also, turn up front-panel output control RV40 (Fig. 6-28). (This might just cure the problem.) If not, continue playing a known-good tape and monitor the audio at pin 7 of IC60. The output should be about 100 mV (for this circuit).

If there is no audio output from IC60, check the circuits from the heads to IC60, including Q61 (off), Q62, and Q63 (on). If there is audio from IC60, check for audio at pin 18 of IC300 (Dolby). If there is no audio from IC300, check the audio path from IC60 to pin 7 of IC300. Make certain that pin 24 of IC500 and pin 11 of IC300 are low (about $-7$ V). If they are not, suspect IC500. If pin 24 of IC500 goes high, IC300 goes into the record mode, rather than the playback mode.

If audio is at pin 18 of IC300, check for audio at RV40 (Fig. 6-28). If it is absent, be sure that Q40 and Q41 (Fig. 6-27) have not been turned on (mute condition) by a signal at pin 27 of IC500. If Q40 and Q41 are not on, check the audio path from IC300 to RV40. Then check the audio from RV40 to the line-out jack and the phones jack. If audio is available at RV40, but not at the jacks, suspect a problem with IC40.

If playback audio is present, but the background noise increases with time, or the playback audio decreases at high frequencies with time, the heads might need to be degaussed. Some technicians never degauss the heads, whereas others degauss at each service (using a commercial head eraser). In between these extremes, some technicians degauss only when certain symptoms occur. The two most common symptoms are increased background noise and decreased high-frequency output—especially when the deck is used over long periods of time. The author has no recommendations on head degaussing except: Never degauss the heads with a cassette in place (particularly a customer's favorite tape or an expensive shop test tape).

If audio is absent or abnormal during recording, but the tape transport appears to be normal (the tape runs properly in both directions), first check the playback circuits as just described. Always clear any playback problems before you check record. This applies to virtually all cassette decks. If there is no audio (or poor audio) during playback, the problem could be common to both recording and playback. However, if the problem is only in the record mode, you have quickly isolated the trouble to a few circuits (bias oscillator, record amp, etc.).

If it is not possible to record any audio (with good playback), start by checking for audio at pin 15 of IC300 (Figs. 6-29 and 6-30). Also try to cure the problem by turning up the front-panel record control (RV01) and by adjusting the record level (RT50) and bias current (RT400) as described in the service literature.

If audio is present at pin 15 of IC300, check the audio path from IC300 to the heads. Also be certain that pin 24 of IC500 and pin 11 of IC300 are high (about $+6$ V) to place IC300 in record mode. If no audio is at the input to IC50, be sure that Q50 and Q51 have not been turned on (mute condition) by a signal at pin 27 of IC500.

If audio is at the heads, but the deck does not record, check the bias oscillator (Fig. 6-30). As a general rule, if the tape can be erased, the oscillator and erase head are good. However, the bias signal might not be reaching the recording/playback heads. Check for an 85-kHz bias signal on both sides of C400 and RT400, as well as the adjustment of RT400.

Remember that each type of tape requires a different amount of bias current. For example, the bias signal measured at C400 on our deck is about 3.4 V for normal tape, 5.2 V for chrome, and more than 10 V for metal tape. This is determined by the tape-type switch (S5) setting.

If audio is at the heads and the bias voltage is correct, suspect that the problem is at the heads. Before pulling the heads (a tedious job) be sure that Q62 is turned on and Q62/Q63 are off, placing the heads in a condition to record. If not suspect Q61, Q62, Q63, or Q70.

If audio is at pin 15 of IC300, trace the audio path from the line-in and microphone jacks (Fig. 6-29) through Q01, Q02, Q300, and IC01. Obviously, if you can record from line-in, but not from the microphone, suspect IC02.

Also remember that Dolby NR switch S1 controls Q300, which, in turn, controls the Dolby filter. However, even if the filter circuit or Q300 and S1 fail, you will probably be able to record—even though the recording will be poor. Before you condemn any Dolby circuit, be certain that the tape is being played back in the same mode as during recording. If Dolby is not used during record, do not use Dolby during playback (in a hopeless attempt to improve quality). If Dolby C is used during recording, play back in Dolby C, not Dolby B, etc. If there is no substantial difference in sound quality, with and without Dolby (of the correct type, B or C), suspect the Dolby noise-reduction (NR) package IC300 or the control signals to IC300.

## 6.14   Typical Audio-Circuit Tests and Adjustments

The procedures described in Chapters 1 through 5 are generally sufficient to test and adjust most present-day audio amplifiers. However, the procedures should be compared to those found in the service literature for the specific amplifier being serviced. (When all else fails, follow the instructions.) The amplifier circuits described in this chapter do not have internal-adjustment controls, as is true for most present-day IC audio amplifiers. However, most audio amplifiers have bass, treble, and balance controls. In the absence of specific test and adjustment procedures in the service literature, here are some points to consider when testing any audio-amplifier circuit.

### 6.14.1   Initial control settings

Operate the volume control for a midrange volume level or as specified in the service literature. Set the bass, treble, and balance controls to their midrange.

Apply a 1-kHz sine-wave signal to each of the various inputs (phono, tuner, CD, tape play, auxiliary, etc.). Adjust the balance control until both channels show the same level indication on the front-panel display. If there is no front-panel display, adjust balance until the outputs across both speakers are identical.

### 6.14.2   Analyzing audio-circuit test results

If the balance control must be set far from the midrange to get equal output (with an identical signal at both inputs), the mismatch is severe. This can be the result of problems in the balance circuit but is not limited to the balance network. For example, the problem can be a mismatch in IC701 (Fig. 6-24) if the problem is evident at all inputs of the amplifier. In such cases, it is necessary to replace IC701 as a package—even though one channel might be good.

Of course, if there is a mismatch at only one input or one output of the amplifier, the problem can be pinned down easily. For example, if the mismatch is at only the phono input, suspect the phono preamp. On the other hand, if the mismatch appears at only the tape-record output, suspect a problem with the tape-record buffer.

If both channels produce essentially the same signal (with an identical signal at both inputs and with the balance control at midrange), the next step is to test the range of both the bass and treble controls. Typically, the bass and treble controls have a ±8- to 10-dB range, at some specific frequency. For example, the bass control of the amplifier described in this chapter has a ±8-dB range at 100 Hz, whereas the treble control has a ±10-dB range at 10 kHz. (50 Hz and 20 kHz are also common bass and treble frequencies.)

Set both the bass and treble controls to midrange. Apply a 100-Hz signal to the inputs of both channels. Set the balance control so that the outputs of both channels are identical. Keep the treble and balance controls at midrange. Vary the bass control from one extreme to the other. Notice that the output of each channel varies from about 8 dB above and below the output existing at the bass midrange setting.

Return both the bass and treble controls to midrange and apply a 10-kHz signal to the inputs of both channels. Leave the balance control set so that both outputs are identical. Keep the bass and balance controls at midrange. Vary the treble control from one extreme to the other. Notice that the output of each channel varies about 8 dB above and below the output existing at the treble midrange setting.

If the audio-amplifier circuit passes the tests described here, it is reasonable to assume that the amplifier is functioning normally and no troubleshooting is required.

### 6.15   Loudspeaker Circuit Tests/Adjustments

In addition to checking amplifier characteristics, audio-equipment loudspeakers should also be checked. This brings up some problems. Although it is possible to test a loudspeaker for such characteristics as *sound pressure level*

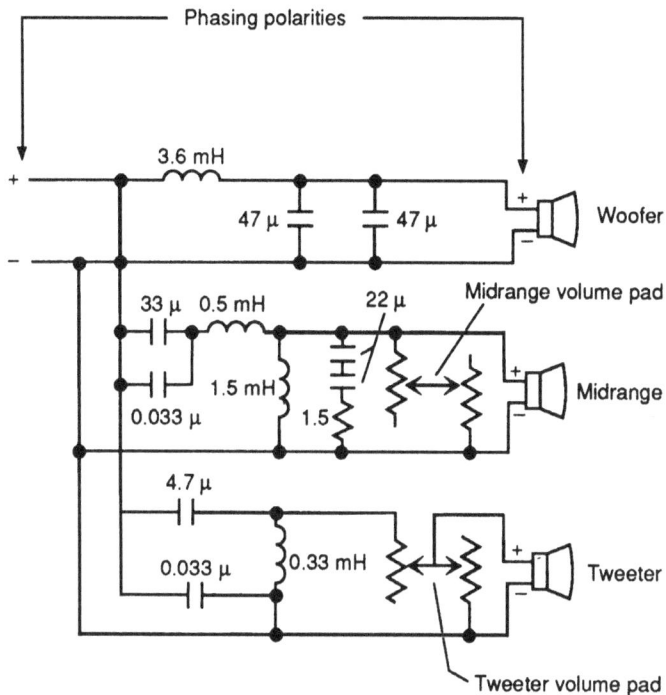

Figure 6.31.   Typical loudspeaker circuits.

(*SPL*) under laboratory conditions, the most practical test is "by ear." Unfortunately, you and the customer have different ears, so the results are uncertain (at best).

To further complicate loudspeaker problems, some speakers are adjustable. For example, the speakers shown in Fig. 6-31 have volume controls (also called *pads*) in both the midrange and tweeters. Although these are not usually customer adjustment controls, the pads are often adjusted to some arbitrary setting "to match the customer's ear."

From a troubleshooting standpoint, be sure that the speaker adjustments (if any) can control volume at the corresponding speaker and that the controls are smooth (no abrupt changes in volume as the control is adjusted).

## 6.16   Example of Audio-Circuit Troubleshooting

This step-by-step troubleshooting problem involves locating the defective part (or improperly connected wiring) in a combination discrete-IC audio amplifier. Figure 6-32 shows the schematic diagram.

This particular circuit was chosen as an example because it combines both IC and discrete components. The CA3094B is a programmable amplifier (where the gain is set by the resistor at pin 5) similar to the OTA ICs described in Chapter 2. No matter what trouble symptom is involved, the actual fault

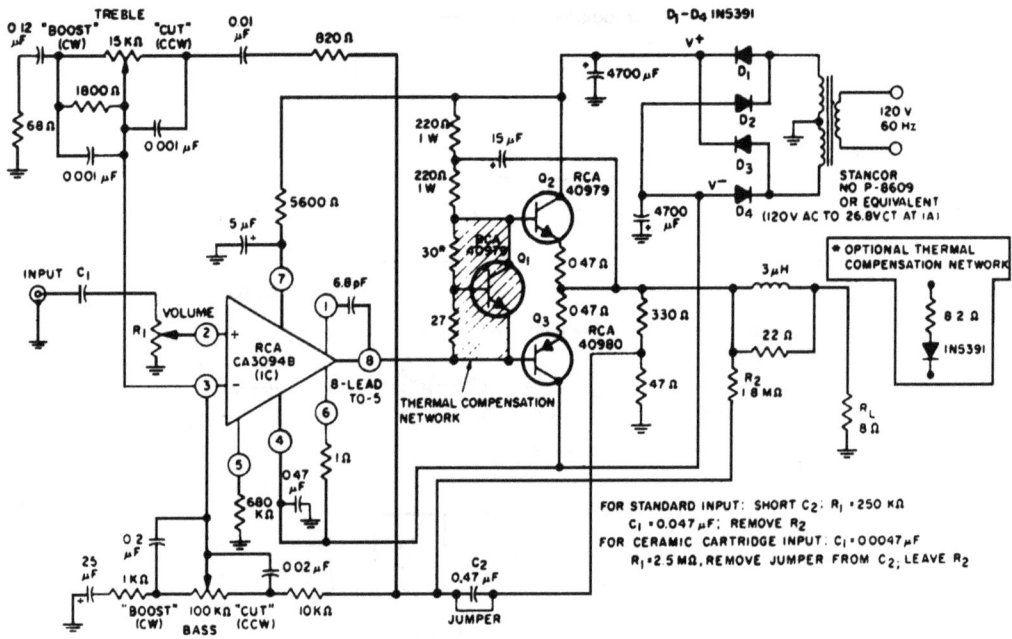

**Figure 6.32.** 12-W audio amplifier with discrete output and IC input (Harris Semiconductor, *Linear & Telecom ICs,* 1994, p. 2-100).

**TYPICAL PERFORMANCE DATA**
**For 12-W Audio Amplifier Circuit**

| | | |
|---|---|---|
| Power Output (8Ω load, Tone Control set at "Flat") | | |
| Music (at 5% THD, regulated supply) . . . . . . . . . . . . . . . . | 15 | W |
| Continuous (at 0.2% IMD, 60 Hz & 2 kHz mixed in a 4:1 ratio, unregulated supply) See Fig. 8 In ICAN-6048 . . . . . . | 12 | W |
| Total Harmonic Distortation | | |
| At 1 W, unregulated supply . . . . . . . . . | 0.05 | % |
| At 12 W, unregulated supply. . . . . . . . . | 0.57 | % |
| Voltage Gain. . . . . . . . . . . . . . . . . . . . . . | 40 | dB |
| Hum and Noise (Below continuous Power Output) . . . . . . . . . | 83 | dB |
| Input Resistance . . . . . . . . . . . . . . . . . . . . . | 250 | kΩ |
| Tone Control Range . . . . . . . . . . . . . . . . See Fig. 9 In ICAN-6048 | | |

can eventually be traced to one or more of the circuit parts (transistors, ICs, diodes, capacitors, etc.), unless, of course, someone has wired the parts incorrectly. Even then, the following waveform, voltage, and resistance checks will indicate which branch within the circuit is at fault.

### 6.16.1   Initial tests

If you are servicing this circuit in an existing piece of equipment, the first step would be to study the literature and test circuit to confirm the trouble. In this example, the only "literature" is Fig. 6-32. The circuit description claims an output of 12-W into an 8-Ω load. Although the load is shown as $R_L$, it can be assumed that the circuit will be used with an 8-Ω speaker. There are no test points or waveforms, the voltage information is incomplete, and there is no resistance-to-ground information. However, with this fragmentary information, you can test the circuit, monitor the signal at various points in the circuit, and localize trouble using the test results.

The first step is to apply a signal at the input and monitor the output. The input can be applied at C1 (as shown). The output is measured at $R_L$, or at an 8-Ω speaker connected in place of $R_L$. Use the resistor or the speaker, but never operate the circuit without a load. Q2 and Q3, and possibly Q1, can be destroyed if operated without a load.

To produce a 12-W output across an 8-Ω load, a signal of about 10 V must be at the speaker, and at the junction of the Q2/Q3 emitters. ($E^2/R$=W, $10^2/8$=12.5 W). If the circuit has a 40-dB gain (as claimed), 0.1 V (100 mV) at the input should be sufficient to fully drive the speaker, depending on the setting of R1.

Connect an audio generator to the input and set the generator to produce 0.1 V at a frequency of 1 kHz (or some other frequency in the audio range). Set both the bass and treble controls to midrange, and adjust R1 until you get a good tone on the speaker and/or readable signal at the Q2/Q3 emitter junctions. (The tone will probably burst your eardrums at this point, so adjust $R_1$ until the tone is reasonable.)

Now vary both the bass and treble controls. Both tone controls should have some effect on the tone, but the bass control should have the most control. Change the generator frequency to 10 kHz and repeat the tone-control test. The treble control should have the most effect with a 10-kHz signal.

If the circuit operates as described thus far, it is reasonable to assume that the circuit is good. Quit while you are ahead! If you have access to distortion meters, check the distortion against the performance data on Fig. 6-32. Also check the actual voltage input (at pin 2 of the IC) when a 10-V signal appears at the output and find the true amplifier voltage gain (which should be 40 dB).

### 6.16.2   Troubleshooting (audio circuit)

If the circuit does not operate as described, set R1 and the bass/treble controls to midrange. Monitor the signal voltages at pins 2 and 8 of the IC. You can monitor all test points with an ac voltmeter, a dc voltmeter with a rectifier probe, or with a scope. If you use the scope, you can see any really abnormal distortion at the test point, together with the signal voltage. However, minor distortion will probably be unnoticed (because you are using a sine wave at the input instead of using square waves).

If a signal is at pin 2 of the IC, but not at pin 8 (or if the signal at pin 8 shows little gain over the pin-2 signal), the problem is at the IC portion of the circuit. Check all voltages at the IC. You do not know the exact values, but there are some hints.

The transformer has a 26.8-V center-tapped secondary, so $V+$ and $V-$ should be about 12 to 15 V (and should be substantially the same). In any event, pins 4 and 6 of the IC should be about $-12$ V and pin 7 should be about $+12$ V (although pin 7 will probably be at a lower voltage than pins 4 and 6 because of the 5600-$\Omega$ resistor at pin 7).

If the IC voltages appear to be good, but the gain at pin 8 is low, it is possible that the IC is bad, that the resistor at pin 5 is not of the correct value (this resistor determines IC gain), or that too much feedback is at pin 3 (from the output through C2 and the tone controls).

If a good signal is at pin 8 of the IC, but not at the output (speaker or $R_L$), the problem is at the discrete portion of the circuit (Q1, Q2, and Q3 or the associated circuit parts). Check the collector voltages of Q2 and Q3. These voltages should be about 12 V and should be substantially the same as at pins 4, 6, and 7 of the IC.

The waveform (signal) and voltage checks that are described here should be sufficient to locate any major defect in the circuit, including bad wiring. Of course, if the circuit operates, but performance is not as claimed, it is possible that the problem is one of poor physical layout, wrong component values, etc. The basic techniques described here can be applied to most audio circuits.

# Linear-Supply Circuit Troubleshooting

This chapter is devoted to troubleshooting linear power supplies. Switching power supplies are covered in Chapter 8. The chapter starts with a review of linear supply basics (half wave, full wave, bridge, regulators, zeners, programmable zeners, IC regulators, current foldback, op-amp regulators, heatsinks, etc.) from a troubleshooting standpoint. This is followed by an in-depth review of linear-supply testing and connections. It then describes an example of linear-supply troubleshooting that covers both discrete and IC circuits.

## 7.1 Linear-Supply Basics

This section describes basic power-supply and linear-regulator circuits. These include half-wave, full-wave, voltage-doubling, and voltage-tripling circuits, as well as zener and adjustable-shunt (or programmable-zener) regulation circuits.

### 7.1.1 Half-wave circuits

Figure 7-1 shows the basic half-wave power-supply circuit. The cathode of diode CR1 is connected to filter capacitor C1 and filter choke L1. Another filter capacitor (C2) is also used to smooth out the pulsating direct current. Resistor R1 is placed across the supply output. (R1 is not used in all circuits; instead, the output is connected directly to the load circuit.) Resistor R1, if used, is known as a *bleeder resistor* because it places a small current drain on the supply and helps stabilize the output. If R1 is made up of a single tapped resistor or a series of several resistors, it is possible to take several different voltages from the bleeder network. For example, if the power-supply output is 30 V and the bleeder consists of six equal-value resistors, the available voltages are 5, 10, 15, 20, 25, and 30 V, as shown.

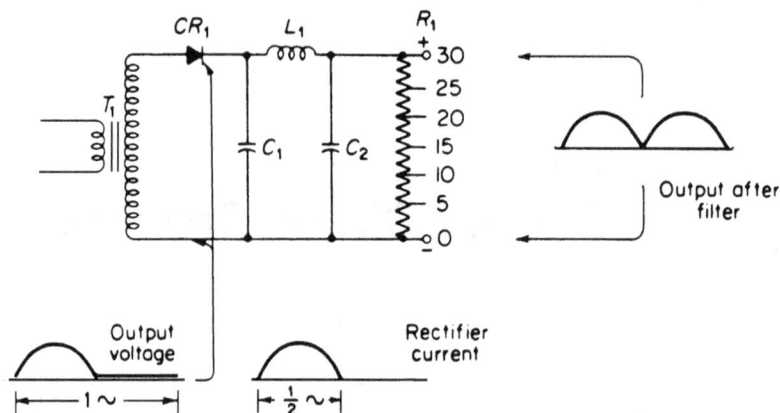

**Figure 7.1.** Basic half-wave power supply.

In troubleshooting any rectifier circuit, the important point to remember is that the circuit output is a direct current, but not necessarily a constant voltage. Practical circuits have some variation in amplitude (known as *ripple*). The output voltage decreases slightly between cycles (the negative peak of the ripple voltage), then increases at the peak of each half cycle (positive peak of the ripple voltage). Peak-to-peak ripple voltage is expressed as a percentage of the total supply output voltage. For example, if the supply produces 100 V of direct current across the bleeder resistor and the ripple is 3 V peak-to-peak (p-p), there is a 3% ripple (which would be totally unacceptable for almost any type of linear supply!) If excess ripple is in any linear supply, the problem is with the filter circuit.

### 7.1.2 Full-wave circuits

Figure 7-2 shows the basic full-wave power-supply circuit. Both the positive and negative alternations of the ac cycle are used. For this reason, the full-wave circuit requires two diodes and a transformer with a center tap. One lead of the transformer secondary is connected to diode CR2. The center tap is connected to the common or ground circuit. The total voltage across the transformer secondary is about twice the voltage that appears at the supply output.

In troubleshooting a full-wave circuit, notice that the filter capacitors charge at a rate twice that of half wave because current flows through the bleeder resistor in the same direction on both alternations (half cycles). This makes the ripple frequency twice that of the half-wave frequency. Because the discharge time between peaks of full-wave pulsating direct current is only half of that found in half-wave circuits (there is a shorter time between peaks), the filter capacitors have less time to discharge. This makes it easier to maintain a relatively high charge and makes the output smoother than that of a half-wave circuit.

### 7.1.3  Bridge circuits

Figure 7-3 shows the basic full-wave bridge power-supply circuit. The bridge circuit makes it possible to have full-wave rectification using a transformer without a center tap. Four rectifiers are used in the basic bridge circuit. Also, some rectifier packages contain four rectifiers connected in the bridge configuration.

When a positive alternation occurs, with the top end of the transformer secondary positive, current flows from the secondary bottom, through CR3,

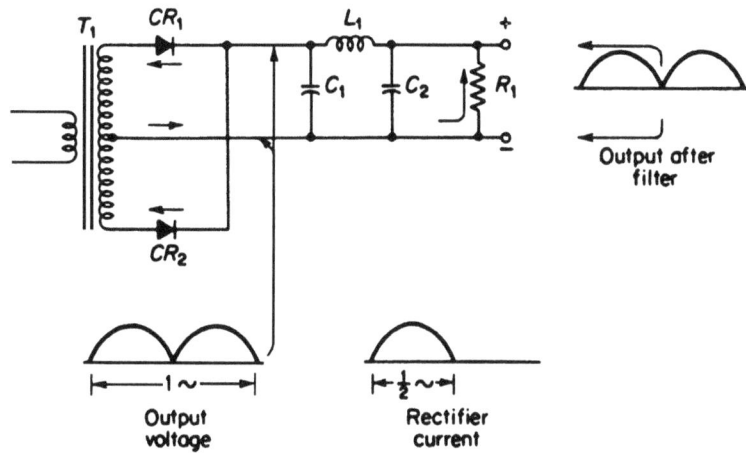

**Figure 7.2.**  Basic full-wave power supply.

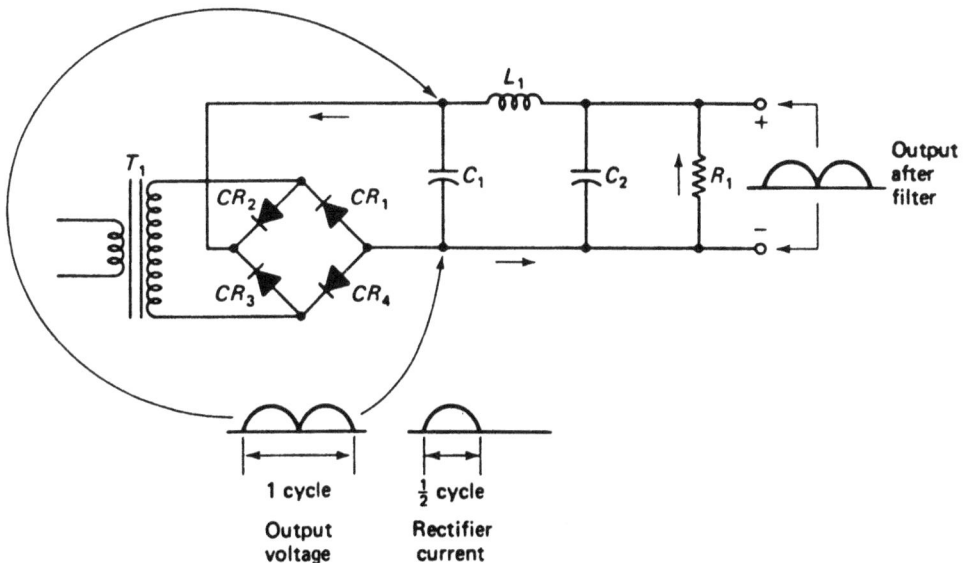

**Figure 7.3.**  Basic full-wave-bridge power supply.

the load or bleeder resistor R1, and diode CR2 to the secondary top. On the next alternation, when the bottom end of the transformer secondary is positive, current flows from the secondary top, through CR1, R1, and CR4 to the secondary bottom. Full-wave rectification occurs through the bleeder or load resistance in the same direction on both half cycles. From a troubleshooting standpoint, this produces a higher ripple frequency and more efficient filtering.

### 7.1.4   Basic linear power supply

Figure 7-4 shows the circuit for a basic linear power supply, using a full-wave bridge with only one filter capacitor. The following is a summary of the components and characteristics that relate to practical troubleshooting.

*Power transformer*   Transformer T1 is usually of the step-down type. A typical primary voltage is 110 to 120 V (or 220 to 240 V), with a secondary voltage between 12 V and 50 V. With a full-wave bridge and a single capacitor as shown, the dc output of the supply is about 1.3 times the transformer secondary voltage (a T1 secondary of 10 V produces about 13 V at the supply output). In addition to voltage, transformers are rated for power capability—usually a volt-ampere (VA) rating, rather than a true ac-power rating. For safety, the transformer power rating should be at least 1.3 times the dc-output power of the complete circuit (if 10 W of dc power is required, the transformer should be capable of at least 13 W).

*Load resistance and power-supply resistance*   These two terms are often confused. Neither value can be measured directly during testing or troubleshooting, but must be calculated on the basis of voltage and current.

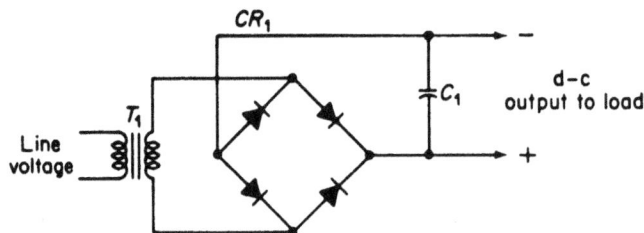

d-c output $\approx 1.3 \times$ secondary voltage
Transformer power (or VA) rating $> 1.3 \times$ d-c output power
Load resistance $=$ output voltage/output current
Power supply resistance $=$ (no-load $V$ − full load $V$)/current
$C_1$ working voltage $> 1.3 \times$ output voltage
$C_1$ capacitance (in $\mu F$) $\approx 200{,}000$/load resistance $\times$ max ripple
Secondary voltage $\approx$ required d-c output/1.3
d-c power output $=$ output voltage $\times$ load current
$CR_1$ ratings $=$ see text

**Figure 7.4.**   Basic linear power supply.

*Load resistance (or effective load impedance)*   This is found when the output voltage across the load is divided by the load current. For example, with a dc output of 10 V and a load current of 2 A, the load impedance is 5 Ω.

*Power-supply resistance (or internal impedance)*   This is found when the change in output voltage is divided by a change in load current. For example, if the no-load or lite-load voltage is 28 V with 0.5 A and the full-load output is 25 V with 2 A, the power-supply resistance is $(28-25)/(2-0.5)=3/1.5=2$ Ω.

*Diode characteristics*   Four basic diode characteristics must be considered during testing or troubleshooting: maximum reverse voltage, forward voltage, reverse current, and forward current.

*Maximum reverse voltage (also known as peak inverse voltage or PIV)*   This is the amount of reverse voltage that a diode can withstand without breakdown. Data sheets often list two values for reverse voltage: average (or normal) and peak (or maximum). From a troubleshooting standpoint, the average reverse-voltage rating is about twice the dc-output voltage for full-wave bridge rectifiers, but must be greater than either the secondary-ouput voltage or the dc-output voltage.

*Forward voltage (or forward-voltage drop)*   This is the amount of drop across the diode in the forward-bias condition (diode conducting). Ideally, the forward-voltage drop should be zero. In troubleshooting a practical circuit, the drop is more like 0.5 to 1.0 V. Also, in a practical circuit, this drop is offset by the voltage buildup across the capacitor.

*Reverse current (or leakage current)*   This is the amount of current flow when the diode is reverse-biased (diode not conducting). Ideally, the reverse current should be zero. In practical troubleshooting, reverse current is less than a few microamperes (at most, a few milliamperes).

*Forward current*   This is the maximum current capacity of the diode in the forward-bias condition. In practical troubleshooting, the forward-current rating of each diode in a full-wave bridge rectifier is typically about twice the dc-output current, and must be greater than the dc-output current.

*Capacitor characteristics*   Capacitors used in full-wave bridge supplies are electrolytics. This is because a high capacitance is required. As shown in Fig. 7-4, the capacitance rating is selected on the basis of allowable ripple and load resistance. For example, assuming a load resistance of 50 Ω and a maximum permissible ripple of 5%, the capacitance values should be 200,000/(50×5), or 800 μF. In practical troubleshooting, never reverse the leads to an electrolytic capacitor! At the very least, this will destroy the capacitor.

## 7.1.5   Voltage-doubling circuits

Figure 7-5 shows the basic voltage-doubling circuit. (Refer to Chapter 9 for additional information on voltage-doubling circuits.) Using such circuits, it is possible to increase an available ac voltage without a transformer. When the ac-line alternation is such that terminal 1 is negative and terminal 2 is positive, as in Fig. 7-5A, current flows from terminal 1 in the direction of the arrows. The current path from terminal 1 includes CR2 and C2 before a return to positive terminal 2. Capacitor C2 is charged during this current flow with a polarity (as shown). During the next alternation, when terminal 1 is positive and terminal 2 is negative, current flows (as shown in Fig. 7-5B). As a result, current flows from terminal 2 toward C1 and charges capacitor C1 to the peak of the ac voltage. Current continues through diode CR1 to positive terminal 1.

From a troubleshooting standpoint, remember that capacitors C1 and C2 are charged to the peak values of the ac-line voltage on alternate half cycles. The dc-output voltage is about double that of the ac voltage. The two capacitors are, in effect, in series (and the polarities add to increase the total voltage).

### 7.1.6   Voltage-tripling circuits

Figure 7-6 shows the basic voltage-tripling circuit. With such circuits, it is possible to triple an available ac voltage without a transformer. When the ac-line alternation is such that terminal 1 is positive and terminal 2 is negative, current flows from terminal 2 through CR1 to terminal 1. Diode CR1 is forward-biased under these conditions, permitting capacitor C1 to be charged to the ac-line voltage. During the next alternation, when terminal 1 is negative and terminal 2 is positive, CR1 is reverse-biased and does not conduct. Instead, CR2 is forward-biased and it permits current flow toward terminal 2 through C1.

When troubleshooting, remember that the charge (already across C1) adds to the ac line voltage and places a charge across C2. This charge is propor-

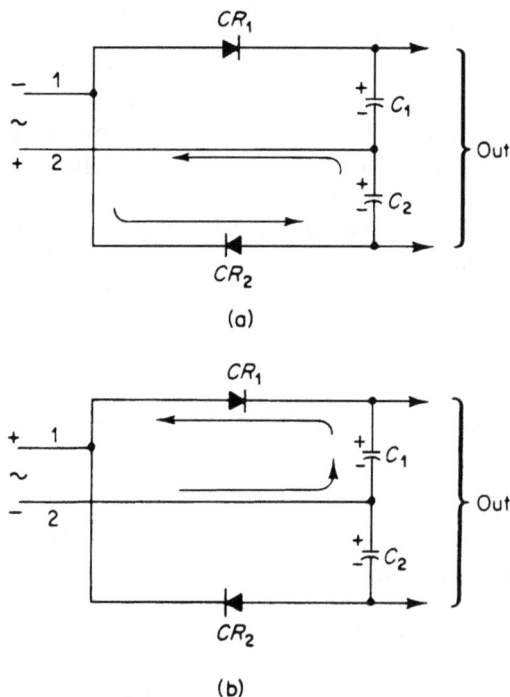

(a)

(b)

Figure 7.5.   Basic voltage-doubling circuit.

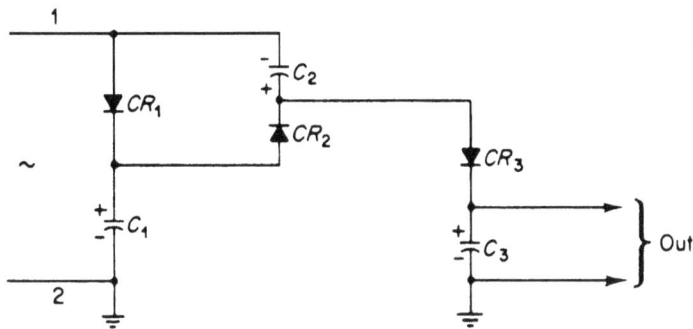

**Figure 7.6.**  Basic voltage-tripling circuit.

tional to the line voltage added to the charge across C1. During the next cycle, when terminal 1 is again positive and terminal 2 is negative, C1 is again charged as before. At the same time, CR3 conducts and charges C3. When CR3 conducts, the charge across C3 is composed of the line voltage plus the existing charge across C2. Because the C2 charge is already double the voltage, this is added to the line voltage, which produces triple the voltage across C3.

### 7.1.7  Basic regulation

The amount of voltage that can be taken from a basic power-supply circuit is determined by the voltage available at the source (line or transformer), less any voltage drop through the rectifiers and filter. This voltage drop is determined by the amount of current flowing through the supply components (rectifiers, etc.). As more current is drawn from the supply, the voltage drop across the components increases and the output voltage is reduced. When less current is drawn from the supply, the output is increased because the components have a minimum voltage drop.

The variation of voltage output with respect to the amount of current drawn from the supply is known as *voltage regulation* or simply *regulation*. Such regulation is expressed as a percentage, with the value being determined as follows:

$$\% \text{ regulation} = \frac{(\text{no-load voltage}) - (\text{full-load voltage})}{\text{full-load voltage}} \times 100$$

This equation applies to voltage regulation and takes into account the proportions of voltage increase and decrease with a change of load on a power supply. The smaller the difference between full-load and no-load voltages, the better the regulation.

Filter capacitors (such as C1 in Fig. 7-4) provide some measure of regulation for the basic supply circuit. If the output voltage drops, the capacitor discharges and serves to keep the voltage constant. The regulation provided by filter capacitors is sufficient for a few applications. However, in most cases, the supply output must be maintained at some crucial voltage or current. This requires some form of linear or switching regulator. (Switching regulators are covered in Chapter 9.)

The most common method for linear voltage regulation involves zener diodes. Zeners maintain the voltage drop across their terminal at a constant value, no matter what current is being drawn (within limits). The basic zener is also used as a voltage reference in other linear-regulator configurations covered in this chapter.

### 7.1.8   Basic zener circuit

Figure 7-7 shows the basic zener circuit. In the simplest form, a zener regulator consists of a series resistance ($R_S$) and a shunt-connected zener (CR1). The value of $R_S$ is set by the load requirements. In troubleshooting any zener circuit, if $R_S$ is too large, the zener is not able to regulate at large values of current. If $R_S$ is too small, the zener dissipation rating (power or wattage rating) might be exceeded at low-load current values.

Crucial zener characteristics are zener (or avalanche) voltage, maximum current, and power capability. The power-dissipation capability of a zener can be considered as the zener voltage multiplied by the maximum current rating. Zeners are rarely rated for greater than a few watts of power dissipation. This rating is a room-temperature (typically 25°C) power dissipation and must be derated for higher temperatures.

In theory, the zener maintains the output at the avalanche voltage (with a tolerance from about 0.1 to 0.9 V), despite any changes in input voltage and load. In practical troubleshooting, if the input falls to less than about 1.4 times the zener voltage, the zener does not go into the avalanche condition and there is no regulation. On the other hand, if the input voltage rises to a point where the maximum current is exceeded, the zener overheats (and possibly is

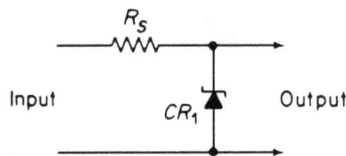

Output voltage = Zener voltage
Zener power dissipation = Zener voltage × Zener current
Minimum input voltage > 1.4 × Zener voltage
Zener voltage ≈ 0.7 × lowest input voltage
Safe Zener power ≈ 3 × load power

$$R_S \text{ (in ohms)} \approx \frac{(\text{max input voltage} - \text{Zener voltage})^2}{\text{safe power dissipation}}$$

Input voltage variation < 30 % of max input voltage
Load current = Zener voltage/load resistance
Load power = Zener voltage × load current
Safe Zener current ≈ 3 × load current

**Figure 7.7.**  Basic zener circuit.

destroyed). Current through the zener is limited by series resistor $R_S$. If the value of $R_S$ is properly calculated during design, the maximum input voltage produces a current that can safely be dissipated by the zener, and the minimum input voltage provides more than enough current to meet the load requirements.

### 7.1.9  Adjustable-shunt regulators

Figure 7-8 shows the schematic and characteristics of an adjustable-shunt regulator (the classic Texas Instruments TI430). Such devices are also known as *programmable zeners* and are similar to three-terminal regulators (Section 7.3) or voltage references. These regulators are a type of zener in IC form. The IC in Fig. 7-8 is a three-terminal programmable shunt (Section 7.2), with a

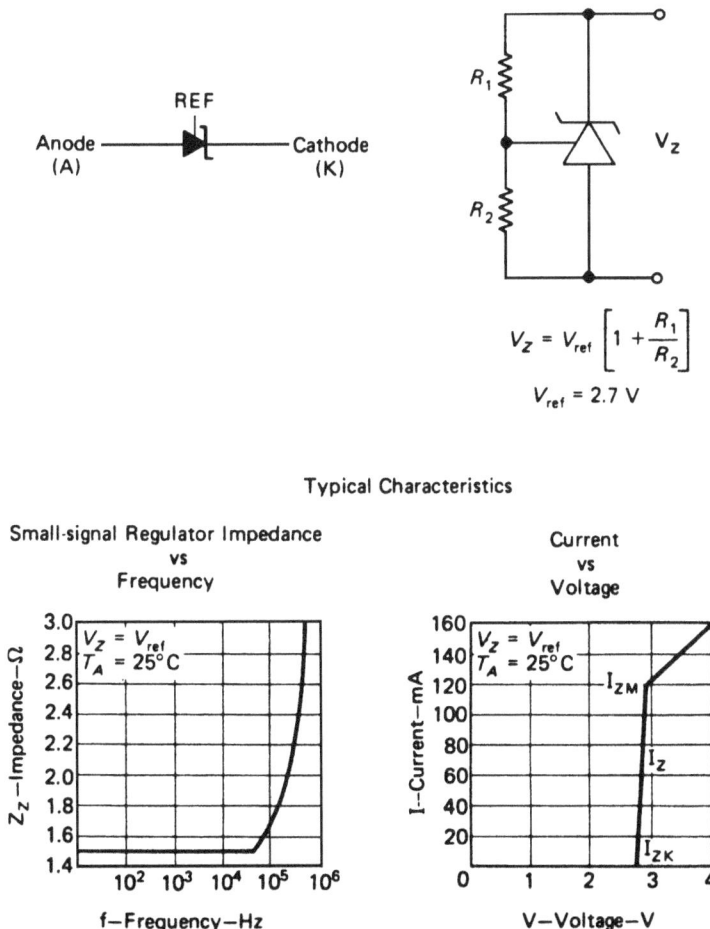

$$V_Z = V_{ref}\left[1 + \frac{R_1}{R_2}\right]$$

$$V_{ref} = 2.7 \text{ V}$$

Typical Characteristics

Figure 7.8.  Adjustable-shunt regulator.

thermal stability (or temperature coefficient, tempco) of 100 100 ppm/°C, and can be used to replace other zeners of similar ratings.

In practical troubleshooting, remember that programmable zeners have a sharp breakdown or avalanche point, as shown by the graphs. This particular device is programmable over a voltage range of about 3 to 30 V and is capable of shunting up to 100 mA. The zener is programmed by an external resistor network connected across the anode and cathode, with the reference terminal connected to the junction of R1/R2. The resistor values are selected so that the voltage developed across R2 is 2.7 V for the desired programmed zener voltage.

The programmable zener is not limited to shunt-regulator functions. Figures 7-9 and 7-10 show the programmable zener in a variety of regulator functions. The figures also show the calculations necessary to find the values of external components in each application.

## 7.2    Discrete Feedback-Regulator Basics

This section describes basic feedback-regulator circuits, including voltage shunt, high-current, constant-current (CC), and voltage series. It also covers overload protection for regulators and feedback regulation. Although the circuits described here are in discrete form, the same basic principles apply to the IC regulators (Section 7.3) and combination discrete/IC regulators described throughout the chapter.

### 7.2.1    Extending zener regulation

The voltage-control ability of a zener can be increased if the zener is used to control the operating point of a transistor or group of transistors. Two basic types of transistor regulators use zeners: shunt (placed across the supply output) and series (in series with the output). Although some zeners can handle heavy currents and dissipate large amounts of power, it is usually more practical to extend the current rating with a power transistor. In either the shunt or series circuit, the power transistor is mounted on a heatsink to dissipate maximum power.

With either shunt or series, the emitter-based junction of the power transistor is held at some fixed value with a low-power zener. The power transistor (typically silicon) emitter remains within 0.5 V of the zener voltage, despite change in input voltage or output (load) current. If a germanium transistor is used (not typical), the emitter remains within about 0.2 V of the zener voltage.

Although popular at one time, the shunt circuit has generally been replaced by the series circuit. The shunt regulator has two disadvantages: more power consumption (which is lost to the load) and a continuous load on the basic power supply. That is, the shunt circuit draws a continuous or quiescent current—even when no current is being delivered to the load. For these reasons, the series linear regulator is concentrated on in this chapter, although some of the basic shunt regulators are described, many of which are still in use.

Series Regulator

Shunt Regulator

$$R_s = \frac{(\text{MAX INPUT } V - 2.7V)^2}{100 \text{ mA}}$$

$$R > \frac{V_{in}}{100 \text{ mA}}$$

$$V_{out} \approx \left(1 + \frac{R_1}{R_2}\right)V_{ref}$$

$$V_{ref} \approx 2.7 \text{ V}$$

$$V_{out} \approx \left(1 + \frac{R_1}{R_2}\right) V_{ref}$$

$$V_{ref} \approx 2.7 \text{ V}$$

Current Limiter

Focus Coil Current Source

$$I_{limit} = \frac{V_{ref}}{R_{CL}}$$

$$V_{ref} \approx 2.7 \text{ V}$$

$$R > \frac{V_{in}}{100 \text{ mA}}$$

$$I_{coil} = \frac{V_{ref}}{R_F}$$

$$V_{ref} \approx 2.7 \text{ V}$$

$$R > \frac{V_{in}}{100 \text{ mA}}$$

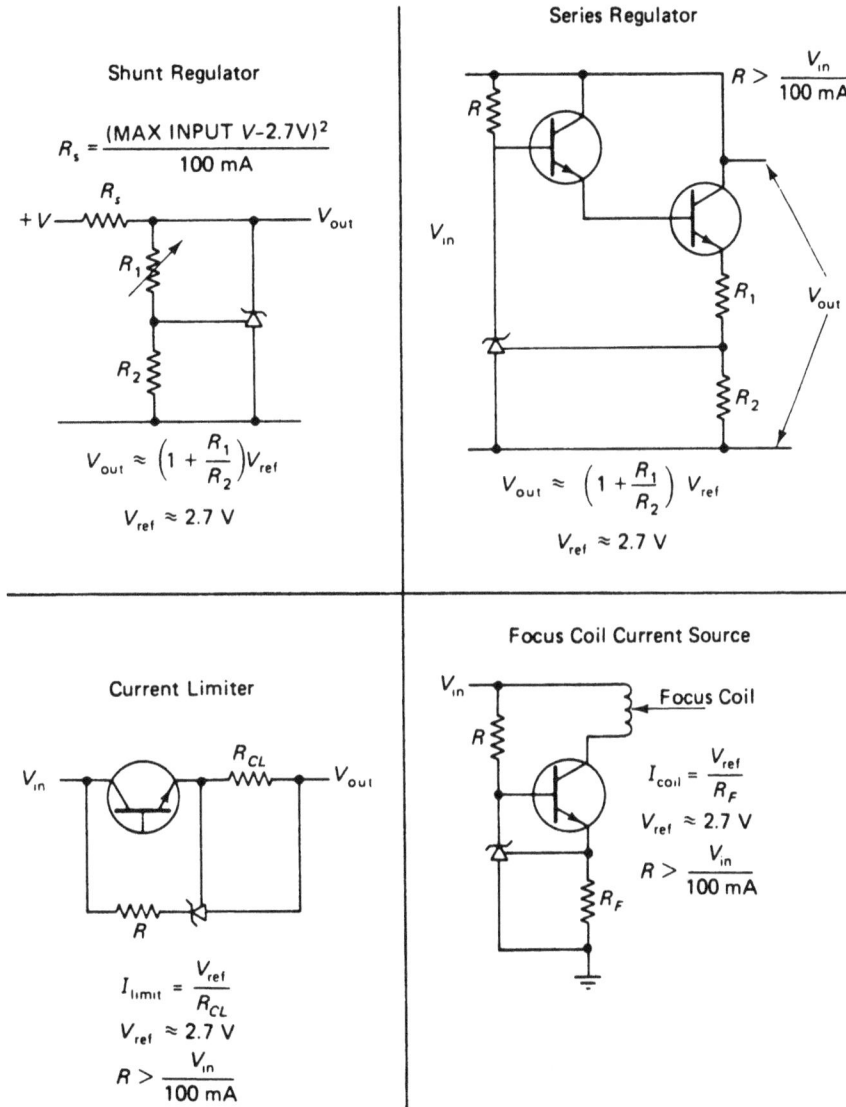

**Figure 7.9.** Applications for programmable zeners.

## 7.2.2  Shunt voltage regulators

Figure 7-11 shows the simplest form of shunt regulator using a transistor. Transistor Q1 appears across the supply output as a variable-bleeder resistor, with current flowing between the emitter and collector. Base current flows through zener CR1. Both of these currents, as well as load current, flow through series resistor R1.

In practical troubleshooting, if the supply load increases, more current is drawn through R1 and the supply output tends to drop. Under these conditions,

$$V_{out} \approx \left(1 + \frac{R_1}{R_2}\right)V_{ref}$$

$$V_{ref} \approx 2.7 \text{ V}$$

| Switch Position | 1 | 2 | 3 | 4 | 5 | 6 |
|---|---|---|---|---|---|---|
| Meter Range | 0-10 | 10-20 | 20-30 | 30-40 | 40-50 | 50-60 |

Fixed Voltage Regulator

$R_1$

$R_2$

Controlling $V_{out}$ of a
Fixed Voltage Regulator

0-10 V
Voltmeter

7.3 K    17.3 K    27.3 K    37.3 K    47.3 K

2.7 K

Voltmeter Scaler

$R_{SC}$

$$I_L = \frac{V_{ref}}{R_S}$$

$$V_{ref} \approx 2.7 \text{ V}$$

$R_S$

LOAD

Current Regulator
With Floating Load

$V_{in}$

$R_S$

$R$

LOAD

$$I_L = \frac{V_{REF}}{R_s}$$

$$V_{ref} \approx 2.7 \text{ V}$$

$$R > \frac{V_{in}}{100 \text{ mA}}$$

Current Regulator
With Grounded Load

Supply Rail Monitors

$V_x$

$R_2$

$R_1$

3.2 K

3.2 K

Led is on for
$V_{mix} < V_x < V_{max}$

$V_x$

$R_2$

$R_1$

3.2 K

3.2 K

$R$

$$R_1 = (V_{min} - 3.2) \text{ k}\Omega \qquad R_2 = (V_{max} - 3.2) \text{ k}\Omega \qquad R = \frac{V_x - (\text{Nominal Led Voltage})}{\text{Nominal Led Current}}$$

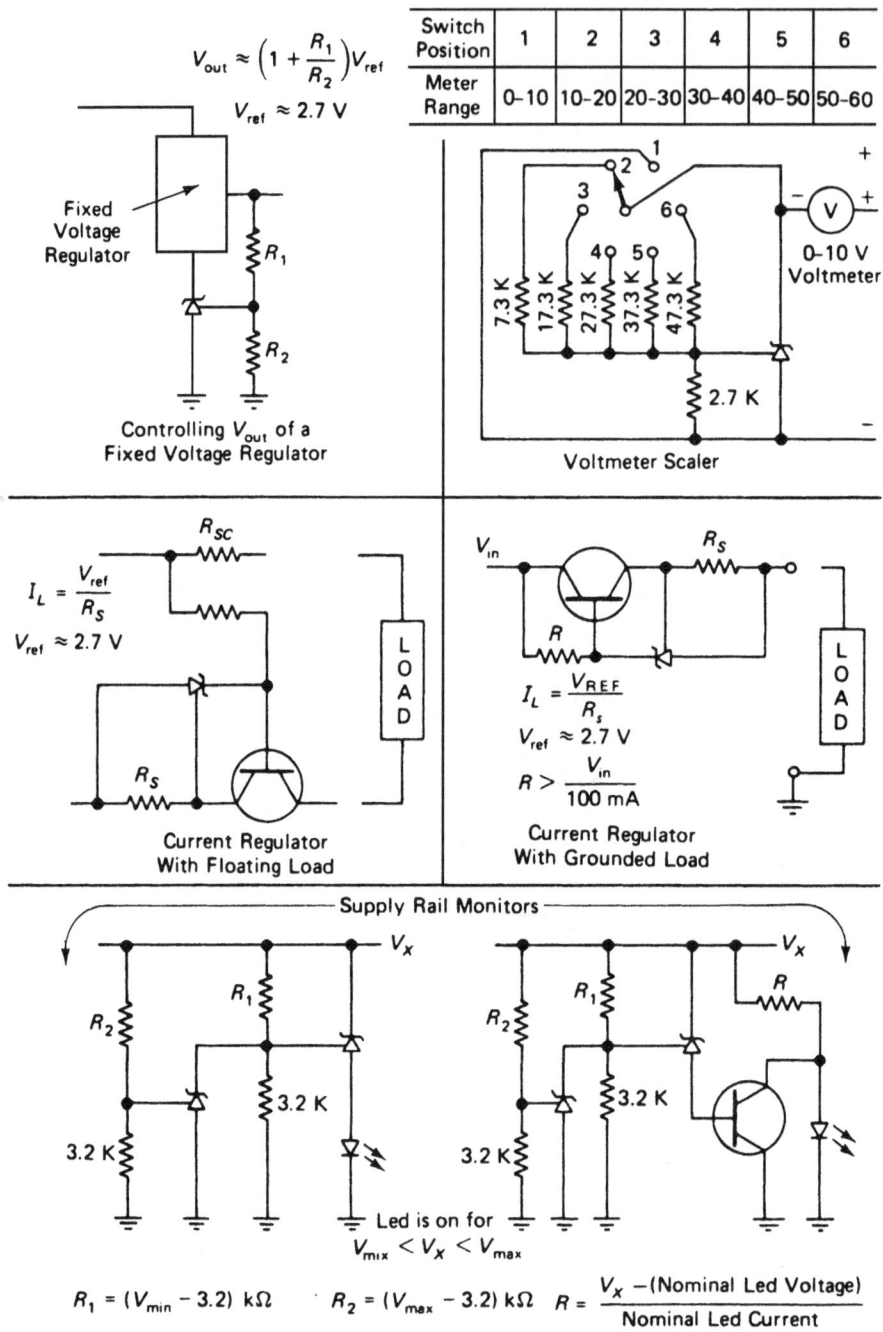

**Figure 7.10.** Additional applications for programmable zeners.

less current is drawn through CR1, the forward bias on Q1 is lowered and less emitter-collector current is drawn from the supply. This causes a lower drop across R1 and tends to increase the supply output voltage, thus offsetting the initial drop in voltage. The opposite occurs if the load decreases.

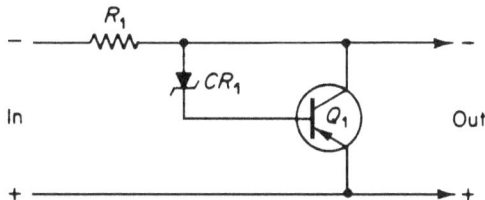

**Figure 7.11.**  Basic shunt regulator.

Figure 7-12 shows the basic *cascade shunt circuit*. Such circuits are used when the supply is subject to large current changes. The cascade circuit increases the effectiveness of the basic shunt circuit. Transistors Q1 and Q2 are placed across the supply output and act as variable resistors. The base current for Q1 flows through CR1, with base current for Q2 flowing through R2. The voltage across R2 is determined by current flowing through R2 and Q1. All of these currents, as well as load current, flow through series resistor R1.

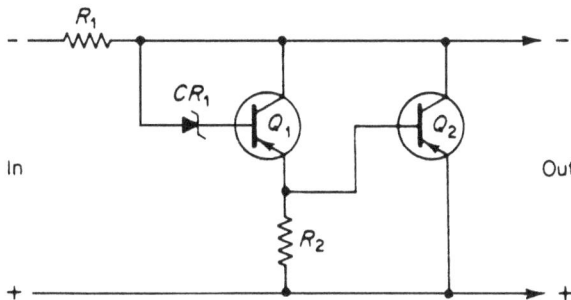

**Figure 7.12.**  Basic cascade shunt regulator.

In practical troubleshooting, if the load decreases, less current is drawn through R1 and the output voltage rises. More current is drawn through CR1 and through the base-emitter of Q1. This raises the forward bias of Q1, causing more emitter-collector current to be drawn from the supply. Also, more current is passed through emitter resistor R2, causing the voltage drop across R2 to increase. When this occurs, the forward bias on Q2 increases and more Q2 emitter-collector current is drawn from the supply. The increased current through Q1 and Q2 causes a larger drop across R1. This decreases the supply output voltage, thus offsetting the initial rise in voltage. The opposite occurs if the load increases.

### 7.2.3  Extending zener regulation

Figure 7-13 shows a shunt-regulated linear supply for outputs larger than the zener voltage. Neglecting the effect of $R_S$ (or assuming that the source voltage is at the junction of $R_S$ and R1), the output voltage is determined by the $R_1/R_2$ ratio: *output voltage* $= (R_1 + R_2)/R_2$. For example, if $R_1$ and $R_2$ are of the same value, the output voltage is twice that of the zener voltage.

In practical troubleshooting, R3 compensates for variations in the supply to the regulator circuit. If the supply appears to be overcompensated, $R_3$ is probably too large. A small $R_3$ resistance produces undercompensation. In an experimental circuit, R3 is often found by trial and error using a variable resistance.

### 7.2.4  Regulating voltages lower than the zener value

Figure 7-14 shows a shunt-regulated supply for outputs less than the zener voltage. Potentiometer R2 acts as a variable voltage divider and sets the regulated output voltage. The forward bias on Q1 is determined by the voltage drop across R1 (or the current through R1). If the supply voltage increases, more current flows through CR1 and R1, causing an increase in forward bias on Q1. The increase in Q1 collector-emitter current passing through $R_S$ causes a greater drop across $R_S$, offsetting the initial rise in supply voltage. The opposite occurs if the supply voltage decreases.

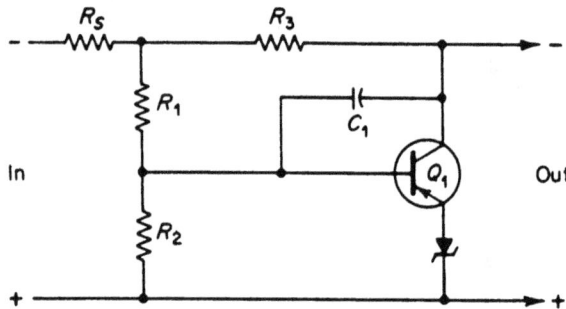

**Figure 7.13.**  Shunt-regulated supply for voltages higher than the zener.

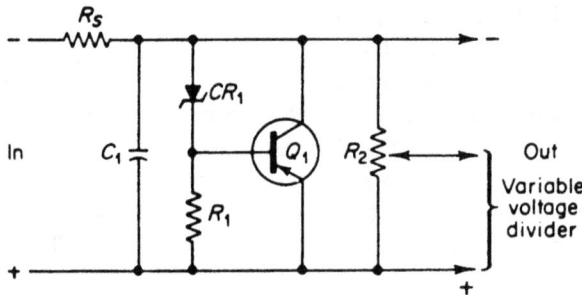

**Figure 7.14.**  Shunt-regulated supply for voltages lower than the zener.

### 7.2.5   Series voltage regulators

Figure 7-15 shows the basic series voltage regulator. Transistor Q1 is in series with the supply output (and possibly a fixed series resistance) and acts as a variable series resistor, with current flowing between the collector and emitter. (Q1 is often called a *series-pass transistor.*) Current also flows through resistor R1 and zener CR1 to establish a voltage at the base of Q1. The base voltage remains fixed in relation to one terminal of the power supply (the negative terminal, in this case), but it varies in relation to the other terminal.

If the supply load increases, more current is drawn through the Q1 emitter-collector resistance and the output voltage drops. Under these conditions, the Q1 emitter voltage drops, but the base remains constant, increasing the forward bias on Q1 (an NPN, in this case). The increase in forward bias on Q1 decreases the Q1 emitter-collector resistance, raising the output voltage to offset the initial drop. The opposite occurs if the load decreases.

The following characteristics should be considered when troubleshooting a practical circuit. First, Q1 will probably require a heatsink if the power dissipated by the transistor exceeds 1 W. The input voltage, current, and power-dissipation characteristics of the series regulator are set by series-pass transistor Q1, not by the zener. For example, maximum input voltage to the circuit is set by the maximum collector voltage of Q1. The maximum current and power ratings of Q1 fix the maximum load-current and power-dissipation capabilities of the circuit.

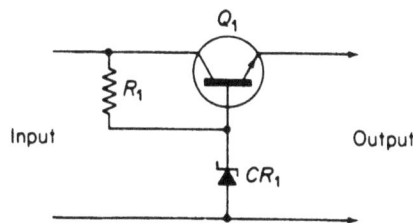

Output voltage :  p-n-p silicon $\approx$ Zener $-0.5$ V
            p-n-p germanium $\approx$ Zener $-0.2$ V
            n-p-n silicon $\approx$ Zener $+0.5$ V
            n-p-n germanium $\approx$ Zener $+0.2$ V

Maximum voltage, power, and current $\approx Q_1$ ratings; see text
Minimum input voltage $\approx 1-2$ V $+$ Zener
Current thru $R_1 \approx 5\%$ of total load current (at minimum input value)
Zener power rating $\approx 0.1 \times$ total power output

$$R_1 (\text{in ohms}) \approx \frac{(\text{minimum input voltage} - \text{Zener voltage})}{(\text{maximum output current} \times 0.5)}$$

$$\text{Maximum current thru } R_1 \approx \frac{(\text{maximum input voltage} - \text{Zener voltage})}{R_1 \text{ resistance}}$$

$R_1$ (in watts) $\approx$ (maximum current thru $R_1$)$^2 \times R_1$ resistance
Zener maximum current rating $\approx$ maximum current thru $R_1$

**Figure 7.15.**   Basic series voltage regulator.

The output voltage is set by the zener and is within about 0.5 V (for silicon transistors) or 0.2 V (for germanium transistors) of the zener voltage. If a PNP is used, the output is 0.2 or 0.5 V below the zener value. With an NPN, the output is 0.2 or 0.5 V above the zener value. The minimum voltage capability of the regulator is also set by the zener voltage. As a troubleshooting guideline, the minimum voltage (at the collector) must be at least 1 V (preferably 2 V) greater than the zener voltage (at the base).

Typically, R1 provides about 5% of the total load current when the input voltage is at minimum. However, the current or power capability of the circuit is set by the zener and series-pass transistor Q1. In practical circuits, it is usually not possible to operate with currents greater than 10 times the maximum rated zener current.

## 7.2.6  High-voltage linear regulation

Figure 7-16 shows a discrete zener/transistor linear-regulator circuit for high voltages. Although solid-state circuits are not usually associated with high voltages, it is possible to use zener/transistor circuits to regulate high-voltage linear supplies. Transistor Q1 is in series with the supply output and acts as a variable series resistor. Current also flows through R1 to establish a voltage at the base of Q1. This voltage is determined by the current flowing through R1 and through the emitter-collector circuit of Q2.

When the forward bias of Q1 is decreased, the Q1 emitter-collector resistance in series with the supply is raised. This causes a higher voltage drop across the emitter-collector resistance and lowers the supply output voltage. The emitter-collector current of Q2 is determined by the forward-bias voltage across R2. This voltage is determined by the emitter-collector current of Q3. The emitter of Q2 is held at a fixed voltage by the action of CR1.

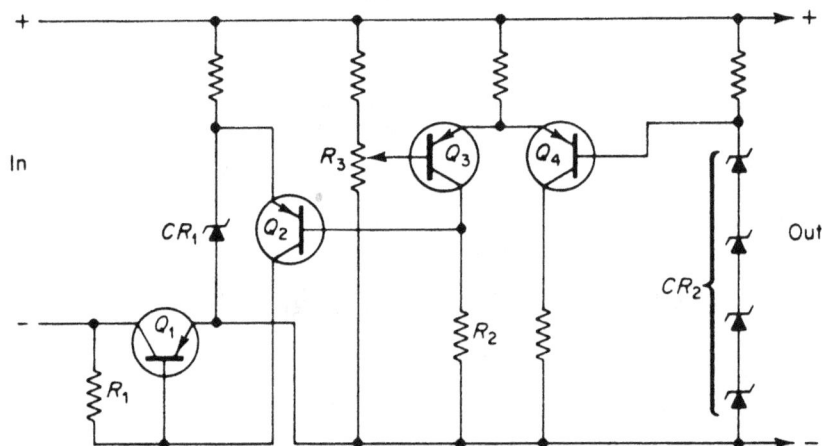

**Figure 7.16.**  Discrete zener/transistor regulator for high voltages.

When troubleshooting this circuit, remember that Q3 and Q4 establish a control voltage for the regulator circuit. The base of Q4 is held at a fixed voltage by CR2. The emitter-collector current of Q4 remains constant, and the emitter of Q4 remains at a fixed voltage. Because the emitter of Q3 is connected directly to the emitter of Q4, the emitter of Q3 also remains at a fixed voltage. However, the voltage at the base of Q3 varies with the supply output voltage.

If the supply output voltage changes, the forward bias on Q3 changes, causing a change in the voltage drop across R2. This changes the forward bias on Q2, causing a change in the emitter-collector current. Because the emitter-collector current of Q2 passes through R1, the voltage drop across R1 changes, making the base of Q1 more or less negative. The change in Q1 base voltage changes the Q1 emitter-collector resistance in series with the supply output. In turn, this causes a higher or lower voltage drop across the emitter-collector resistance and changes the output voltage to offset the initial rise or fall in voltage. The supply output voltage is set by adjustment of R3, which determines the bias voltage on the base of Q3. The supply is adjusted by connecting a voltmeter across the supply output terminals (usually with the load connected) and setting R1 for the desired voltage.

## 7.2.7   High-current linear regulation

Figure 7-17 shows a discrete zener/transistor linear-regulator circuit for high currents. If a very high current must be regulated and all of this current passes through the series-pass transistor, the transistor might not be able to dissipate all of the heat satisfactorily (even though heatsinks are used). This can result in damage to the transistor. To overcome the problem, several transistors can be connected in parallel with each other. Then the string of transistors is connected in series with the supply output (as shown). Transistors Q3 through Q6 are connected in series with the supply output and in parallel with each other, to act as

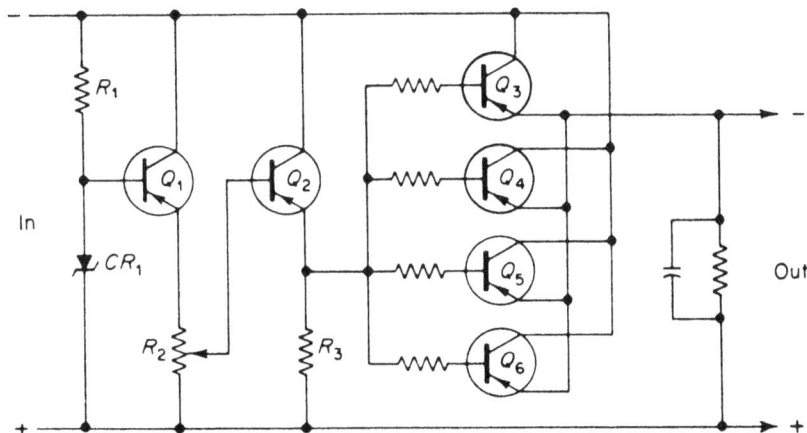

**Figure 7.17.**   Discrete zener/transistor regulator for high currents.

variable series resistors. The current is divided equally among all four so that each transistor dissipates 25% of the total power (and heat).

In troubleshooting this circuit, remember that the base of Q1 is held at a fixed voltage (in relation to the emitter) by zener CR1. However, the voltage across the emitter-collector circuit of Q1 varies with the supply voltage. If the supply voltage changes for any reason, the voltage across R2 varies and changes the forward bias on Q2. In turn, this changes the current through R3. Because the voltage across R3 determines the forward bias on all four transistors (Q3 through Q6), the voltage across R3 determines the amount of resistance offered by Q3 through Q6. A change in the Q3 through Q6 voltage drop changes the supply output voltage to offset the initial change in voltage. The output voltage is set by R2 by connecting a voltmeter across the supply output terminals (usually with the load connected).

### 7.2.8  Constant-current linear regulation

Figure 7-18 shows a discrete zener/transistor regulator circuit for constant currents. In some applications, it is necessary to produce a constant current from a supply (rather than a constant voltage). Notice that a constant-current regulator is sometimes called a *CC regulator* (in contrast to a constant-voltage, CV, regulator). Most of the regulators described in this book are CV.

In troubleshooting this circuit, remember that transistor Q1 still functions as a variable series resistor in the supply output. However, there are two parallel circuit paths: one through zener CR1 in series with bias resistor R1, and the other through R1 and Q1. If there is any variation in current at the supply output, the current through R3 varies and produces a change of forward bias on Q1. In turn, the emitter-collector resistance of Q1 varies to correct the current flow. The net result is that for every change in R3 current, there is an equal and opposite change in Q1 current. Current output is set by R1. Unlike a CV supply, the current remains constant (within limits) despite any load changes, but the CC supply output voltage varies with any load changes.

### 7.2.9  Overload protection

Figure 7-19 shows an overload-protection circuit for a series regulator. (The regulator or portion of the circuit is the same as shown in Fig. 7-15.) A series

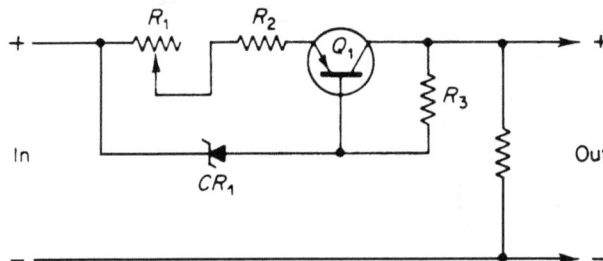

**Figure 7.18.**  Discrete zener/transistor regulator for constant currents.

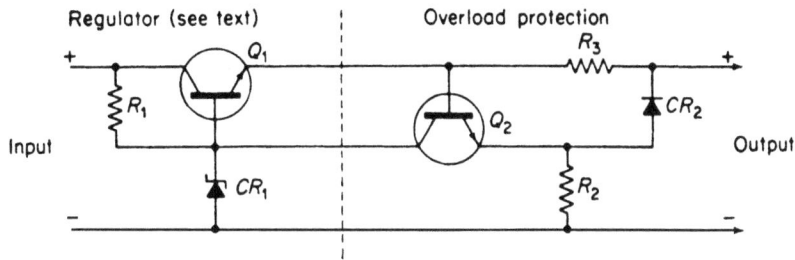

With $Q_2$ and $CR_1$ silicon, drop across $R_3 \approx 1.1$ V, at maximum current output

Output voltage $\approx$ output from regulator $-1$ V

Output current and power $\approx$ same as regulator

$R_2 \approx 10 \times$ load resistance, at maximum current

Load resistance $\approx$ output voltage / load current

$R_3$ (in ohms) $\approx 1.1$ / maximum load current

$R_3$ (in watts) $\approx$ (maximum load current)$^2 \times R_3$ resistance

Figure 7.19. Overload-protection circuit for series regulator.

linear regulator is subject to damage caused by overloads. (This is usually not a problem with shunt regulators.) The overload can be in the form of a high input voltage or an excessive output load. Either way, the series-pass transistor can burn out. Overload caused by high input voltage is not likely, because any prolonged period of high input voltage usually burns out the basic supply components first (transformer, rectifier diodes, etc.). A possible exception is where the series-regulator design is on the borderline of safety. On the other hand, an excessive output load (low load resistance and high input current) is quite common. (A screwdriver shorted across the regulator output can do the job nicely!) For these reasons, any series regulator that might be subject to overloads should be provided with an overload-protection circuit.

In troubleshooting, remember that the circuit is controlled by the voltage across R3. All of the load current passes through R3 and develops a corresponding voltage. When the load current is below a certain value (at a safe level below the maximum rating of the series-pass transistor), the voltage drop across R3 is not sufficient to forward-bias Q2. So, Q2 remains nonconducting as long as the load current is at a safe level.

If diode CR2 is silicon, there is about a 0.5-V drop from the output terminal to the emitter of Q2. Also, if Q2 is silicon, another 0.5-V drop is required before Q2 is turned on. As a result, the drop across R3 must be about 1 V or more (typically 1.1 V) before Q2 turns on. The value of $R_3$ is chosen so that the drop is about 1.1 V when the output load current is at the maximum safe level. When Q2 is turned on, part of the current through R1 is passed through Q2, drawing current away from the base of Q1. Transistor Q1 is cut off (or partially cut off), preventing (or limiting) current flow to the output load. When the load is returned to normal, the drop across R3 is below 1.1 V and transistor Q2 is turned off.

### 7.2.10    Classic feedback linear regulator

Figure 7-20 shows the basic discrete linear regulator with feedback. Such circuits have generally been replaced by IC linear regulators (Section 7.3) or combination IC/op-amp regulators (Section 7.4). For this reason, the circuit is not dwelled upon. However, the following paragraphs describe circuit operation to provide a basis for understanding the feedback principle found in all linear regulators (IC, op amp, combination, etc.).

The output voltage remains within about 0.5 V of the voltage at R3. Any variation in load (which might result in an undesired change in voltage across the load) causes corresponding changes in current through the base-emitter of Q3. Transistors Q1 and Q2 are connected as a Darlington pair or compound. Any variation in output current (caused by load changes) changes the base current of Q2. These changes are amplified through Q1 and Q2. The change in the emitter-collector current of Q1 opposes any change in output current. Q1

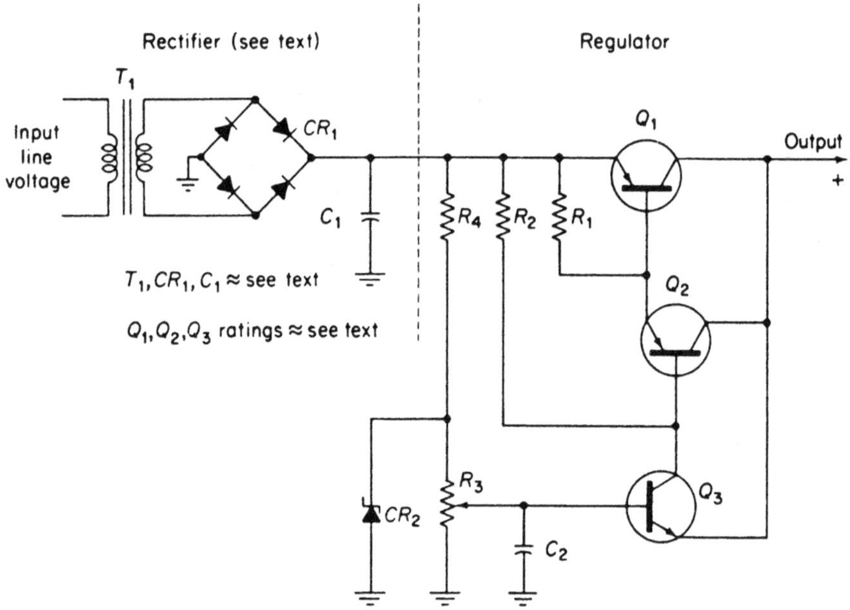

Output voltage ≈ voltage at $R_3$ (± 0.5 V)

Max output voltage ≈ Zener voltage of $CR_2$ ≈ minimum output voltage from rectifier − 2 or 3 V

$R_3$ ≈ max output voltage /(max load current × 0.05)

Current thru $R_3$ < 5% max load current

$R_4$ (in ohms) ≈ $\dfrac{\text{minimum drop across } R_4}{1.2 \times \text{current thru } R_3}$

**Figure 7.20.**    Basic discrete linear regulator with feedback.

is a series-pass transistor. CR2 provides a constant voltage across R2, despite any change in input voltage. As a result, the circuit provides a constant output (in the presence of input-voltage and output-current changes).

## 7.3   IC Linear-Regulator Basics

This section describes basic IC linear-regulator circuits, including voltage regulators, current regulators, short-circuit and overload protectors, current boosting, current feedback, shutdown, and three-terminal regulators.

### 7.3.1   IC voltage regulators

IC voltage regulators provide an output voltage that remains constant (within limits) despite changes in load, input voltage, temperature, etc. In many cases, IC voltage regulators function as the control element in a voltage-regulator circuit (Section 7.4). However, many IC regulators operate entirely without external components. The following is a brief summary of IC voltage-regulator types.

*Fixed-output voltage regulators* maintain the output of a power supply at a fixed voltage and provide positive and/or negative regulation at various current levels. Fixed-output regulators normally require no external add-on components. However, some manufacturers recommend that input/output capacitors be used if the regulator is located some distance from the supply filter.

*Variable-output voltage regulators* can be tailored to regulate at any specific output voltage (with a given range) that is available directly from the IC regulator. Increased output current is available with external current-boosting circuits.

*Current regulators* provide a constant current rather than a constant voltage. In many cases, the basic IC voltage regulator is converted to provide constant-current (CC) operation with a few added external components.

*Floating regulators* for high-voltage operation can deliver hundreds of volts and are limited only by the breakdown voltage of the associated external series-pass transistor.

*Dual-tracking regulators* provide balanced positive and negative output voltages. Generally, the dual-tracking regulators are set to provide a specific output voltage (such as ±15 V), but an external adjustment can change one or both outputs simultaneously (for example, from 8 to 20 V).

Finally, *low-temperature-drift regulators* are usually used for low-voltage operation, where precision regulation is required.

### 7.3.2   Basic IC voltage regulator operation

Figure 7-21 shows a combined block diagram and simplified schematic of a basic IC voltage regulator, which has been divided into four parts or sections: control, bias, dc-level shift, and output.

*The control section* provides for starting and stopping the regulator action. Not all IC regulators have a control section.

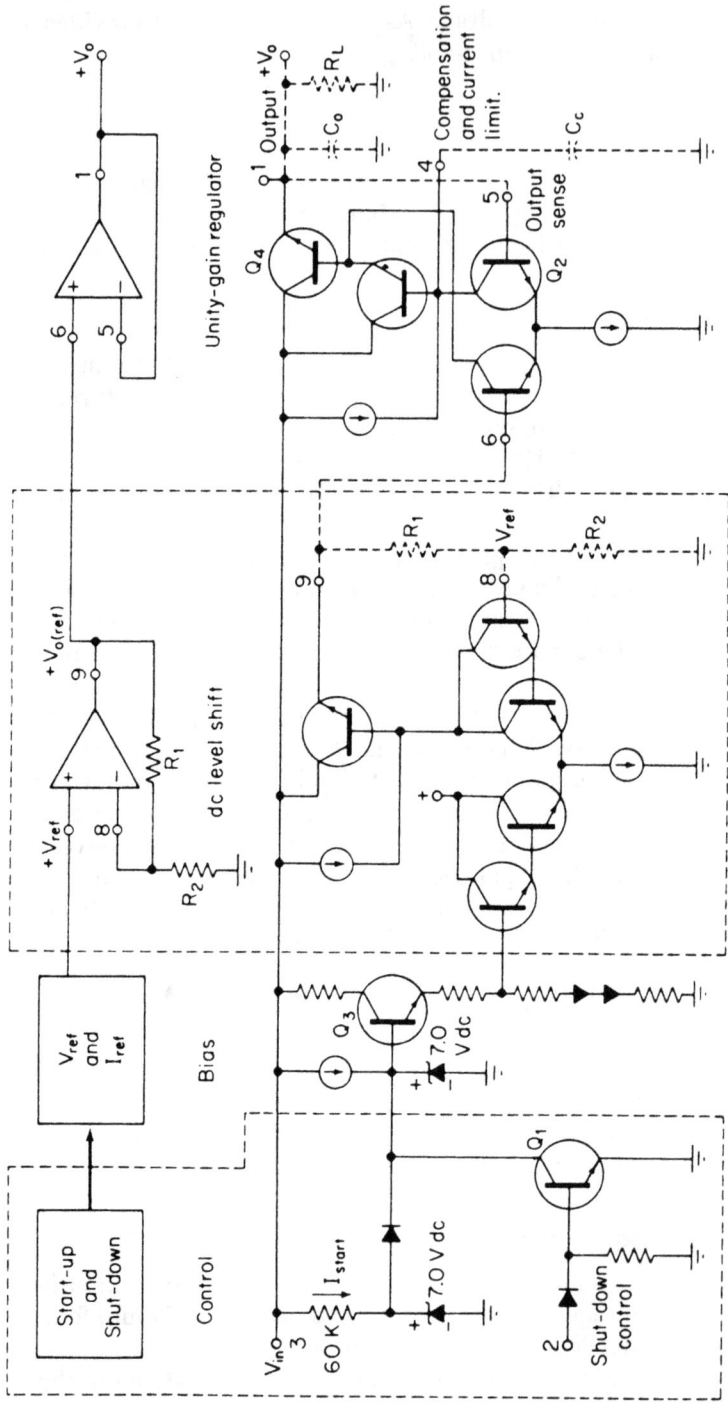

**Figure 7.21.** Basic IC voltage regulator.

*The bias section* provides a fixed, temperature-compensated reference voltage, $V_{REF}$, typically about 3 or 4 V. Because the desired output voltage is rarely an exact value between 3 and 4 V, it is necessary to "multiply" the reference voltage to some higher exact value.

*The dc-level shift section* performs the necessary "multiplication" of the reference voltage. The available output-voltage reference $V_{O(REF)}$ at pins 6 and 9 is determined by the ratio of resistors R1 and R2, which set the feedback (and thus the gain) of the reference amplifier. The $V_{O(REF)}$ can be higher or lower than the fixed $V_{REF}$ supplied by the bias section.

*The output section* is essentially a unity-gain, differential amplifier, followed by a pair of NPN transistors. The differential-amplifier inputs consist of $V_{O(REF)}$ at pins 6 and 9 and the output voltage $(V_O)$ at pins 1 and 5. $V_{O(REF)}$ and $V_O$ are the same when the regulator is functioning normally (at the selected voltage).

Any change in output voltage unbalances the differential amplifier and changes the bias of series-pass transistor Q4. The bias change increases or decreases the current through Q4, as necessary to offset the initial change in output voltage. The maximum output voltage is determined by the minimum input-output differential (the smallest amount of voltage across Q4 that allows normal operation).

### 7.3.3   Short-circuit and overload protection

Figure 7-22 shows the basic IC regulator (Fig. 7-21) connected for short-circuit and overload protection. These circuits provide for current limiting and are not to be confused with thermal-overload circuits found on some IC regulators. The current-limiting circuits shut down the IC when the output current exceeds a desired value. (Thermal-overload circuits provide shutdown if the IC junction temperature exceeds a safe level.)

When troubleshooting either configuration in Fig. 7-22, the transistor (or diode string) is forward-biased when the load current creates a drop across the

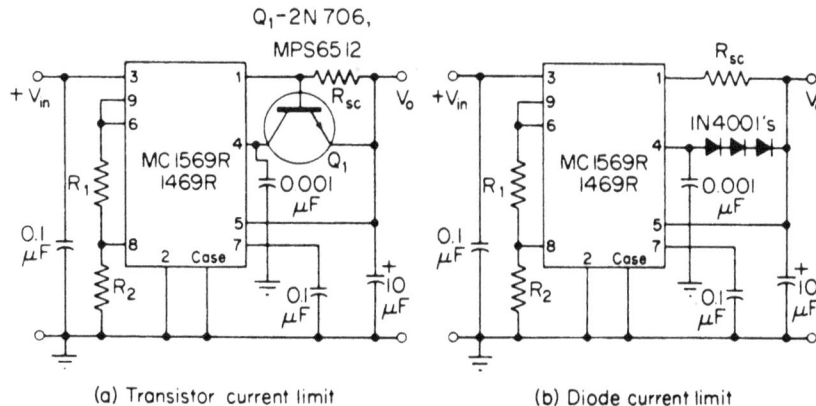

**Figure 7.22.**  Short-circuit and overload-protection circuits.

series resistor $R_{SC}$ equal to the base-emitter drop of Q1 (or equal to the total diode voltage drops). When $R_{SC}$ is of the proper value, the drop across $R_{SC}$ equals the base-emitter drop (or total diode voltage drops) only if the IC regulator maximum-current limit is reached or approached. When the threshold point is reached, the transistor or diodes draw base current away from the NPN transistor pair (at pin 4 of the IC, as shown in Fig. 7-21) and thus limit output current to a desired maximum.

### 7.3.4  Current boosting

Figure 7-23 shows a typical boosted-current circuit (with current limit) using NPN transistors. Current boosting is often used for higher load current or better regulation. Even though the IC regulator (Fig. 7-21) might be capable of delivering the desired current, boosting reduces junction-temperature rise because the series-pass transistor (which dissipates most of the heat) is separate from the IC. So, any temperature-related problems (such as reference-voltage drift caused by temperature increases) are minimized.

In troubleshooting the circuit of Fig. 7-23, the load current is passed through the series-pass power transistor. In turn, the series-pass transistor is controlled by the IC regulator output (pin 1). Similar circuits are often used to increase efficiency at low output voltage because the external series-pass transistor requires a volt or so to remain active.

### 7.3.5  Current regulators

Figure 7-24 shows the basic IC regulator (Fig. 7-21) used as a current regulator, both with and without current boosting. In troubleshooting both configurations, remember that the output current passes through the load ($R_L$), and through resistor R1. The voltage drop across R1 is returned to the IC regulator at pin 5. This voltage is compared with the reference voltage (at pin 6) set

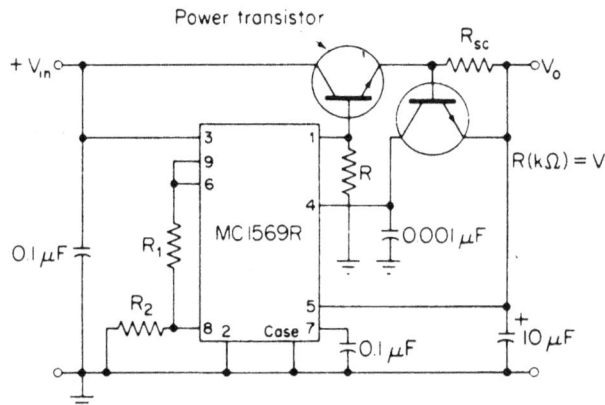

**Figure 7.23.**  Boosted-current circuit with current limit.

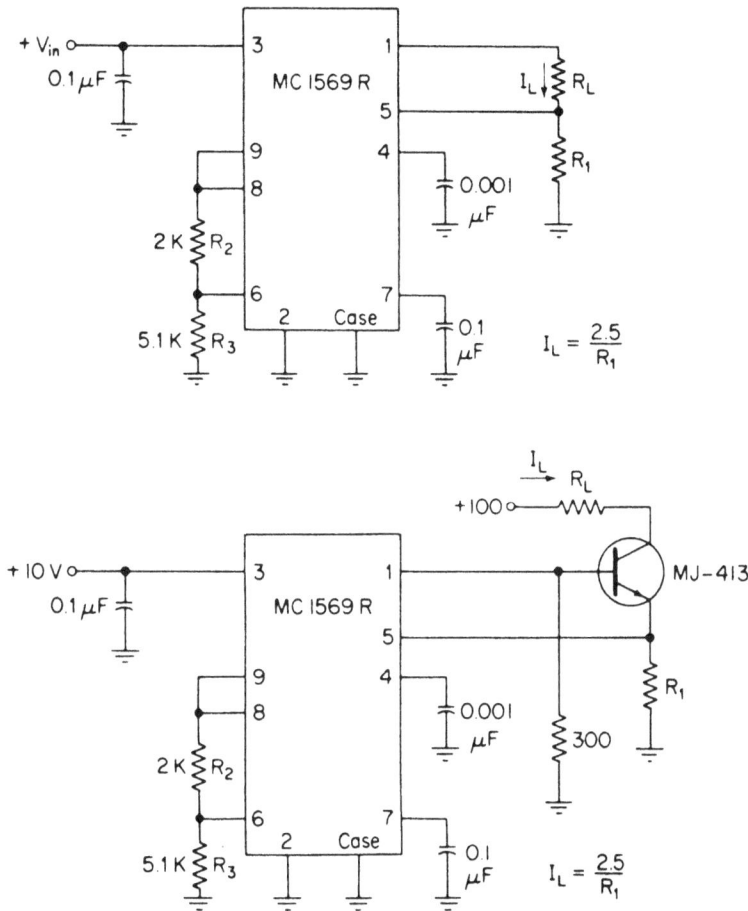

**Figure 7.24.**  Basic IC voltage regulators used as current regulators.

by the $R_2/R_3$ ratio. Any deviations in current are amplified and corrected to maintain load current constant to within less than 1% (typically 0.05%).

## 7.3.6  Shutdown circuits

Many linear IC regulators have some means for shutdown and startup. In the circuit of Fig. 7-21, the IC is controlled by transistor Q1 and the associated circuit. In normal operation, when the IC regulator is functioning, the shutdown-control input at pin 2 is open or connected to a voltage near ground. In this condition, Q1 is nonconducting and has no effect on normal operation of the IC.

Transistor Q1 shunts the internal 7-V reference zener. If Q1 is turned on and saturated, the zener is shorted to ground. Because all internal current sources are biased to the internal reference, the sources cannot function and the IC regulator stops operation (series-pass transistor Q4 is shut off). The IC is shutdown by applying a voltage at pin 2. The voltage must be large enough to drive Q1 into saturation. The IC returns to normal operation when the voltage is removed. The following paragraphs describe the external circuits required to shut down the IC in a variety of operating configurations.

### 7.3.7  Logic-signal shutdown

Figure 7-25 shows how shutdown of the IC regulator can be controlled by a logic gate. It is assumed that the IC regulator is operating in the normal mode (as a positive regulator, referenced to ground) and that the logic gate is of the saturating type, operating from a positive supply. The gate can be TTL (or even the ancient DTL/RTL) as long as the gate output uses an active pull-up resistor.

A typical application for the circuit of Fig. 7-25 is for remote systems where power is conserved until a particular subsystem is needed to function. At such time, the IC regulator is turned on via a control signal (the voltage at pin 2 is removed or goes low), the regulator performs the required task and the subsystem is then shut down (the voltage at pin 2 goes high).

### 7.3.8  Temperature control or shutdown

Figure 7-26 shows how the shutdown of the IC regulator can be controlled (externally) by temperature. This is not to be confused with the internal thermal-overload circuits that are used in many IC regulators.

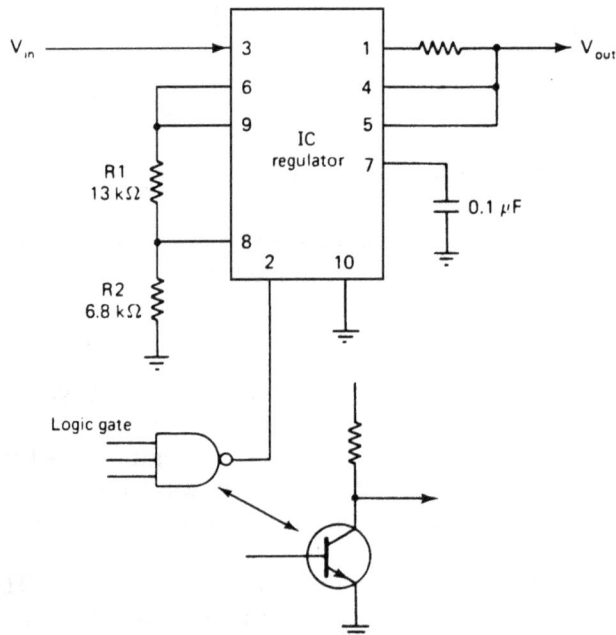

**Figure 7.25.**  Shutdown of IC regulator with logic gate.

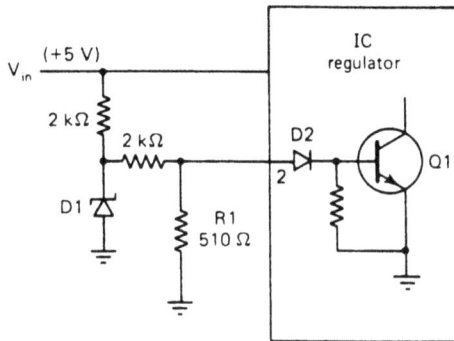

(a) Junction temperature T$_J$

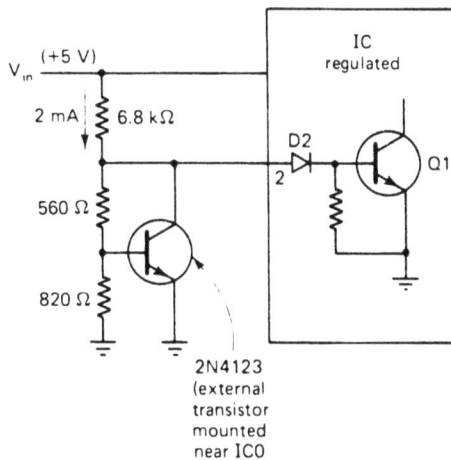

(b) Ambient temperature T$_A$

**Figure 7.26.** Shutdown of IC regulator with temperature control.

In troubleshooting the Fig. 7-26 circuit, remember that most of the IC regulator chip is occupied by the series-pass transistor (because this transistor handles most of the current). This is not true of the discrete-component linear regulators covered in Section 7.2. In an IC regulator, the temperature of the entire IC is set by the series-pass transistor. Variations in load cause the transistor temperature to change, thus changing the temperature of the remaining components on the IC. This "thermal feedback" or "temperature feedback" can cause thermal runaway. An increase in temperature causes an increase in current, which, in turn, causes a further increase in temperature, and so on, until the IC burns out.

It is possible to use thermal feedback to control the IC (externally) and to initiate shutdown if an unsafe temperature is reached. Two methods are shown in Fig. 7-26. One method uses junction temperature ($T_J$). The other method uses ambient temperature ($T_A$).

In troubleshooting a circuit using $T_J$, the temperature of the D2 and Q1 junctions can be considered the same as that of the series-pass transistor. As temperature increases, the voltage required to turn on D2 and Q1 (and thus turn off the regulator) drops. Thus, if a fixed voltage is applied, D2 and Q1 turn on (the regulator turns off) when a predetermined temperature is reached. External zener D1 provides the fixed voltage (which can be set by varying R1). Zener D1 is usually mounted away from the IC so that heat from the IC does not affect D1.

In troubleshooting a circuit using $T_A$, the fixed reference for control is set by the drop across the junction of an external transistor (a 2N4123, in this case). Because the voltage drop is controlled by ambient temperature, the control voltage (and turn-on/turn-off point) is set by the ambient temperature. The external transistor is usually mounted near the IC so that a large increase in IC heat changes the ambient temperature around the transistor.

### 7.3.9   Combined temperature and short-circuit shutdown

The temperature-control or shutdown systems described here are often used in conjunction with the short-circuit and overload circuits described in Section 7.3.3. When a short circuit or overload occurs, increased current is drawn through the ICs and the IC can heat up, possibly to a dangerous level. Using both overload and temperature, the IC turns off when the temperature reaches a predetermined point during overload. When the IC is cool enough, the regulator returns to the ON condition. If the short or overload still exists, the heat process begins again and the IC continues turning on and off as long as the output is shorted.

The technique of combining both temperature and short-circuit protection provides control of both the maximum short-circuit current and the maximum junction temperature. Such complete protection is often considered superior to the current-foldback system described in Section 7.3.12.

### 7.3.10   Current-boost shutdown

Figure 7-27 shows a method of temperature-controlled shutdown used when the IC regulator is operating in a current-boost circuit (Section 7.3.4). The external series-pass transistor (being controlled by the IC regulator) is mounted on a common heatsink with a monitor transistor.

In troubleshooting this circuit, remember that the external series-pass transistor (Q2) heats when the output is shorted or overloaded, but it is the monitor transistor (Q1) that senses the heatsink temperature. Q1 conducts when the heatsink exceeds the normal ambient temperature by a predetermined amount. This applies a signal to pin 2 and shuts off the regulator. The point at which the monitor transistor conducts is set by the values of $R_2/R_3$. Resistor R4 limits the drive into pin 2.

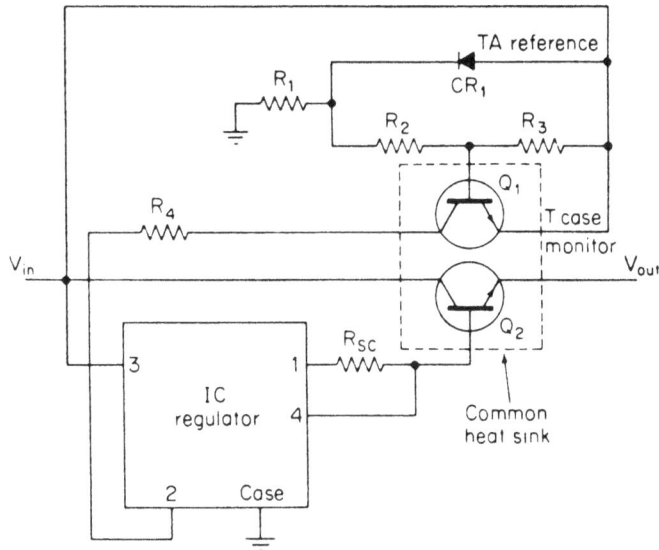

**Figure 7.27.**  Temperature-controlled shutdown with IC regulators. Used in current-boost circuit.

### 7.3.11  Output short-circuit shutdown

Figure 7-28 shows an IC regulator shutdown circuit using an output short circuit for control. External transistor Q1 is normally saturated because of the base drive supplied through resistor-divider R1 and R2. With Q1 in saturation, the voltage on pin 2 of the IC is below the threshold of the IC control circuit.

In troubleshooting this circuit, if the IC output is short-circuited, Q1 turns off, the voltage at pin 2 rises, and the regulator shuts down. The values of $R_1$ and $R_2$ are selected to saturate Q1, as well as to sink minimum regulator load current (typically about 1 mA). Capacitor C1 provides an RC time constant that prevents shutdown when $V_{IN}$ is initially applied (before Q1 can saturate). When the output short circuit is removed, the regulator must be manually reset before regulation is restored.

### 7.3.12  Current foldback

Figure 7-29 shows a short-circuit shutdown circuit for IC regulators that do not have a control section. This circuit reduces the regulator output voltage and current to zero (or near zero) when the load current reaches a certain limit. The technique is called *current foldback* because both the voltage and current drop back (or fold back) to some low (safe) level when the output current reaches a certain limit (as shown by the graph in Fig. 7-29).

In troubleshooting this circuit, the voltage drop across R2 is set to about twice the base-emitter drop of Q1 (somewhere between 1 and 1.4 V) when the output voltage is normal. For Q1 to conduct, the voltage across R3 (that senses

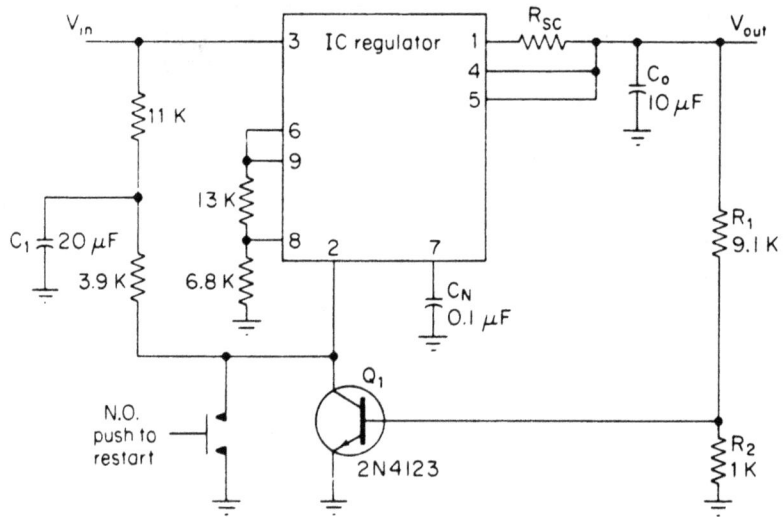

**Figure 7.28.** IC regulator shutdown using an output short circuit.

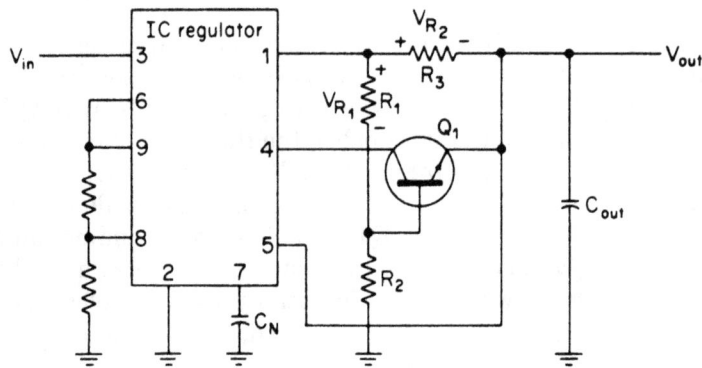

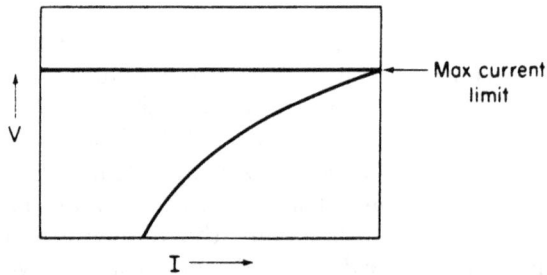

**Figure 7.29.** Short-circuit shutdown for IC regulators without a control section (current foldback).

the output load) must be about equal to the drop across R1, plus the base-emitter voltage. By properly selecting the resistance values, this occurs only if the current limit is reached. When Q1 conducts, the collector connected to pin 4 diverts current drive from the series-pass transistor (Q4 in Fig. 7-21), producing current foldback.

### 7.3.13 Three-terminal regulators

Figures 7-30 and 7-31 show two classic off-line linear power supplies using three-terminal regulators. The three-terminal regulator is the simplest form of linear IC voltage regulator. As the name implies, the IC has only three active terminals: input, output, and common (ground). The input terminal is connected to an unregulated voltage source (typically at the output terminal of a rectifier package). The voltage at the output terminal (connected to the load) remains constant despite variations in load current (within limits). The data sheets for three-terminal regulators specify input/output voltage and current limits and show what external components (if any) are required. The component types and values shown in Figs. 7-30 and 7-31 can be used as first trial values for these classic linear supplies (which provide a variety of voltages for troubleshooting). Section 7.6 describes troubleshooting for power supplies using three-terminal regulators.

## 7.4   Op-Amp Linear-Regulator Basics

This section is devoted to IC operational amplifiers (op amps) used as linear voltage regulators, which are not to be confused with IC packages that function as complete regulator circuits (Section 7.3). Here, the coverage is limited to the use of an IC op amp as the active or gain element in a linear voltage-regulator circuit. The present trend is to combine the gain of an op amp with the stability of an IC regulator to form a precision linear regulator. Such circuits usually show much better regulation than the packaged IC regulators. This is because of the higher available feedback-loop gain in the op amp.

### 7.4.1   Typical IC op-amp voltage regulator

Figure 7-32 shows a typical op-amp regulator circuit that incorporates both the IC regulator (Section 7.3) and an IC op amp. Figure 7-33 shows a typical regulation curve (the variation in output voltage at various load currents). The output voltage drops less than 0.25 V with a load variation from 0 to 300 mA.

The regulator operates from a +20-V source and is adjusted for a +15-V output by R1. The drop across Q1 is 5 V with a 300-mA load. Q1 must be capable of dissipating at least 1.5 W (preferably more). The beta of Q1 must be at least 20. The reference voltage supplied by the IC regulator to the op amp is equal to the ratio of $R_1/R_2$ times a constant of 3.5. For best results, R1, R2, and Q1 should be mounted close together so that the components are all at the same temperature. This minimizes the change in percentage of regulation with

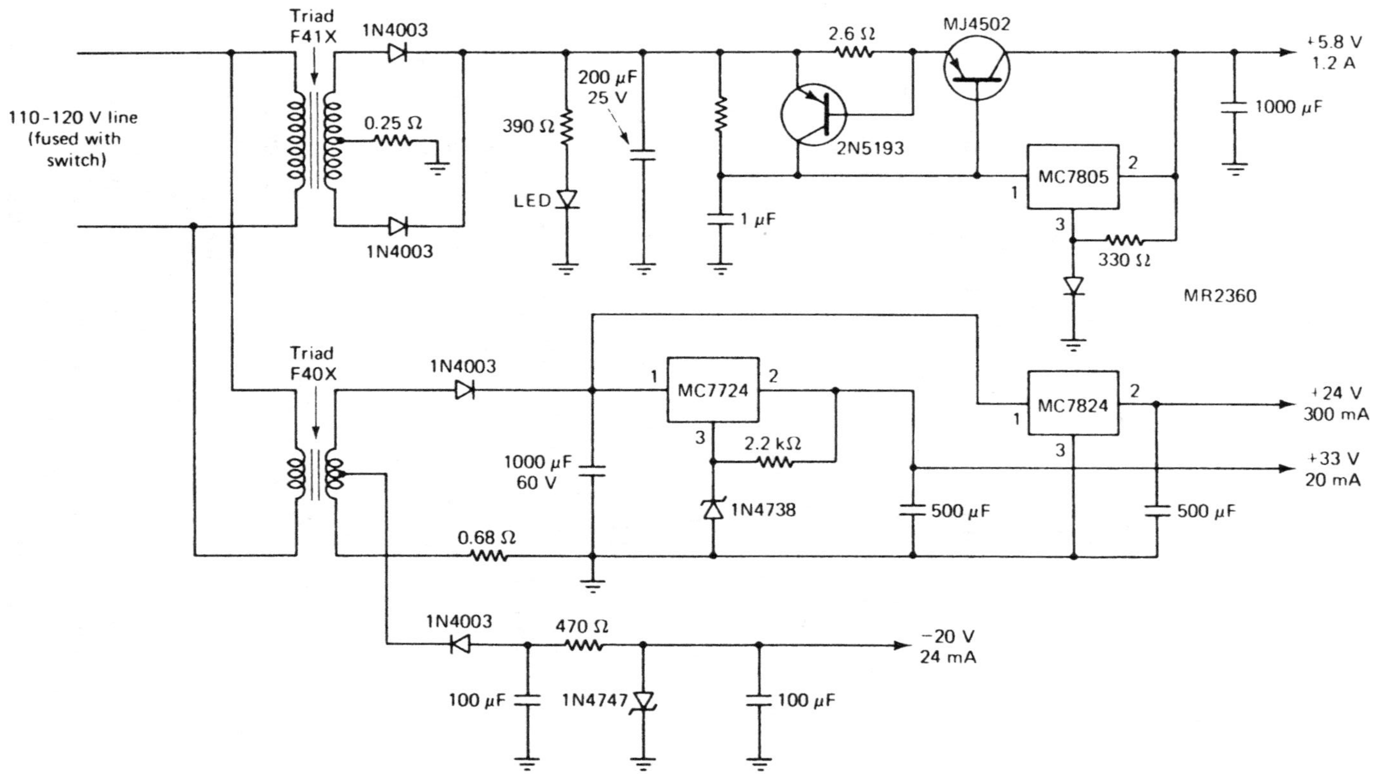

**Figure 7.30.** Typical linear supply with three-terminal regulators.

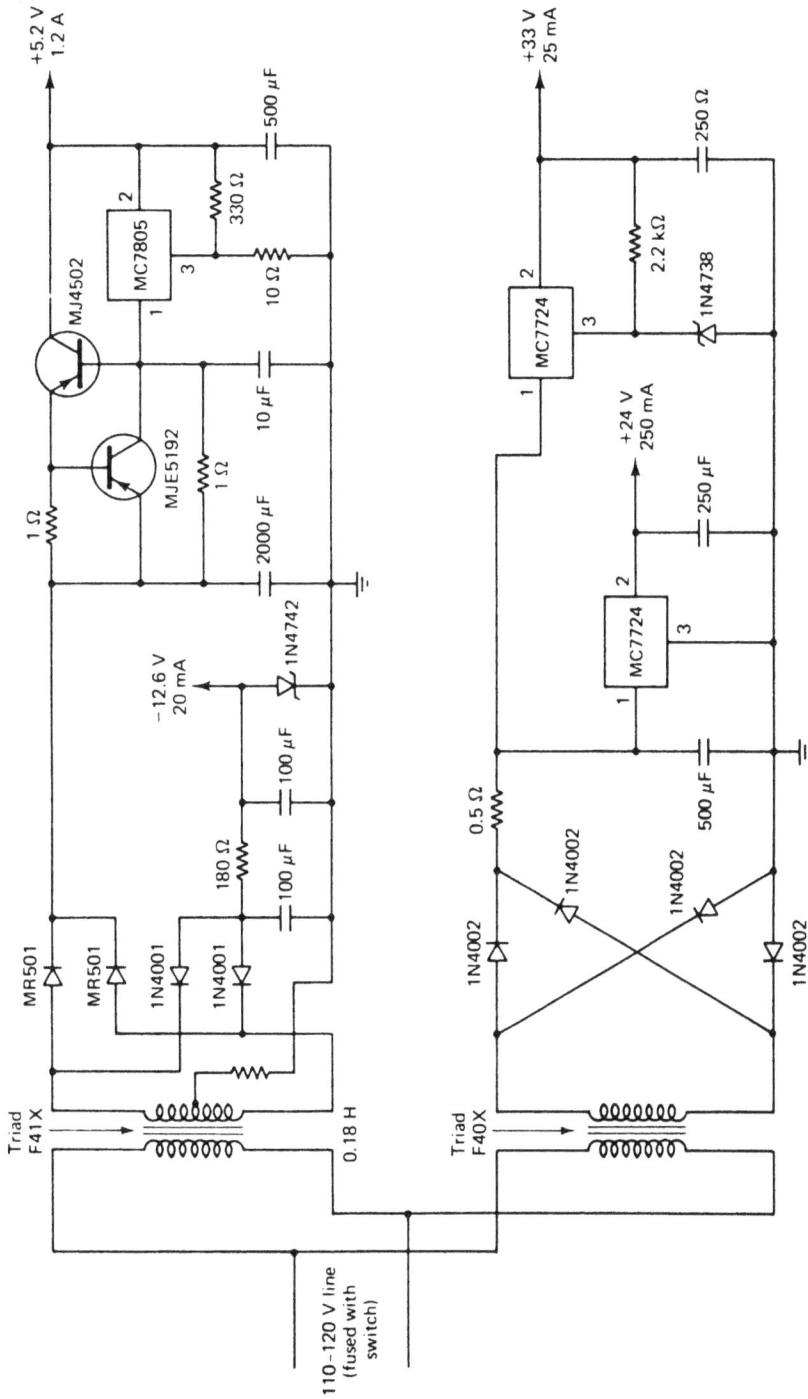

**Figure 7.31.** Alternate linear supply with three-terminal regulators.

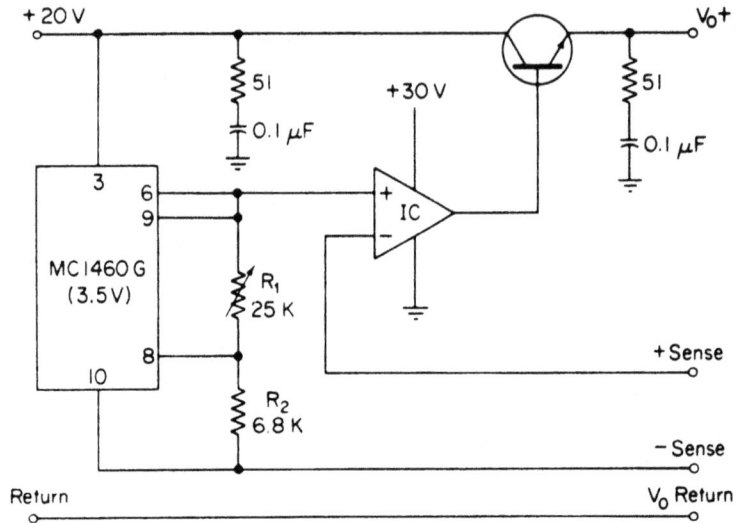

IC op-amp = MC 1539 G, or following characteristics
$A_{Vol}$ = 50,000 (min)
$Z_O$ = 4 K
CMRR = 100 dB
Offset voltage = −4 mV max
$TC_{Vio}$ = 5 $\mu$V/°C
Power supply sensitivity = 150 $\mu$V/V (max)
$Q_1$ = 2N4921 with beta of 20 (min)

Figure 7.32.  Typical op-amp and IC regulator.

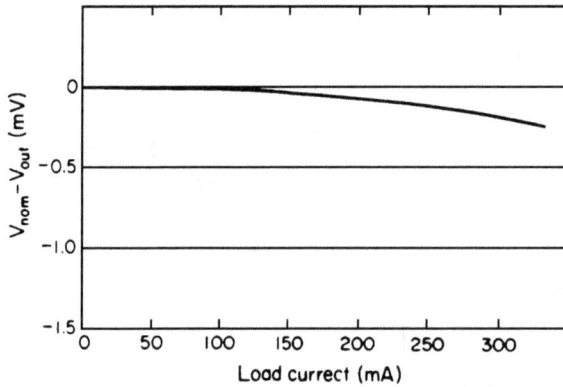

Figure 7.33.  Typical load-regulation curve for op-amp
regulator.

changes in temperature. The temperature characteristics of R1 and R2 should also be matched to the IC regulator and to each other.

### 7.4.2  Remote sensing and ground loops

Notice that the circuit of Fig. 7-32 provides for remote sensing. One of the purposes of remote sensing is to minimize the effects of ground loops. In practical circuits, a well-designed regulator can perform very poorly if layout and current paths have not been considered carefully (particularly ground-current paths or ground loops).

In a typical troubleshooting situation, the regulator circuit elements are mounted on the same board or card as the power supply. However, the regulator output can just as typically be delivered to a load at a remote location. (In this sense, the term *remote* can mean another board or card only a few inches away from the regulator.) In any event, the load current must pass through wires that produce some voltage drop.

Figure 7-34 shows the basic ground-loop model of an op-amp regulator. The crucial load-current resistance paths are shown as RW1 and RW2 (remote wires 1 and 2). These resistances are the most important simply because the paths carry the most current and thus drop the most voltage. Even if the output voltage ($V_O$) equals $V_{REF}$ and the op-amp gain is infinite (although neither condition is practical), the actual voltage across the load ($V_L$) is still a fraction of $V_O$.

This might seem insignificant, but when you consider that #20 wire shows about 10 mΩ of resistance per foot, there is also about 1 mV per 100 mA of load current per foot. Assuming a 100-mA load (which is quite small for a typical power transistor used as a series-pass transistor, Q1), a $V_{REF}$ of 10 V and a distance of 6 inches from the regulator card to the load card (12 inches total), the actual $V_O$ is offset by 1 mV from $V_{REF}$.

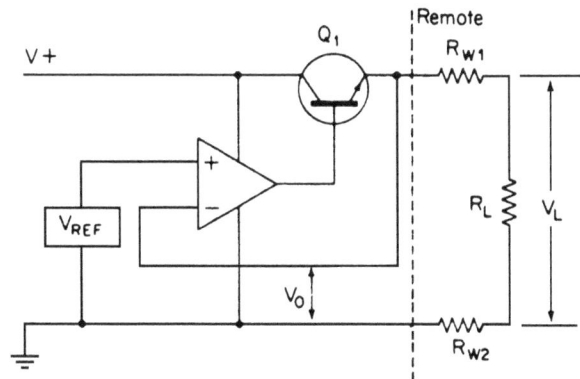

$$V_L \approx V_O \times \frac{R_L}{R_{W1} + R_{W2} + R_L}$$

**Figure 7.34.**   Ground-loop model of an op-amp regulator.

The problem is compounded further if the regulator output is connected to the load by binding posts or plug-in terminals, rather than being soldered. Even solder joints can result in a loss of a few millivolts, if not properly made.

Figure 7-35 shows how the circuit of Fig. 7-34 can be converted to minimize (but not totally eliminate) ground-loop problems with remote sensing. In Fig. 7-35, it is assumed that the sense lines are the same length as the load lines and that the voltages across the inverting and noninverting inputs to the op amp are as shown by the equations. In effect, both inputs are offset by the same amount.

In the circuit of Fig. 7-35, the common-mode voltage is increased by the drop across $R_{W2}$. If the common-mode rejection of the op amp is good (at least 90 dB), the added common-mode voltage does not significantly affect performance of the regulator. The sense lines should be #20 wire or smaller. (The lines need not be larger because they carry very little current.) Resistance $R_{W1}$ increases the open-loop output impedance of the regulator. However, the additional 10 to 20 mΩ (for a load-wire length of 1 foot) does not significantly affect performance.

## 7.5    Linear-supply tests

The basic function of most linear supplies is to convert alternating current into regulated direct current. Ac is first converted to dc by rectifiers (the basic power-

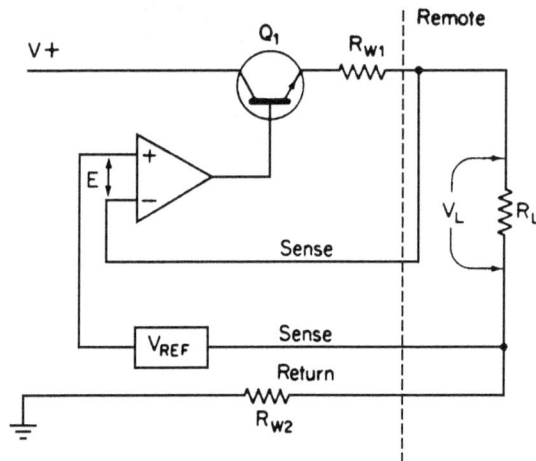

$$V_E^- = V_L + I_L R_{W2}$$

$$V_E^+ = V_{REF} + I_L R_{W2}$$

$$V_E^+ - V_E^- = E = V_{REF} - V_L, \text{ and}$$

$$\frac{V_E^+ + V_E^-}{2} = \frac{V_L + V_{REF}}{2} + I_L R_{W2}$$

**Figure 7.35.** Op-amp regulator with remote sensing.

supply function), then the dc is maintained at a desired voltage (or current) output (the basic regulator function). These functions can be checked by measuring the output voltage or current without a load, with a load, and possibly with a partial load. If the supply delivers the full-rated output voltage (or current) into a full-rated load, the basic function is met. In practical troubleshooting, it is also helpful to measure the regulating effect of a supply and the amplitude of any ripple at the output, as a minimum. Additional, more sophisticated tests are described in this section.

## 7.5.1 Basic output tests

Figure 7-36 shows the basic linear-supply test circuit. This simple circuit permits the supply to be tested at no load, half load, and full load, depending on the position of switch S1. With S1 is position 1, no load is on the supply. At positions 2 and 3, there is half load and full load, respectively.

In practical troubleshooting, the load resistors (R1 and R2) must be noninductive so that a reactance is not placed on the supply output. Generally, the output load is considered as pure resistance for testing purposes. Of course, if the supply must be operated with a reactive load, the supply can be so tested with a simulated inductance (capacitance or inductance) in addition to the pure resistance. The values of $R_1$ and $R_2$ are chosen on the basis of output voltage and load current (maximum or half load). For example, if the linear supply is designed for an output of 25 V at 500 mA full load, the value of $R_2$ is $25/0.5 = 50\ \Omega$. The value of $R_1$ is $25/0.25 = 100\ \Omega$. When supplies with various

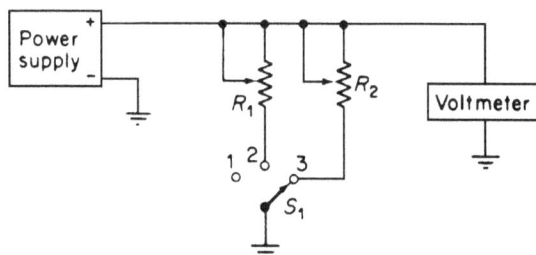

$$R_1\ \text{(in ohms)} = \text{half}-\text{load} = \frac{\text{rated output voltage}}{\frac{1}{2}\ \text{max rated current}}$$

$$R_2\ \text{(in ohms)} = \text{full}-\text{load} = \frac{\text{rated output voltage}}{\text{max rated current}}$$

$$\text{Actual load current} = \frac{\text{voltage readout}}{\text{value of } R_1 \text{ or } R_2}$$

$$\%\ \text{regulation} = \frac{\text{no}-\text{load voltage}-\text{full}-\text{load voltage}}{\text{full}-\text{load voltage}} \times 100$$

$$\text{Power supply internal resistance} = \frac{\text{no}-\text{load voltage}-\text{full}-\text{load voltage}}{\text{current (amperes)}}$$

**Figure 7.36.** Basic linear-supply test circuit.

outputs are to be tested, R1 and R2 should be variable and adjusted to the correct resistance value before test (using an ohmmeter, with the supply turned off). R1 and R2 must also be capable of dissipating the rated power without overheating. For example, using the previous values for $R_2$ and $R_1$, the power dissipation of R2 is 25 times $0.5 = 12.5$ W (use at least 15 W), and the dissipation for R1 is 25 times $0.25 = 6.25$ W (use at least 10 W).

Use the following procedure to measure linear-supply output:

1. Connect the equipment as shown in Fig. 7-36.

2. If R1 and R2 are adjustable, set both to correct values, using the equations.

3. Apply power. Set the input voltage to the correct value. Use the midrange value for the input voltage if a variable transformer or variac is available. For example, if the supply is designed to operate at 110 to 120 V, set the input to 115 V. (In most practical applications, simply use the available input supply voltage.)

4. Measure the output voltage at each position of S.

5. Calculate the current at positions 2 and 3 of S1, using the equations for actual load current. For example, assume that R1 is 100 Ω and the meter indicates 22 V at position 2 of S1. The actual load current is $22/100 = 220$ mA. If the supply output is 25 V at position 1 and it drops to 22 V at position 2, this means that the supply does not produce full output with a load. This is an indication of poor regulation, possibly resulting from poor design (in the case of an experimental supply) or component failure (when discovered as part of troubleshooting).

### 7.5.2 Basic output-regulation tests

Linear supply output regulation is usually expressed as a percentage and is determined by the following equation:

$$\% \text{ regulation} = \frac{(\text{no-load voltage}) - (\text{full-load voltage})}{\text{full-load voltage}} \times 100$$

A low percentage-of-regulation figure is desired because this indicates that the output voltage changes very little with load. Use the following steps when measuring linear-supply output regulation.

1. Connect the equipment as shown in Fig. 7-36.

2. If R2 is adjustable, set R2 to the correct value, using the equation in Fig. 7-36.

3. Apply power. Measure the output voltage at position 1 (no load) and position 3 (full load).

4. Using the equation, calculate the percentage of regulation. For example, the no-load voltage is 5 V and the full-load voltage is 4.750 V, the percentage of regulation is $(5 - 4.750)/4.750 \times 100 = 5.3\%$ (extremely poor regulation!).

### 7.5.3  Basic input-regulation tests

Linear-supply input regulation is usually expressed as a percentage that represents the maximum allowable output variation (with a given load) for maximum-rated input variation. For example, if the supply is designed to operate with an ac input from 110 to 120 V, and the dc output is 5 V, the output is measured (1) with an input of 120 V and (2) with an input of 110 V. If the output does not change, input regulation is perfect (and probably impossible). If the output varies by 0.250 V, the output variation is $(0.250/5) \times 100 = 5\%$ (again, poor regulation).

The input regulation can be measured at full load and/or at half load, as desired, using the same test connections shown in Fig. 7-36. However, the input voltage must be varied from maximum to minimum rated values (with a variable transformer or variac) and monitored with an accurate voltmeter.

### 7.5.4  Basic ripple tests

Any linear supply, no matter how well regulated and filtered, has some ripple. This ripple can be measured with a meter or scope. Usually, the factor of greatest concern is the ratio between the ripple and full output voltage. For example, if 0.05 V of ripple is measured with a 5-V output, the ratio is $(0.05/5) \times 100 = 1\%$.

### 7.5.5  Measuring ripple with a meter

Use the following procedure to measure output ripple with a meter.

1. Connect the equipment as shown in Fig. 7-36.
2. If R2 is adjustable, set R2 to the correct value using the equation in Fig. 7-36.
3. Apply power. Measure the dc-output voltage at position 3 (full load).
4. Set the meter to measure ac. Any voltage measured under these conditions is ripple.
5. Find the percentage of ripple as a ratio between the two voltages (ac/dc).

One problem often overlooked when measuring ripple during troubleshooting is that the voltage is probably not a pure sine wave. Most meters provide accurate ac-voltage indications only for pure sine waves. A better method is to measure ripple with a scope, where the peak value can be measured directly.

### 7.5.6  Measuring ripple with a scope

A scope can be used to display the ripple waveform, from which the amplitude, frequency, and nature of the ripple voltage can be determined. Usually, the scope is set to the ac mode when measuring ripple because this blocks the dc output of the supply. Normally, ripple voltage is small in relation to supply voltage. If the scope gain is set to display the ripple, the supply dc voltage drives the display off-screen.

1. Connect the equipment as shown in Fig. 7-37.

2. Apply power. If the test is to be made under load conditions, close switch S1 and adjust R1 for the desired load. Open S1 for a no-load test.

3. Adjust the scope controls to produce two or three stationary cycles of each wave on the screen.

4. Measure the peak amplitude of the ripple on the voltage-calibrated vertical scale.

5. Measure the frequency of the ripple on the horizontal scale. (The ripple frequency is probably the same as the line frequency.) When measuring ripple frequency, notice that a full wave produces two ripple "humps" per cycle. A half-wave supply produces one hump per cycle.

6. If the supply dc output must be measured simultaneously with the ripple, set the scope for the dc mode. The baseline should be deflected upward to the dc level and the ripple should be displayed at that level. If this drives the display off-screen, reduce the vertical gain and measure the dc output.

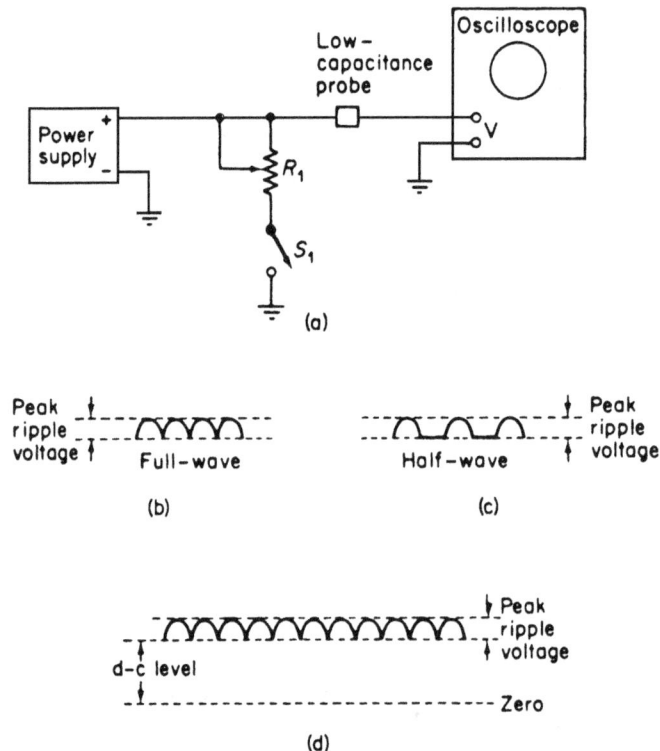

**Figure 7.37.** Measuring ripple with a scope (basic).

Then return the vertical gain to a level where the ripple can be measured and use the vertical-position control to bring the display back on the screen for ripple measurement. (It is generally easier to measure the dc voltage with a meter!)

7. Study the ripple waveform for defects in the linear supply (which can arise from poor design, poor layout, or component failure).

If the supply is unbalanced (one rectifier passes a different current than the others), the ripple humps will be unequal in amplitude.

If noise or fluctuations are in the supply (particularly if the supply includes zeners), the ripple humps will vary in amplitude or shape.

If the ripple varies in frequency, the ac source is probably varying.

If a full-wave supply produces a half-wave output ripple, one rectifier is not passing current.

### 7.5.7   Linear-supply advanced tests

The tests described thus far for linear supplies are generally sufficient as experimental circuits used by the hobbyist. However, a number of other tests can be applied to commercial supplies and to supplies found in more sophisticated equipment. Figure 7-38 shows test connections suitable for measuring the six most important operating specifications of a constant-voltage (CV) supply. These include source effect, load effect, periodic and random deviation (PARD), load-effect transient-recovery time, drift, and temperature coefficient.

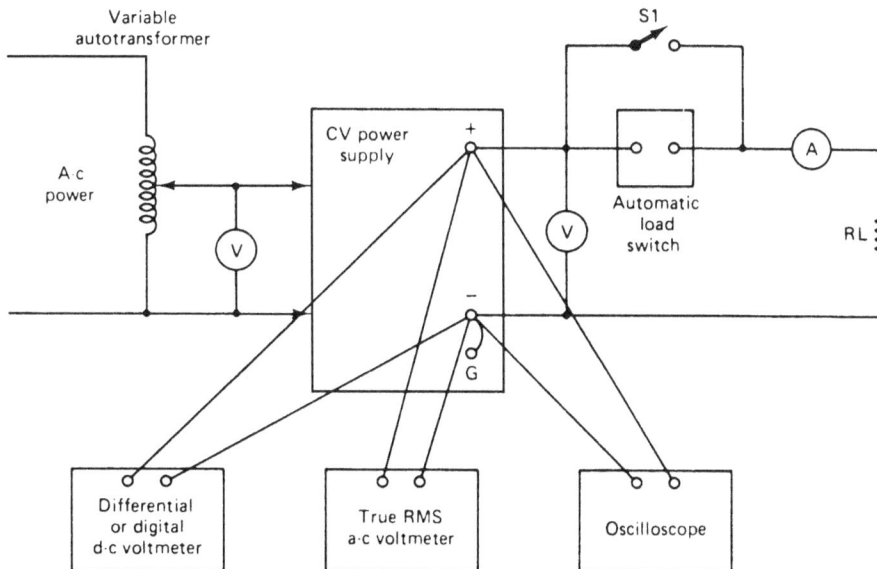

**Figure 7.38.**   Advanced linear-supply test circuit.

### 7.5.8  Test precautions

The following precautions apply primarily to completely assembled and packaged linear supplies, but can also pertain to supplies in the design/experimental stage. Most of the tests described here can be performed with only four test instruments: a variable autotransformer, a differential or digital dc meter, a true ac voltmeter, and an oscilloscope.

Be certain that the autotransformer has an adequate current rating. If not, the input ac voltage applied to the supply might be severely distorted and the supply's rectifying/regulating circuit might operate improperly. The dc voltmeter should have a resolution of 1 mV (or better). The ac voltmeter should have a resolution of 1 mV full-scale. (Hewlett-Packard HP3420B, HP3455A, and HP3400A meet these requirements.) The scope should have a sensitivity of 100 μV/cm and a bandwidth of 400 Hz for all measurements, except noise spike. A 5-mV sensitivity and a 20-MHz bandwidth are required for noise-spike measurements. (The HP180C with a 1821A time base and a 1806A vertical plug-in meet all requirements, except for noise spike. Use an 1803A plug-in with the HP180C for noise spike.)

### 7.5.9  Test connections

The connections shown in Fig. 7-38 apply to CV linear supplies with no remote-sensing leads. When remote sensing is involved, use the connections of Fig. 7-39. Either way, test connections should be permanent (not using clip leads) and should be at the closest point to the supply (so that the monitoring device sees the same performance as the feedback circuit in the supply). Clip-lead connections often produce measurement errors because the contact resistance between the clip leads and supply terminals is typically greater than the supply output impedance.

All measurement instruments must be connected directly by separate pairs of leads. Twisted pairs or shielded cable should be used to avoid pickup on the leads. Here is a simple test for pickup and/or ground-loop effects during testing/troubleshooting. Turn off the supply and observe the scope for any unwanted signals (particularly at the line frequency) with the leads connected between the measurement points (+S and −S, Fig. 7-39, for example). Then connect both scope leads to whichever terminal is grounded to the supply. If any noise is in either test condition with the supply off, you have possible pickup and/or ground-loop effects.

### 7.5.10  Source effect or line regulation

Source effect and line regulation are sometimes interchanged. No matter what the test is called, the measurement is made by turning the variable autotransformer throughout the specified range, from low line to high line, and noting the change in voltage at the output. The test is performed with all other parameters constant. The supply should stay within the specifications for any rated output voltage, combined with any rated output current. A typical spec-

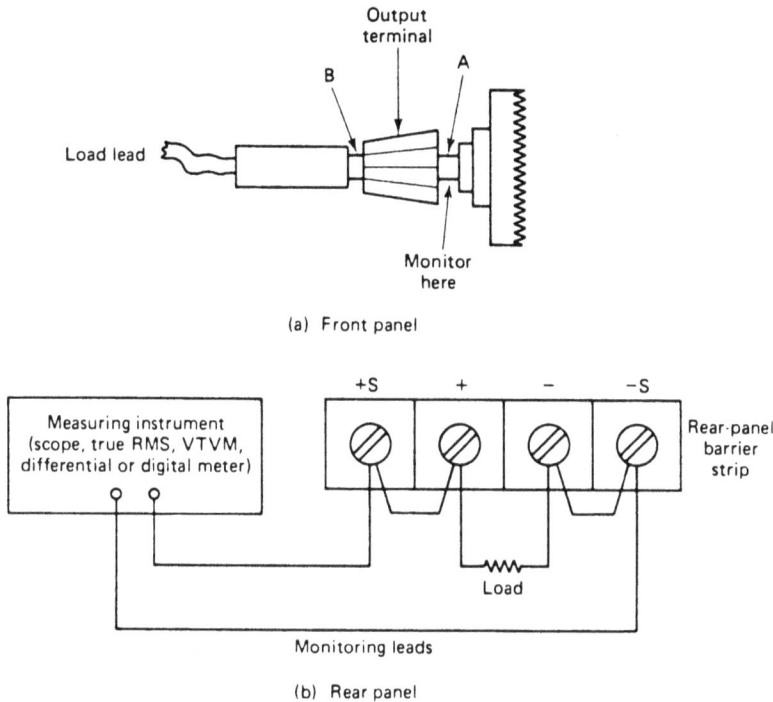

(a) Front panel

(b) Rear panel

**Figure 7.39.** Test connections for remote sensing.

ification might be "no more than a 5-mV change in output when the supply is delivering 5 V at 500 mA, as the line input is varied from 110 to 120 V." The extreme source-effect test is taken with maximum output voltage and maximum output current.

### 7.5.11 Load effect or load regulation

These terms are also often interchanged. The test is performed with all other parameters constant. The supply should stay within specifications for any rated output voltage, combined with any rated input line voltage. A typical specification might be "no more than a 5-mV change in output when the supply is delivering 5 V at 500 mA with the line at 120 V." The measurement is made by closing and opening switch S1 (Fig. 7-38) and noting the resulting static change in output voltage. This extreme load-effect test is with the maximum output voltage and maximum output current.

### 7.5.12 PARD (or noise and ripple)

In some commercial supplies, the term *PARD* (*periodic and random deviation*) has replaced the terms *ripple* and *noise*. When used, PARD represents deviation of the dc-output voltage from the average value, over a specified bandwidth, and with all other parameters constant.

Figure 7-40 shows some typical PARD (or noise and ripple) measurement circuits, together with some typical waveshapes. Notice that there are three possible waveshapes, each with the equation necessary to convert from one to the other. This is because the output ripple of a linear supply might resemble a square wave or a sawtooth wave more than a sine wave—even though the ripple is locked to the source or line frequency.

The connections shown in Fig. 7-40A produce a continuous ground-loop condition. Any current circulating in this loop causes a drop in series with the scope input. The resulting signal can easily be much greater than the true supply ripple and can completely invalidate the measurement.

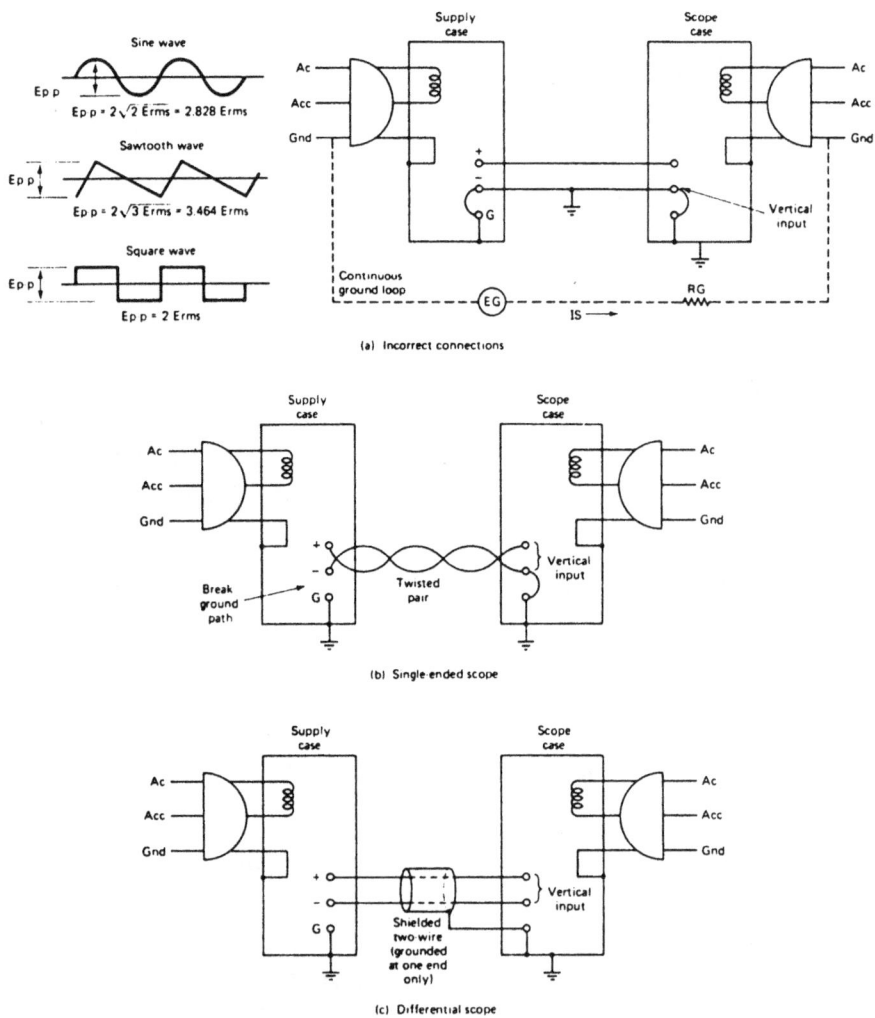

**Figure 7.40.**  Typical PARD (or noise and ripple) measurement circuits.

Figure 7-40B shows the correct method of measuring output ripple using a single-ended scope. The ground-loop path is broken by floating the supply output (breaking the ground path at the supply end). Use either a twisted pair or (preferably) a shielded two-wire cable. (Ground the shield at one end only to prevent ground currents from generating noise signals in the shielded leads.) When troubleshooting, be sure that the scope is not displaying ripple induced in the leads or picked up from grounds, short both scope leads together and touch both output terminals. If the ripple magnitude of the "shorted" test approaches that of the measurement ripple, the test is not reliable. Look for ground-loop or pick-up problems.

In most troubleshooting situations, the single-ended scope is satisfactory for all ripple and noise measurements. In certain cases, or for measurements where both the supply and scope cases are connected to ground (for example, if both are rack mounted), the differential scope with a floating input is required. (As an alternate, two single-wire shielded cables can be substituted for the shielded two-wire cable.)

In testing or troubleshooting, the equations in Fig. 7-40 can be helpful in converting the peak-to-peak scope display to the values given in the supply specification. For example, if a supply is rated at a maximum of 1 mV (RMS) of ripple, the maximum peak-to-peak ripple is 2.828 mV, assuming that the ripple is a sine wave. If the ripple is closer to a sawtooth, the maximum peak-to-peak is 3.464 mV. For a square-wave ripple, the maximum peak-to-peak is 2 mV. Of course, all of these values are approximations and depend on the precise waveform.

## 7.5.13    Regulator ripple rejection and output resistance

*Ripple rejection* is a term applied to linear IC regulators (Section 7.3), but not to the entire linear supply. Figure 7-41 shows typical test connections for measuring ripple rejection. Although this circuit is designed specifically for particular IC regulator (the Harris Semiconductor CA3085), variations of the circuit can be used with other regulators. The circuit also provides a means of measuring output resistance for the regulator. The procedures for measuring both ripple rejection and output resistance are given in Fig. 7-41.

## 7.5.14    Regulator noise voltage

Figure 7-42 shows the typical test connections for measurement of noise voltage. Again, these measurements apply to the IC regulator and not to the entire linear supply. Also, the circuit is designed specifically for the Harris Semiconductor CA3085, but it can be used with other IC regulators (with changes in component values). In practical troubleshooting, remember that both the input (pin 3) and booster transistor (pin 2) are connected to a fixed 25 V. Any noise appearing at the output is being generated in the IC.

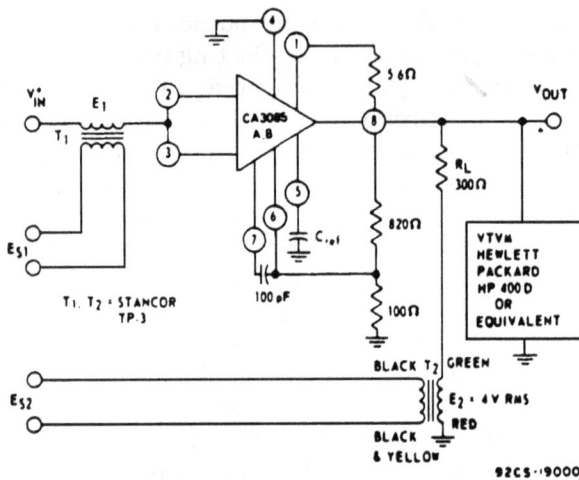

**TEST PROCEDURES FOR TEST CIRCUIT FOR
RIPPLE REJECTION AND OUTPUT RESISTANCE**

**Output Resistance**

Conditions

1.  $V_{IN}$ = +25V, $C_{REF}$ = 0. Short $E_1$
2.  Set $E_{S2}$ at 1 kHz so that $E_2$ = 4V rms
3.  Read $V_{OUT}$ on a VTVM, such as a Hewlett-Packard, HP400D or equivalent
4.  Calculate $R_{OUT}$ from $R_{OUT}$ = $V_{OUT}$ ($R_L/E_2$)

**Ripple Rejection — I**

Conditions

1.  $V_{IN}$ = +25V, $C_{REF}$ = 0. Short $E_2$
2.  Set $E_{S1}$ at 1 kHz so that $E_1$ = 3V rms
3.  Read $V_{OUT}$ on a VTVM, such as a Hewlett-Packard, HP400D or equivalent
4.  Calculate Ripple Rejection from 20 log ($E_1/V_{OUT}$)

**Ripple Rejection — II**

Conditions

1.  Repeat Ripple Rejection I with $C_{REF}$ = 2 μF

**Figure 7.41.**   Measuring ripple rejection and output resistance (Harris Semiconductor, *Linear & Telecom ICs*, 1991, p. 2-43).

However, any noise from the 25-V source (ripple, noise spike, etc.) can appear at the output.

### 7.5.15  Transient recovery time

Figure 7-43 shows the waveforms involved with transient recovery of a linear supply. If a step change in load current is imposed on the output, the output voltage shows a transient similar to that in Fig. 7-43. The output impedance of any supply rises at high frequencies, producing an *equivalent output inductance*. If the load current is stepped or switched rapidly enough that the high frequencies associated with the leading edge of the step change can react with this inductance, a spike occurs on the output terminals. So, it is essential that rise time is established before specifying an amplitude for transient recovery.

As an example, a supply with an effective output inductance of 0.1 μH will show a load-transient spike of about 0.16 V if the load is switched with a rise time of 1 A/μs. However, the spike amplitude is only 160 μV if the load is switched at 1 A/ms. In the latter case, the output might not show up because the spike is small compared to the static change in output voltage associated with the full-load change.

Notice that Y shown in Fig. 7-43 is usually specified separately for each type of supply, but it is usually the same as the load-regulation specification (Section 7.5.11). Z is the specified load-current change, but it is usually equal to the full-load current of the supply. The nominal output voltage is the dc level halfway between the steady-state output voltage, both before and after the imposed load change. Unless otherwise specified, transient-recovery time can be measured at any input line voltage, combined with any output voltage and load current (within ratings). Also, transient-recovery time is not always defined in the same way. Sometimes the specification is omitted completely. In

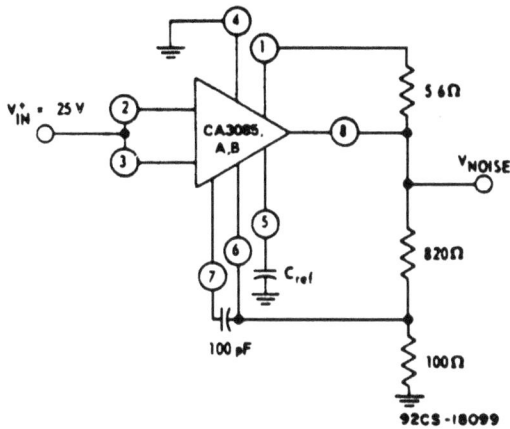

**Figure 7.42.** Measuring regulator noise voltage (Harris Semiconductor, *Linear & Telecom ICs,* 1991, p. 2-43)

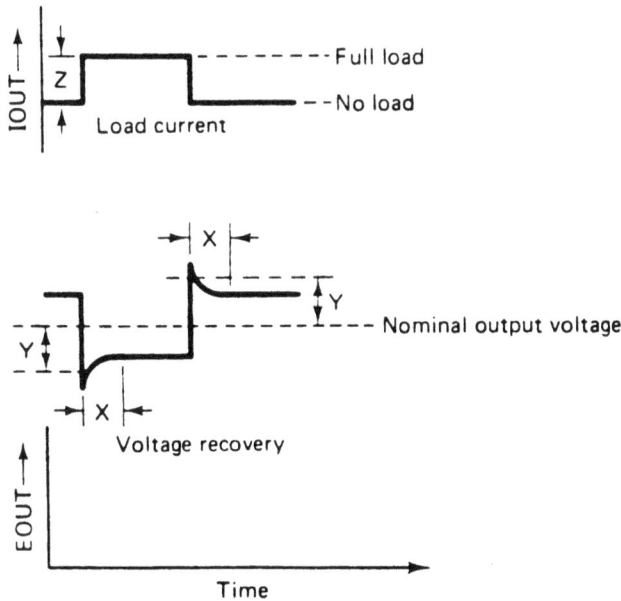

**Figure 7.43.** Waveforms involved with transient recovery of a linear supply.

other cases, a given time is specified, but not the level or percentage of output-voltage recovery.

Figure 7-44 shows an automatic-load switch used by Hewlett-Packard to measure the load-effect transient-recovery time. The switch uses a mercury-wetted relay operated by a 60-Hz line voltage. (Use of the 60-Hz voltage makes it easy to observe on the scope.) The load-current duty cycle is controlled by R1. The auto-load switch is connected into the test setup of Fig. 7-38. Do not exceed the maximum load ratings shown in Fig. 7-44 (to prevent damage to the relay contacts).

Figure 7-45 shows an alternate test circuit (including waveforms) for transient-recovery time, developed by Harris Semiconductor. Instead of a mechanical relay, the load is applied and removed from the supply output by pulses applied to a transistor. Again, this circuit applies specifically to the IC regulator, but a similar circuit can be developed to test a complete supply. Notice the similarity of the theoretical waveforms shown in Fig. 7-43 and the actual scope display of Fig. 7-45 (both the input-current pulse and the output-voltage pulse).

### 7.5.16   Drift (stability)

Drift is measured by monitoring the supply output on a differential or digital voltmeter over a stated measurement interval (typically 8 hours after a 30-minute warmup). In some cases, a strip chart is used to provide a permanent record. A thermometer should be placed near the supply to verify that the ambient temperature remains constant during the period of measurement. The supply should be at a location immune from stray air currents (away from open doors or windows, and from air-conditioning vents). If possible, place the supply in an oven and hold it at a constant temperature. For a thorough test, measure both drift and PARD (or noise and ripple) over an 8-hour interval. (Be

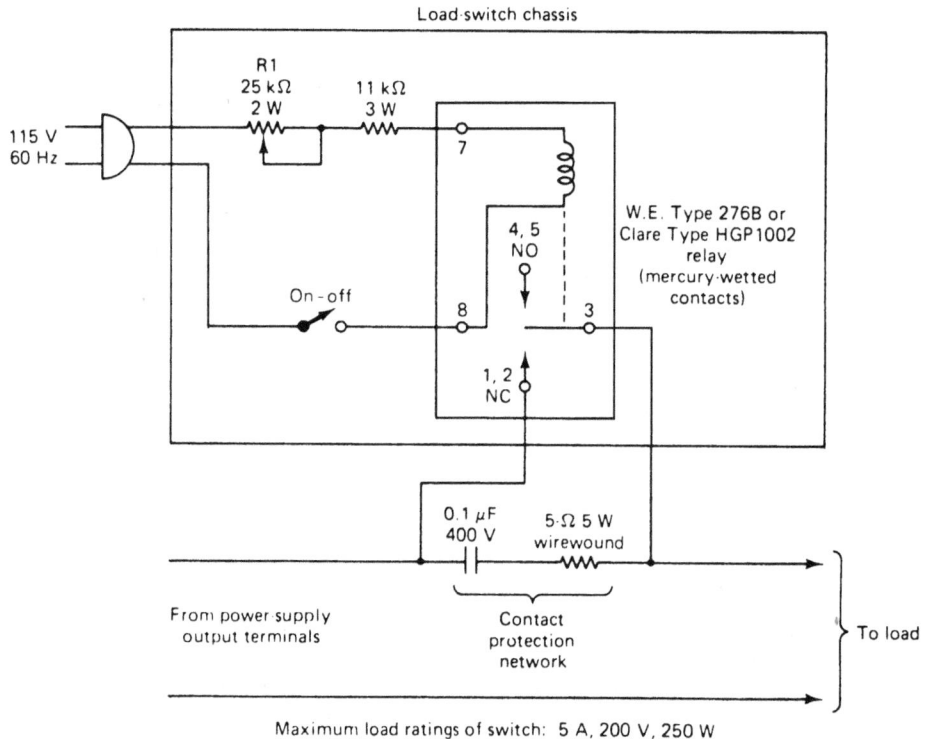

**Figure 7.44.**   Automatic-load switch used to measure load-effect transient-recovery time.

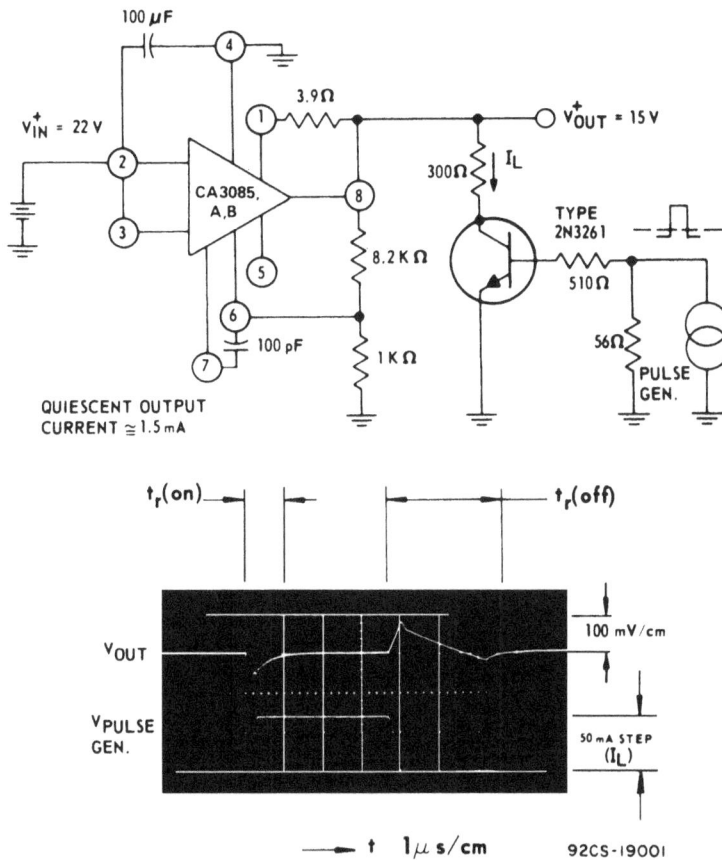

**Figure 7.45.**  Alternate test circuit for transient-recovery time (Harris Semiconductor, *Linear & Telecom ICs,* 1991, p. 2-44).

certain that the measuring instrument has a stability at least an order of magnitude better than the stability specification of the supply.) Typically, a linear supply should drift less over the 8-hour period than during the first 30-minute warmup period.

## 7.5.17  Temperature coefficient

Temperature-coefficient measurements are made by placing the supply in an oven and varying the temperature over a given range, following a 30-minute warmup. The supply is allowed to stabilize at each measurement temperature. Although there is no standardization, a typical temperature coefficient specification is the change in output voltage per °C in ambient temperature, from 0°C to 55°C. The measuring instrument (digital or differential voltmeter) should be placed outside the oven and must have a long-term stability adequate to ensure that any voltmeter drift does not affect measurement accuracy.

### 7.5.18    Constant-current-supply advanced tests

Figure 7-46 shows test connections suitable for measuring the most important operating specifications of a constant-current (CC) linear supply. The automatic load switch shown in Fig. 7-46 is used to interrupt the load periodically, when measuring transient-recovery time. The procedures for CC supplies are similar to those of CV supplies. By comparing Fig. 7-46 to Fig. 7-38, you can see that CC and CV measurements have three major differences. First, the load switch is connected in parallel, rather than in series with the supply load. This is because the supply performance is checked between short-circuit and full-load conditions, rather than between open-circuit and full-load conditions.

Second, a current-monitoring resistor $R_M$ is inserted between the supply output and the load. To simplify grounding problems, connect one end of $R_M$ to the grounded output terminal of the supply.

Third, all CC measurements are made in terms of change in voltage across $R_M$. The CC performance is calculated by dividing these voltage changes by the resistance value of $R_M$.

## 7.6    Linear-Supply Troubleshooting

This section describes the troubleshooting procedures for two linear-supply circuits. One is an adjustable off-line supply and the other is a dual pre-regulated supply circuit.

### 7.6.1    Adjustable off-line supply troubleshooting

Figure 7-47 shows a complete off-line supply with an adjustable output and an IC regulator. The first step in troubleshooting the circuit is to test the supply (as described in this chapter), using both the basic and advanced tests. With a 90-mA load, you should be able to set the output voltage between 3.5 and 20 V, using the 10-k$\Omega$ pot at pin 6 of the CA3085.

If you find no voltage, check the neon lamp when the power switch is closed (right after you are sure that the power cord is plugged in!). If the lamp is off, suspect that the fuse is blown.

If the lamp is on, check for ac at the transformer secondary (about 24 V across the diodes). If ac is absent, suspect a problem with the transformer. If ac is present, check for dc between pin 3 of the CA3085 and ground. If it is absent, suspect problems with the diodes. It is also possible that the 500-$\mu$F capacitor is shorted or leaking badly.

If dc is at pins 2 and 3 (about 25 V), but there is no output (pin 8) or you cannot adjust the output across the range, suspect a problem with the 5-$\mu$F output capacitor, the 100-pF compensation capacitor, and the CA3085 (in that order).

If you get the correct voltage output (across the range) but there is excessive ripple, or if line/load regulation is out of tolerance (alleged to be 0.2%), suspect a problem with the CA3085 (although leaking filter capacitors are always a possibility).

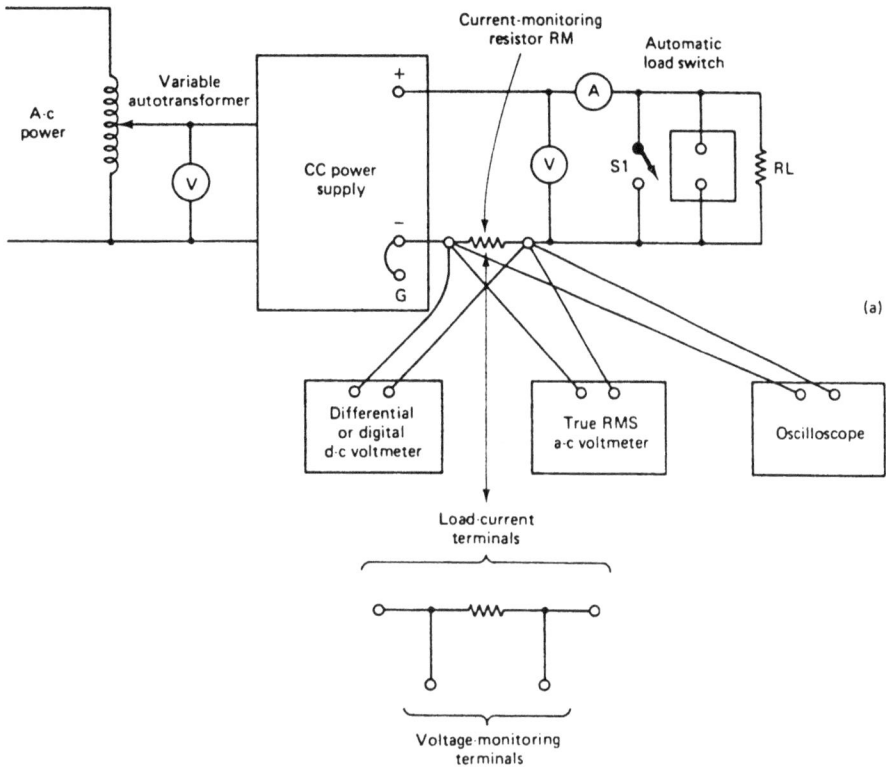

**Figure 7.46.**   Constrant-current linear-supply test circuit.

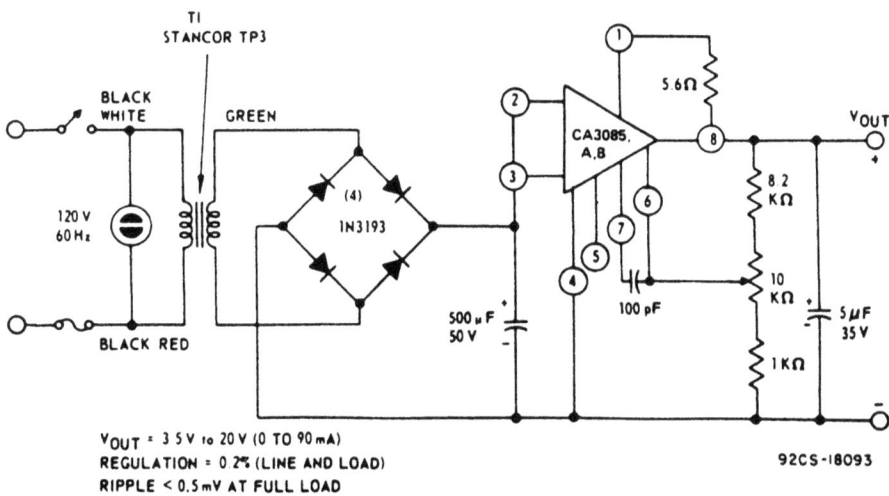

$V_{OUT} = 3.5\,V$ to $20\,V$ (0 TO 90 mA)
REGULATION = 0.2% (LINE AND LOAD)
RIPPLE < 0.5 mV AT FULL LOAD

92CS-18093

**Figure 7.47.**   Offline supply with adjustable output (Harris Semiconductor, *Linear & Telecom ICs,* 1991, p. 2-41).

### 7.6.2    Dual pre-regulated supply troubleshooting

Figure 7-48 shows a complete off-line supply with ±12-V outputs developed by Linear Technology. This circuit includes packaged diodes (the MDA201s), three-terminal regulators (the LT1086s), and IC regulators (the LT1011s) in both sections.

Notice that neither the +12-V nor the −12-V outputs are adjustable. The adjustment pin on the three-terminal regulators is connected to a fixed voltage-divider network in both cases. If you get an output voltage, but not at the correct level, check the values of the 124-Ω and 1.07-kΩ resistors.

If you get +12 V, but not −12 V (or vice versa), you have eliminated half of the circuits as suspects. The same is true if you get excessive ripple or if the two supplies do not show substantially the same line/load regulation. If both supplies are equally bad, suspect a problem with the transformer.

Assume that the +12-V supply is bad, but the −12-V supply is good (−12-V output with a 1.5-A load, when the line is varied between 90 and 130 V ac). If there is no +12 V, check for ac at the input of the MDA201 and dc at the output. If the ac is absent or is abnormal, suspect a problem with the transformer winding. If you find ac, but not dc, suspect that the MDA201 (or the 4700-μF capacitor) has failed.

If dc is coming from the MDA201, compare this to the dc from the MDA201 in the −12-V supply. While you are at it, compare the voltage at $V_{IN}$ of both

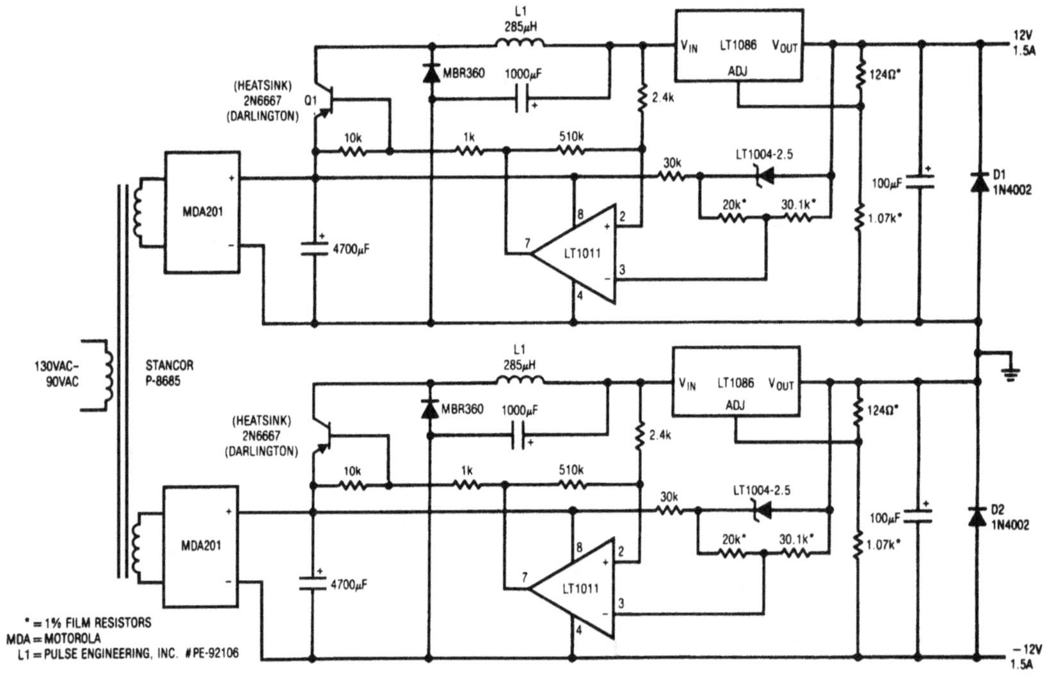

**Figure 7.48.**   Dual pre-regulated off-line linear supply.

LT1086 regulators. If $V_{IN}$ for the bad +12-V LT1086 is not substantially the same as for the −12-V regulator, suspect a problem in Q1 or the associated parts (such as L1, the LT1011, the MRB360, and the 1000-µF capacitor).

If the dc is the same for both $V_{IN}$ terminals, but the +12-V output is absent or abnormal, suspect a problem with the 100-µF output capacitor, diode D1, the LT1086, or the LT1004 zener (in that order). If the zener is leaking badly (or is completely dead), you will get a different voltage at the $V_{IN}$ terminals of the LT1086.

# Switching-Supply Circuit Troubleshooting

This chapter is devoted to troubleshooting for switching power supplies. It starts with a review of switching supply basics (duty cycle, frequency, transistor and diode characteristics, PWM versus PFM, boost, buck-boost or inverting, continuous versus discontinuous, buck, flyback, forward, current-boost, Cuk, inductor problems, EMI problems, etc.) from a troubleshooting standpoint. This is followed by an in-depth review of switching-supply testing and connections. Then examples of switching-supply troubleshooting are described.

## 8.1 Switching-Supply Basics

At one time, switching power supplies (also known as *switch-mode supplies*) were constructed in discrete form, including the oscillator required for a switching regulator. In present-day circuits, the switching regulator is in IC form. Figure 8.1 shows the block diagram of the basic switching regulator. The function of this circuit is to convert an unregulated dc input to a regulated dc output. For this reason, switching regulators are often referred to as *dc-dc converters*.

In a switching regulator, the power transistor is used in a switching (or on/off) mode, rather than in the continuous mode of a linear supply (Chapter 7). As a result, switching regulator efficiency is usually in the 70 to 95% range, which is more than double that of linear regulators. In addition to increased efficiency, the switching regulator can provide outputs that are greater than the input, if desired. The output of linear regulators is always lower than the input. Switching regulators can also invert the input (produce a positive output for a negative input, and vice versa), unlike the conventional linear regulator. High-frequency switching regulators offer considerable weight and size reductions and better efficiency at high power than linear supplies.

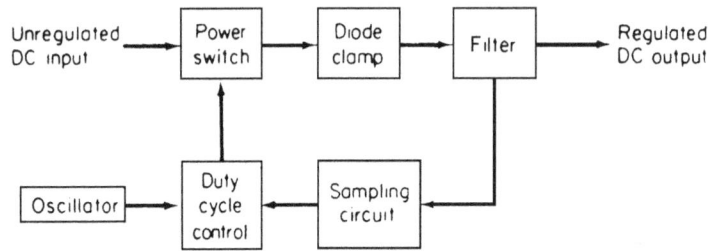

**Figure 8.1.** Basic switching-regulator functions.

### 8.1.1  Switching-regulator circuit problems

Switching regulators are not without problems that affect troubleshooting. In addition to requiring more complex circuits, switching regulators produce electromagnetic interference (EMI). However, with proper design, EMI can be reduced to acceptable levels. Such design techniques involve the use of low-loss ferrite cores for transformers and chokes or inductors, of high-permeability magnetic alloys for shielding, and of miniature semiconductors and IC devices for switching and regulation circuits.

### 8.1.2  Switching duty cycle

The circuit of Fig. 8.1 regulates by switching the series transistor (power switch) to either the ON or OFF condition. The duty cycle of the series transistor determines the average dc output. In turn, duty cycle is adjusted in accordance with a feedback that is proportional to the difference between the dc output and a reference voltage.

### 8.1.3  Switching frequency

Switching is usually at a constant frequency just above the audible range, although some switching regulators use a variable frequency with changing line and load. With some switching-regulator ICs, it is possible to set or change the switching frequency with an external capacitor. One of the first troubleshooting guides to remember is that higher frequencies are generally less efficient because transistor switching losses and ferrite-core losses increase. On the other hand, lower switching frequencies in the audible range might cause certain components to "sing" or might produce interference in audio circuits being powered by the regulator.

### 8.1.4  Transistor and diode characteristics

Switching regulators must use transistors with a gain-bandwidth product ($f_T$) of at least 4 MHz to operate efficiently (an $f_T$ of 30 MHz is even better). Darlington transistors and MOSFETs are also used in switching regulators. A fast-recovery rectifier or a Schottky barrier diode is used as a free-wheeling clamp to keep the switching-transistor loadline within safe operating limits and to increase efficiency. Other devices used in some switching regulators include gates, flip-flops (FFs), op-amp comparators, timers, and rectifiers.

## 8.2 Switching-Regulator Basics

Figure 8.2 shows a theoretical switching-regulator circuit (in the buck or step-down configuration) and the related waveforms. The high efficiency of switching regulators is the result of operating the series transistor in a switching mode. When the transistor is switched on, the full input voltage is applied to the dc filter. When the transistor is switched off, the input voltage is zero. With the transistor turned on and off for equal amounts of time (50% duty cycle), the dc load voltage is half of the input voltage. The output voltage ($V_o$) is always equal to the input voltage ($V_{in}$) times the duty cycle ($D$). In other words, $V_o = DV_{in}$.

Varying the duty cycle sometimes compensates for changes in the input voltage. This technique is used to produce the regulated output voltage as follows. Repetitive operation of the switching transistor at a fixed duty cycle produces the steady-state waveforms shown in Fig. 8.2. With the switch closed, the inductor current ($I_L$) flows from the input voltage ($V_{in}$) to the load. The difference between the input and output voltage ($V_{in} - V_o$) is applied across the inductor. This causes $I_L$ to increase during the "switch-closed" period.

With the switch open, stored energy in the inductor forces $I_L$ to continue to flow to the load and to return through the diode. The inductor voltage is then reversed, and is approximately equal to $V_o$. $I_L$ decreases during the "switch-open" period. The average current through the inductor equals the load current. Because the capacitor keeps $V_o$ constant, load current $I_o$ is also constant. When $I_L$ increases above $I_o$, the capacitor charges, and when $I_L$ drops below $I_o$, the capacitor discharges (as shown by the theoretical waveforms).

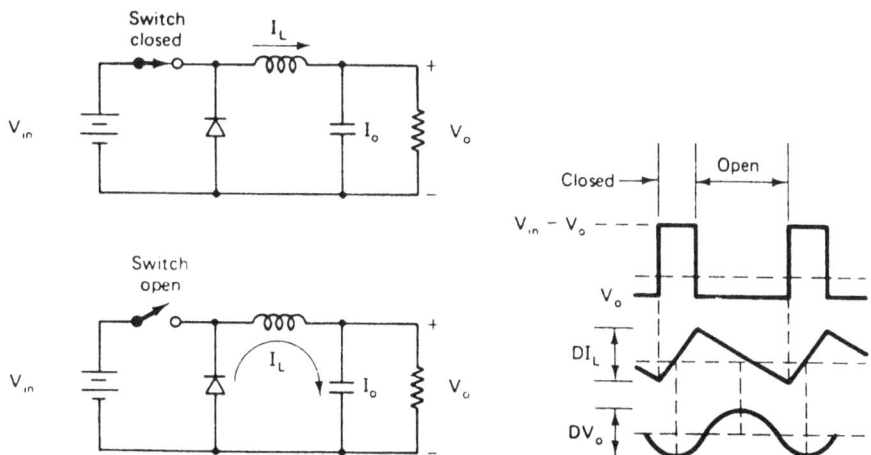

**Figure 8.2.**   Theoretical switching-regulator circuit (buck type).

From a troubleshooting standpoint, the end results of steady-state operation are:

- The average inductor voltage is zero, but a wide variation from $(V_{in} - V_o)$ is experienced.
- The direct current flowing through the inductor equals the load current. A small amount of sawtooth ripple is also present.
- The dc voltage on the capacitor is equal to the load voltage. A small amount of ripple is also present at the capacitor.

To be effective, a regulator must provide compensation for changes in both $V_{in}$ and $I_o$. In a switching regulator, the $V_{in}$ changes are automatically compensated for by duty-cycle variation in a closed-loop feedback system (Fig. 8.1). Input regulation and ripple rejection depend on loop gain. Changes in $I_o$ are more difficult to offset and load transient response is generally poor (rapid changes in load are difficult to regulate). From a troubleshooting standpoint, changes in $I_o$ are offset with temporary duty-cycle changes. For example, a change in load from zero to full-load results in the following:

- The duty cycle increases to maximum (the transistor might simply stay on).
- The inductor current takes many cycles to increase to the new dc level.
- The duty cycle returns to the original value.

## 8.3   PWM versus PFM

Switching regulators (or converters) are sometimes classified as to how they control output voltage. The two most common approaches are *pulse-width modulation* (*PWM*) and *pulse-frequency modulation* (*PFM*). Both approaches control the output by varying the duty cycle.

In troubleshooting, remember that with PWM regulators the frequency is held constant and the width of each pulse is varied. PWM regulators are often used in high-power switching circuits that operate from the ac line. With PFM regulators, the pulse width is held constant and the duty cycle is controlled by changing the pulse repetition rate.

The two basic approaches have many variations. For example, the *current-mode control* is a refinement of PFM. Likewise, *pulse skipping* is a refinement of PFM. These and other variations of PWM/PFM are covered throughout this chapter.

## 8.4   Common Switching-Regulator Configurations

Many switching-regulator configurations are possible. The choice of which configurations to use generally narrows down to such factors as voltage, polarity, voltage ratio, and fault condition. For example, if the output voltage must be larger than the input, a buck converter cannot be used. If the input voltage is negative and the output must be positive, some form of inverter is required. If the regulator must be current-limited, the basic boost circuit is of no value.

Even with these obvious limitations, most applications still have many choices of configuration. For example, to convert +28 V to +5 V, the buck, flyback, forward, and current-boosted configurations could be used. The following section covers most of the configurations, describing both the capabilities and limitations from a troubleshooting standpoint.

### 8.4.1  Boost or step-up

Figure 8.3 shows the theoretical boost or step-up configuration. Figure 8.4 shows a typical IC switching regulator (the Raytheon RC4190) connected in a practical step-up converter configuration. Figure 8.5 shows the corresponding waveforms found during circuit troubleshooting. The following paragraphs describe how both theoretical and practical functions are performed and how the functions relate to troubleshooting. Notice that switch S in the theoretical circuit is replaced by transistor Q1 in the practical circuit. Capacitor C, diode D, and inductor L in Fig. 8.3 are replaced by C1, D1, and $L_x$ in Fig. 8.4.

As shown in Fig. 8.3, when switch S is closed, the battery voltage is applied across inductor L. The charging current flows through L, building

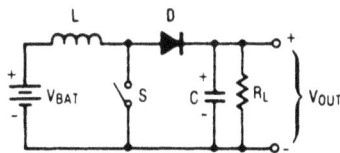

**Figure 8.3.**  Theoretical boost or step-up configuration (*Raytheon Linear Integrated Circuits,* 1989, p. 9-7).

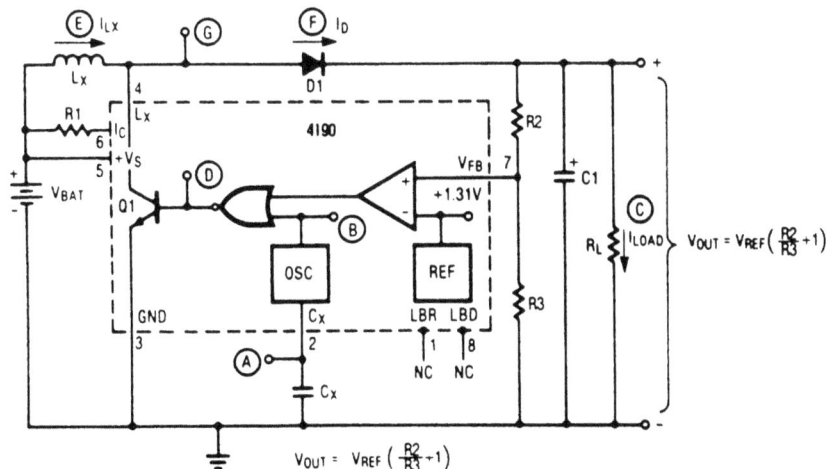

**Figure 8.4.**  Practical step-up converter (*Raytheon Linear Integrated Circuits,* 1989, p. 9-8).

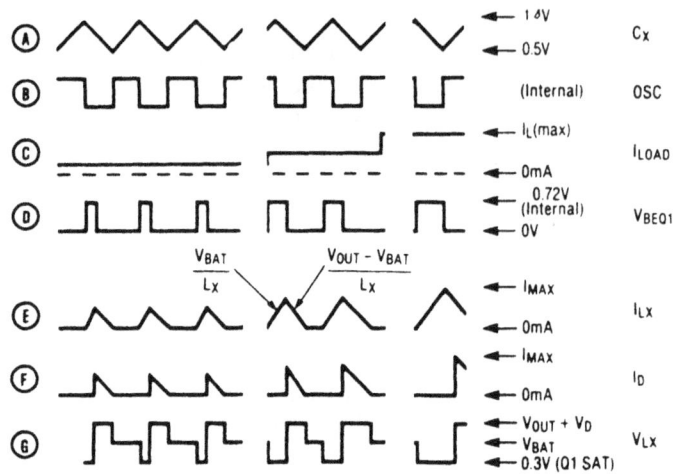

**Figure 8.5.** Step-up converter waveforms (*Raytheon Linear Integrated Circuits,* 1989, p. 9-8).

up a magnetic field, increasing as the switch is held closed. While S is closed, diode D is reverse-biased (open circuit) and current is supplied to the load by capacitor C. Until S is opened, the current through L increases linearly to a maximum value determined by the battery voltage, inductor value, and amount of time S is held closed, $I_{\text{peak}} = V_{\text{batt}} / (L \times T_{\text{on}})$.

When S is opened, the magnetic field collapses and the energy stored in the field is converted into a discharge current that flows through L in the same direction as the charging current. Because current has no path to flow through S, it must flow through D to supply the load and charge output capacitor C. If the switch is opened and closed repeatedly (by some form of oscillator), at a rate greater than the time constant of the output RC, then a constant dc voltage is produced at the output.

An output voltage higher than the input voltage is possible because of the high voltage produced by a rapid change of current in the inductor. When S is opened, the inductor voltage rises high enough (instantly) to forward-bias D and it adds to the battery voltage. In a practical IC regulator, a feedback-control system adjusts the on-time of switch S (controlling the level of the inductor current) so that the average inductor discharge current equals the load current, thus regulating the output voltage.

When power is first applied to the practical IC (Fig. 8.4) during troubleshooting, the current in R1 supplies bias current to pin 6 of the IC. This current is stabilized by a unity-gain, current-source amplifier and then used as a bias current for the 1.31-V (a bandgap-type reference, in this case). The stable bias current generated by the reference is used to bias the remainder of the ICs.

At the same time that the IC is starting up, current flows through $L_x$ and D1 to charge C1 to the battery voltage, less the drop across D1 (for $V_{\text{batt}} - V_{\text{p}}$). At this point, the feedback pin (pin 7) senses that the output voltage ($V_{\text{out}}$) is too low, then the comparator output changes to a logic zero. The NOR gate then combines the oscillator or square wave with the comparator signal. If the comparator output is zero and the oscillator output is low, then the NOR-gate output is high and switch transistor Q1 is forced on. When the oscillator goes high again, the NOR-gate output goes low, and Q1 turns off.

The turning on and off of the switch transistor (Q1) performs the same function as the opening and closing of switch S in Fig. 8.3. That is, energy is stored in the inductor during the on-time and is released into output capacitor C1 during the off time. The comparator continues to allow the oscillator to turn Q1 on and off until enough charge is delivered to C1 to raise the feedback voltage above 1.31 V. When the feedback is above the reference, the feedback system varies the duration of the on time in response to changes in load current or battery voltage, as shown by the waveforms in Fig. 8.5. If the load current increases (waveform C), then Q1 remains on (waveform D) for a longer period of the oscillator cycle, thus allowing the inductor current (waveform E) to build up a higher peak value. The duty cycle of Q1 varies in response to changes in both load and line.

In any switching regulator, the inductor value and oscillator frequency are (or should be) carefully tailored to the battery voltage, output current, and ripple requirements of the application. From a troubleshooting standpoint, if the inductor value is too high or the oscillator frequency is too high, then the inductor current never reaches a value high enough to meet the load-current drain and the output voltage drops (or collapses). If the inductor value or oscillator frequency is too low, then the inductor current builds up too high, causing excessive output-voltage ripple. It is also possible for the switch transistor to be overdriven, or for the inductor to be saturated, with low inductor values and/or low oscillator frequency.

## 8.4.2 Continuous versus discontinuous

The operation of switching regulators can be continuous or discontinuous. With continuous operation, current through the inductor never drops to zero during the switch (transistor) off time. With the discontinuous mode, if the load current is low enough, the inductor current can drop to zero in some cases.

This brings up two problems when troubleshooting a switching-regulator circuit. First, at light load currents, it is possible that the on-time of the switch transistor cannot be reduced to a low enough level to prevent the lightly loaded output from drifting high. If this occurs, most switching regulators will begin "dropping cycles" where the switch transistor does not turn on for one or more cycles. Although the output control is maintained, the condition produces subharmonic frequencies which might not be acceptable in some applications. (If the frequencies are in the audio range, one or more components might "sing.")

The main problem in troubleshooting discontinuous regulators is when heavy load currents are involved. This requires a high ratio of switch current to output current, and results in switching "spikes." Such spikes are shown in Fig. 8.6, which illustrates the voltage and current waveforms for the switch

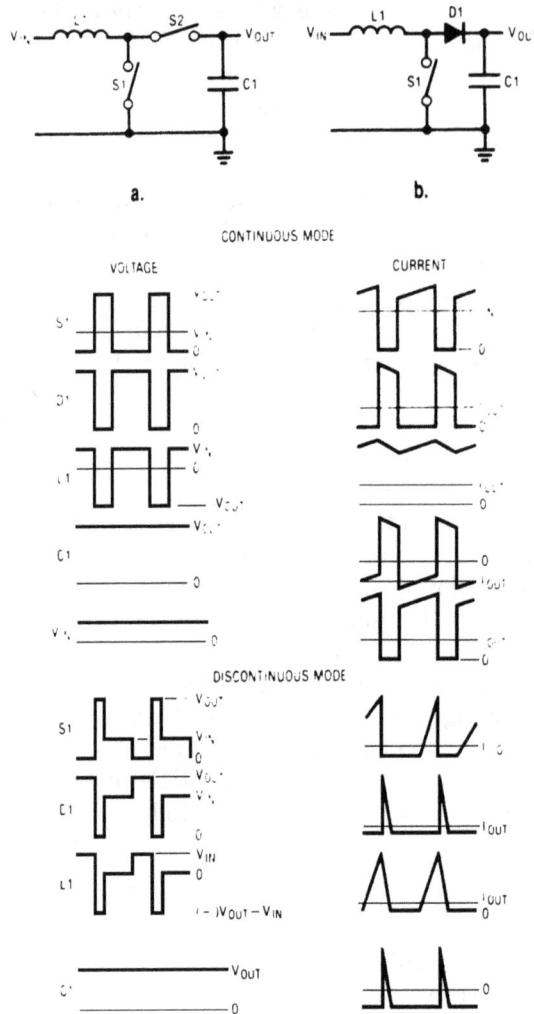

**Figure 8.6.** Waveforms for switch, diode, and output capacitor of a boost regulator (Linear Technology, Application Note 19, p. 14).

(transistor), diode, and output capacitor of a boost regulator. As a general rule, with either continuous or discontinuous, the diode and output capacitor must be capable of handling peak currents, as well as average currents. Of course, the current limits for the switch (Q1 in Fig. 8.4) are part of the IC specifications.

### 8.4.3   Buck-boost or inverting

Figure 8.7 shows the theoretical buck-boost or inverting configuration. Figure 8.8 shows a typical IC switching regulator (the Raytheon RC4391) connected in a practical inverting converter (or inverter) configuration. Figure 8.9 shows the corresponding waveforms found during circuit troubleshooting. The following paragraphs describe how both theoretical and practical functions are performed and how the functions relate to troubleshooting. Notice that switch S in the theoretical circuit is replaced by transistor Q1 in the practical circuit. Capacitor C, diode D, and inductor L in Fig. 8.7 are replaced by $C_F$, D1, and $L_x$ in Fig. 8.8.

As shown in Fig. 8.9, when switch S is closed, charging current from the battery flows through inductor L, which builds up a magnetic field that increases as S is held closed. When S is opened, the magnetic field collapses, and energy stored in the magnetic field is converted into a current that flows through L in the same direction as the charging current. Because this current has no path to flow through the switch, the current must flow through diode D to charge capacitor C. The key to inversion is the ability of the inductor to become a source when the charging current is removed.

In troubleshooting the practical circuit of Fig. 8.8, remember that the feedback circuit and the output capacitor decrease the output voltage across the inductor to a regulated fixed value. When power is first applied, the ground-sensing comparator (pin 8) compares the output voltage to the +1.25-V reference. Because CF is initially discharged, a positive voltage is applied to the comparator and the output of this comparator gates the square-wave oscillator.

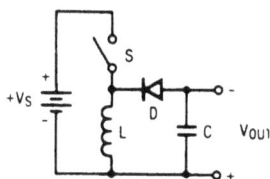

**Figure 8.7.** Theoretical buck-boost or inverting configuration (*Raytheon Linear Integrated Circuits*, 1989, p. 9-54).

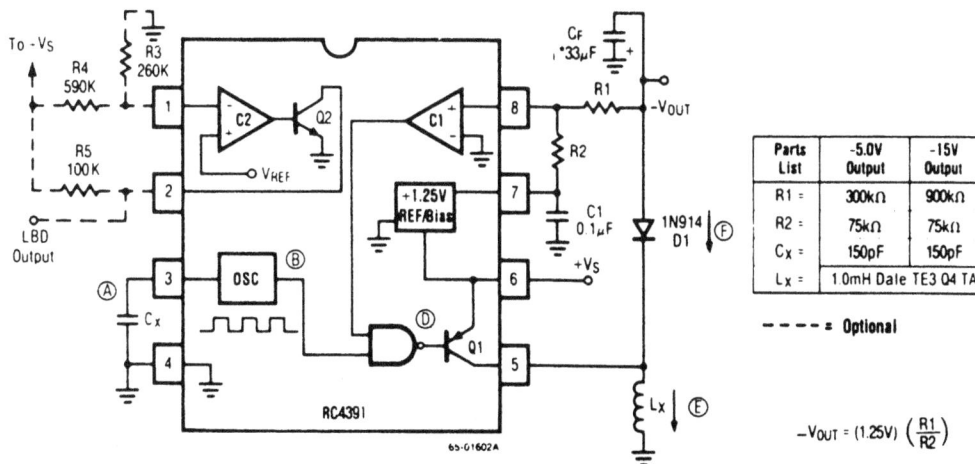

| Parts List | -5.0V Output | -15V Output |
|---|---|---|
| R1 = | 300kΩ | 900kΩ |
| R2 = | 75kΩ | 75kΩ |
| Cx = | 150pF | 150pF |
| Lx = | 1.0mH Dale TE3 Q4 TA | |

- - - - = Optional

$$-V_{OUT} = (1.25V)\left(\frac{R1}{R2}\right)$$

*Caution: Use current limiting protection circuit for high values of CF

**Figure 8.8.** Practical inverting converter (*Raytheon Linear Integrated Circuits*, 1989, p. 9-54).

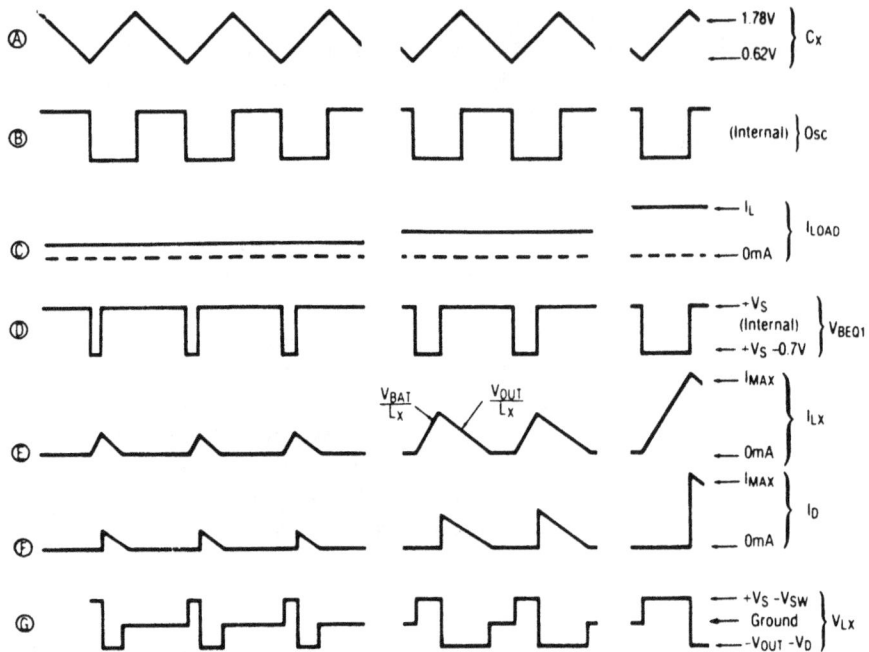

**Figure 8.9.**  Inverter waveforms (*Raytheon Linear Integrated Circuits*, 1989, p. 9-55).

The turning on and off of the switch (transistor Q1) performs the same function as opening and closing the switch (S in Fig. 8.7). Energy is stored in the inductor during the on-time and is released into output capacitor $C_F$ during the off time. The comparator continues to gate the oscillator square wave to Q1 until enough energy is stored in $C_F$ to make the comparator input voltage decrease to less than 0 V. The voltage applied to the comparator is set by the output voltage, the reference voltage, and the ratio of $R_1$ to $R_2$.

### 8.4.4    Buck or step-down

Figure 8.10 shows the theoretical buck or step-down configuration. Figure 8.11 shows a typical IC switching regulator (the Raytheon RC4391) connected in a practical step-down converter configuration. Figure 8.12 shows the corresponding waveforms found during circuit troubleshooting. The following paragraphs describe how both theoretical and practical functions are performed and how the functions relate to troubleshooting. Notice that switch S in the theoretical circuit is replaced by a transistor (within the IC) in the practical IC circuit. Capacitor C, diode D, and inductor L in Fig. 8.10 are replaced by $C_F$, D1, and $L_x$ in Fig. 8.11. Also notice that the ground lead of the IC (pin 4) is not connected to circuit ground. Instead, pin 4 is tied to the output voltage. Using this rearrangement of the feedback system, it is possible to regulate a non-negative output voltage (because of the feedback system senses voltage more negative than the ground lead).

As shown in Fig. 8.10, when switch S is closed, current flows from the battery, through the inductor L, and through the load resistance to ground. After S is opened, stored energy in L causes current to keep flowing through the load. The circuit is completed by catch diode D. Because current flows to the load during the charge and discharge, the average load current is greater than in an inverting circuit. The significance in troubleshooting is that for equal load currents, the step-down circuit requires less peak inductor current than for an inverting circuit. As a result, the inductor of a step-down circuit can be smaller than for inverting, and the switch transistor in a step-down IC is not stressed as heavily for equal load currents.

In troubleshooting the practical circuit of Fig. 8.11, remember that the output filter capacitor ($C_F$) is discharged so that the ground lead (pin 4) potential

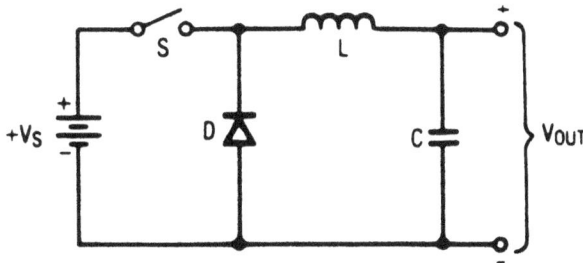

Figure 8.10.   Theoretical buck or step-down configuration (*Raytheon Linear Integrated Circuits*, 1989, p. 9-55).

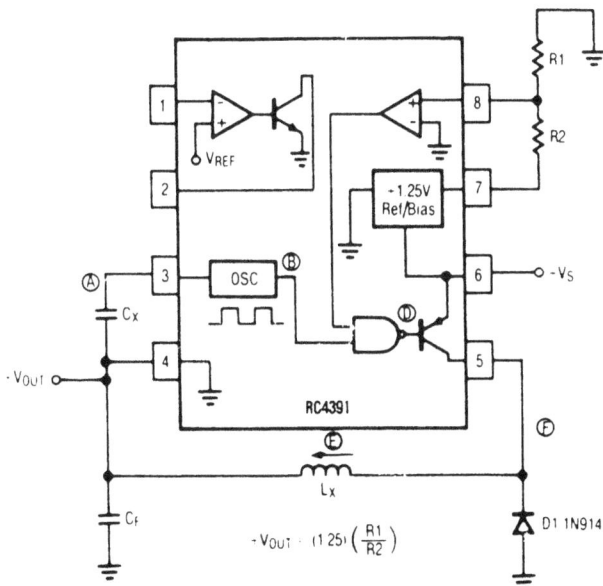

Important Note. This circuit must have a minimum load $\geq 1$ mA always connected

Figure 8.11.   Practical step-down converter configuration (*Raytheon Linear Integrated Circuits*, 1989, p. 9-56).

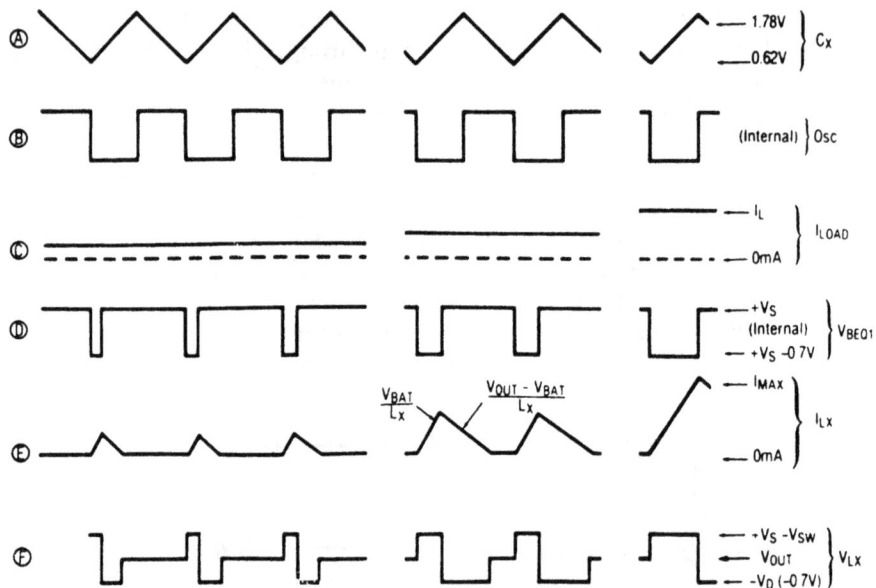

**Figure 8.12.** Step-down converter waveforms (*Raytheon Linear Integrated Circuits,* 1989, p. 9-57).

starts at 0 V. The reference voltage is forced to +1.25 V above the ground lead, pulling the feedback input (pin 8) more positive than the ground lead. This positive voltage forces the control network to start pulsing the output transistor. When the switching action pumps up the output voltage, the ground lead rises with the output until the voltage on the ground lead is equal to the feedback voltage. At that point, the control network reduces the on-time of the switch to maintain a constant output.

### 8.4.5  Flyback

Figures 8.13 and 8.14 show the theoretical flyback configuration, using both an inductor and a transformer, respectively. (Although some flyback converters use a simple inductor, a transformer is generally more common.) Figure 8.15 shows a typical IC switching regulator (the Raytheon RC 4292) connected in a practical flyback configuration. The following paragraphs describe how both theoretical and practical functions are performed, and how the functions relate to troubleshooting. Notice that switch S1 in the theoretical circuits is replaced by external transistor M1 in the practical circuit. Also, M1 is connected to ground, not to the negative-supply circuit. As a result, a simple inductor (as shown in Fig. 8.13) cannot be used in a practical circuit to supply a positive output. By replacing L1 in Fig. 8.13 with a transformer (as shown in Figs. 8.14 and 8.15), a positive output ($V_{OUT}$) can be used as an inverter (Section 8.4.3).

Flyback switching regulators are based on a two-cycle energy transfer. First, energy is stored in an inductor or transformer. Second, the energy is transferred to a load capacitor. If a transformer is used, the circuit can be operated in two input-to-output modes. When the input-current flow and the output-current flow alternate, one preceding the other, the function is a true flyback operation. If the input and output current flow occur at the same time, the function is called a *feed-forward* or simply a *forward-converter operation* (Section 8.4.6).

In the simple flyback (inductor) circuit of Fig. 8.13, when switch S1 is closed, charging current from the battery flows through inductor L1, which builds up a magnetic field that increases as S1 is held closed. When S1 is opened, the magnetic field collapses, and energy stored in the magnetic field is converted into a current that flows through L1 in the same direction as the charging current. Because this current has no path to flow through S1, the current must flow through diode D1 to charge capacitor C1.

The key to inversion in a flyback circuit is the ability of the inductor to become a source when the charging current is removed. (This is the same as with an inverting or buck-boost configuration, Section 8.3.3.) During discharge, the current in inductor L1 decreases. When the current reaches zero, diode D1 stops conducting. The rate of change (with time) of the current in an inductor is proportional to the voltage across the inductor and is inversely proportional to the inductance. Also, the load voltage and/or current can be regulated by controlling the on time of switch S1. As in the case of other regulators, the load capacitor (C1) stores the energy until used by the load.

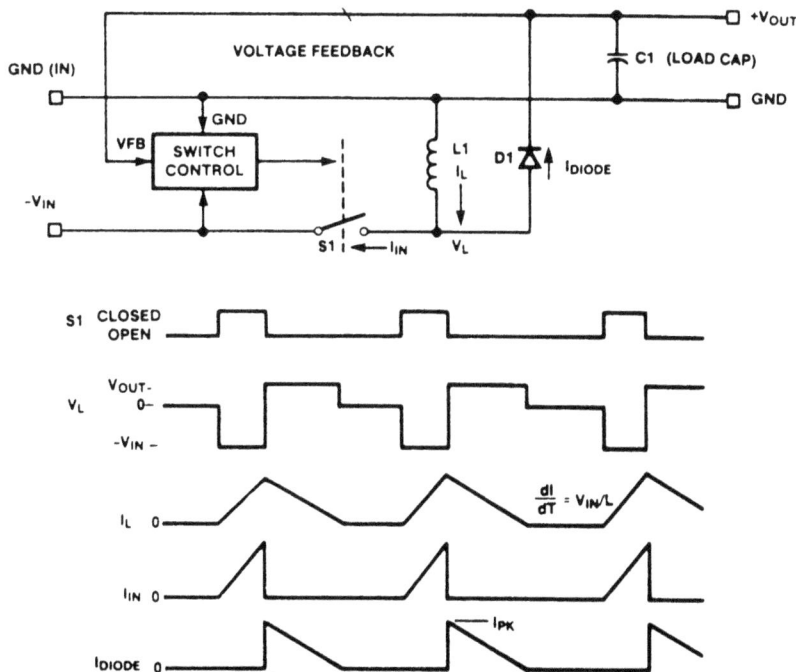

**Figure 8.13.** Theoretical flyback converter with inductor (*Raytheon Linear Integrated Circuits,* 1989, p. 9-40).

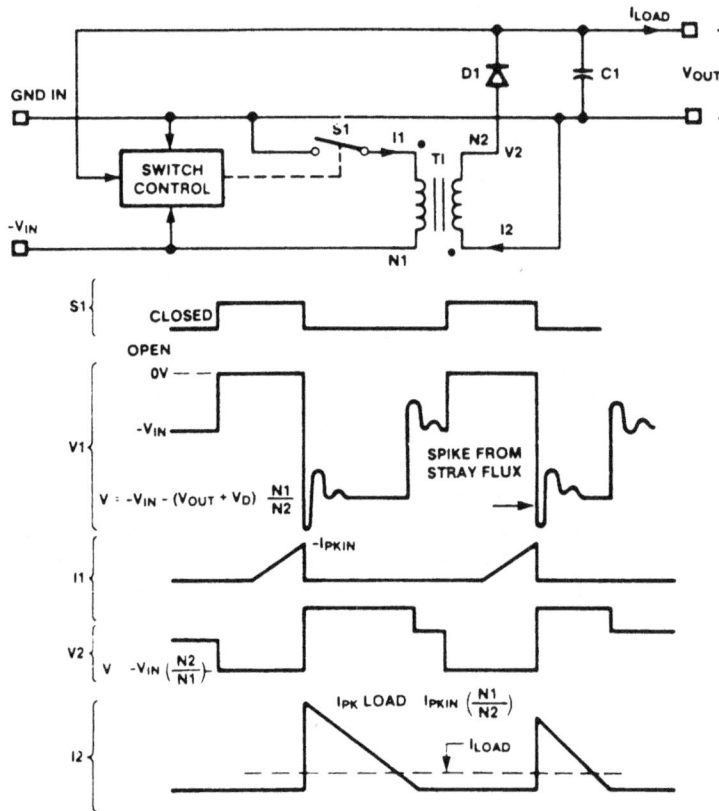

**Figure 8.14.** Theoretical flyback converter with transformer (*Raytheon Linear Integrated Circuits,* 1989, p. 9-41).

The circuit for the transformer flyback configuration (Fig. 8.14) is similar to that of the simple flyback (inductor) configuration (although the waveforms are substantially different). In effect, flyback transformer T1 stores energy with one winding and removes energy with the other winding. The first cycle starts with the closing of S1. This pulls $V_{IN}$ up to ground. Current starts from zero and ramps up to the N1 winding. This stores energy in the magnetic flux of the transformer core. After a controlled time, switch S1 opens, and energy is transferred from the core to the N2 secondary, then to the output.

The circuit for the practical IC flyback regulator in Fig. 8.15 is similar to those of other switching regulators in that the IC contains an oscillator, comparator, error amplifier, reference, and control logic. However, the circuits external to the IC are somewhat more complex (in this particular configuration). The following section briefly covers the IC and its external-component functions as they relate to troubleshooting.

The oscillator at pin 1 of the IC generates a time base for the drive pulse at pin 6. The oscillator frequency is set by an external capacitor ($C_x$) connected to pin 1. The error amplifier compares the feedback and reference signals at pins 2 and 3, and an amplified error signal proportional to the input difference. The current comparator compares the error-amplifier output to a signal that is proportional to the current in the transformer (measured by the voltage across R4).

If the feedback signal at pin 7 is greater than the error signal, the control logic (an FF and an output driver) turns the external transistor (M1) off. The control logic uses an FF to be sure that M1 receives only one pulse for each oscillator cycle. The output driver amplifies the FF output to provide a fast switching signal to M1. The voltage at pin 4 provides $-5$ V for power to the ICs, as well as a reference for the error amplifier. The shunt regulator at pin 5 acts like a zener to clamp the IC, thus preregulating the supply within safe limits.

When power is first applied, the error amplifier senses that the output voltage is lower than required, and sends an error signal through the current comparator to the control logic. In turn, the control logic pulses M1 to increase the output voltage. When the output voltage reaches the desired value, the control logic changes the M1 drive so that the transformer current is maintained at a constant level. The ratio of $R_1$ and $R_2$ determines the value of $V_{OUT}$. Typically, the equivalent resistance of this combination should be in the 25-k$\Omega$ to 100-k$\Omega$ range. This minimizes input-bias current and input-noise errors. (The manufacturer recommends an RN55 metal-film resistor for R1 and R2.)

$R_3$ sets the shunt-regulator (pin 6) current. $R_4$ sets the maximum switch current through M1. Resistor R5 holds M1 off during start-up or any time that

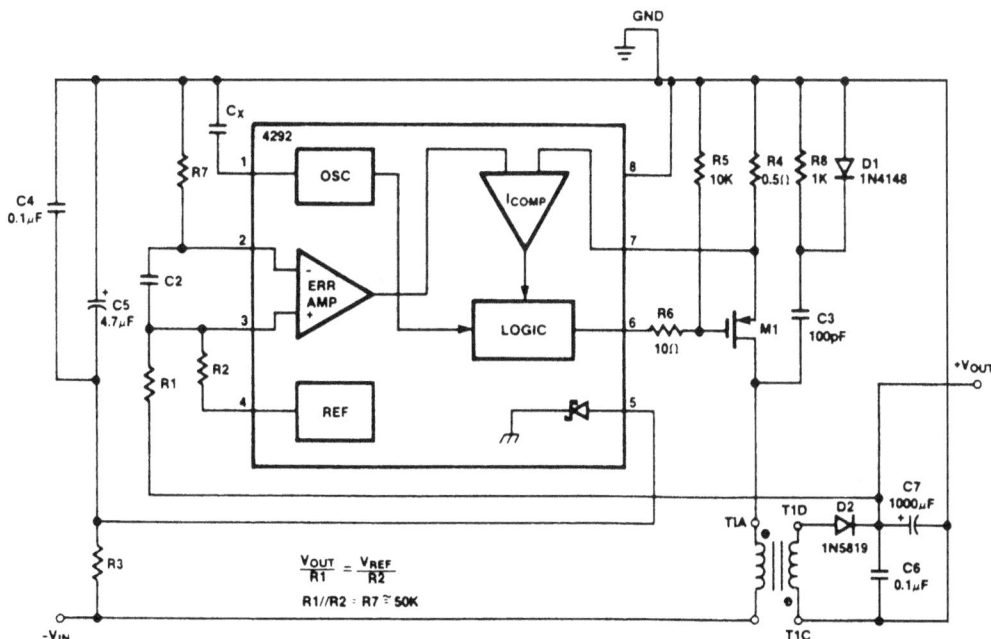

**Figure 8.15.** Practical flyback converter (*Raytheon Linear Integrated Circuits,* 1989, p. 9-43).

the IC is inactive. Resistor R6 provides for a signal loss in the gate drive to M1, thus preventing possible oscillation. Resistor R7 cancels input-bias current errors at the error amplifier inputs.

Resistor R8, capacitor C3, and diode D1 form a "scrubber" network that dampens ringing on the M1 drain and T1 primary, thus reducing voltage spikes that might potentially damage M1. The R8/C3 combination is some-times omitted, depending on the type of M1, the supply voltage, and the T1 characteristics. Capacitor C2 acts as a filter for the feedback signal. Capacitors C4 and C5 filter the shunt regulator (pin 5) voltage. If the shunt current goes too low to supply the IC properly, the output will alternately shut down and turn on at a low frequency ("motor boating"). The frequency is deter-mined by C5. Capacitor C4 must have a low impedance at high frequencies to filter out switching noise.

Capacitors C6 and C7 filter the output voltage. Diode D2 rectifies the output voltage. The basic limitation of the output load power that can be extracted from the circuit is determined by the gate-to-drain capacitance of M1. Although specifically designed to drive capacitance loads, the drive output at pin 6 will not switch large FETs, where drain current exceeds about 10 A. The maximum M1 size is also affected by the ratio of input to output voltage. The manufacturer recommends an International Rectifier IRF9633 or Motorola MTP5P18 for M1.

A disadvantage of any flyback converter is the large amount of energy that must be stored in the form of direct current in the transformer windings. This requires larger cores than would be necessary with pure alternating current in the windings. The problem can be overcome with a forward converter, cov-ered next.

### 8.4.6  Forward

A *forward converter* (Fig. 8.16) avoids the problem of large amounts of stored energy in the transformer core. However, the forward converter does so at the expense of an extra winding on the transformer, two more diodes, and an addi-tional output filter inductor.

When comparing the forward circuit to other switching regulators in trou-bleshooting, remember that power is transferred from input to load through D1 during the switch-on time. When the switch turns off (opens), D1 is reverse biased and the L1 current flows through D2. The additional winding and D3 are required to define the switch voltage during switch-off time. Without this clamp, the switch voltage would jump all the way to breakdown at the moment when the switch is opened (because of the magnetizing current flowing in the transformer primary).

The additional ("reset") winding usually has a 1:1 turns ratio to the prima-ry. This limits switch duty cycle to 50% maximum. Greater than 50%, switch current rises uncontrolled (even with no load) because the primary winding cannot maintain zero dc voltage. Reducing the number of turns on the reset winding allows higher switch duty cycles at the expense of a higher switch voltage. Output voltage ripple of forward converters tends to be low because of

L1, but input ripple current is high because of the lower duty cycles normally used. A smaller core can be used for T1 (compared to the flyback) because there is no net direct current to saturate the core.

### 8.4.7  Current-boosted boost

The circuit shown in Fig. 8.17 is an extension of the standard boost converter (Section 8.4.1). A tapped inductor is used to decrease the switch current for a given load current. This allows higher load currents at the expense of higher switch voltage. However, you must be careful to ensure that the maximum switch voltage is not exceeded.

### 8.4.8  Current-boosted buck

The current-boosted buck converter in Fig. 8.18 uses a transformer to increase output current greater than the maximum current rating of the switch (which

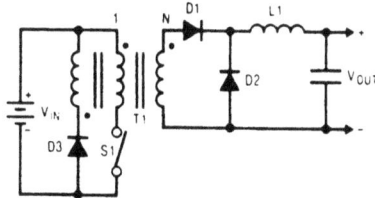

**Figure 8.16.**  Forward converter (Linear Technology, Application Note 19, p. 16).

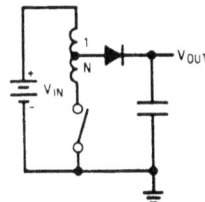

**Figure 8.17.**  Current-boosted boost converter (Linear Technology, Application Note 19, p. 16).

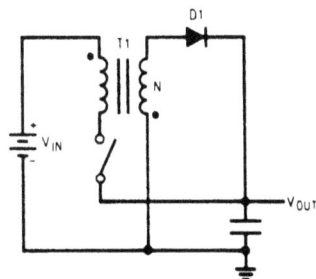

**Figure 8.18.**  Current-boosted buck converter (Linear Technology, Application Note 19, p. 17).

is a transistor in a practical circuit). The current-boosted buck circuit does so at the expense of increased switch voltage during switch-off time. The increase in maximum output current over a standard buck converter (Section 8.4.4) is equal to the input voltage divided by the output voltage, plus the turns ratio times the input-output differential.

For example, in a 15-V to 5-V current-boosted buck converter with a 1:4 turns ratio, the increase in output current is double: $15 / (5 + 1/4 \times 15 - 5) = 2$. This is a 100% increase in output current. However, the maximum switch voltage for a current-boosted buck is increased from input voltage to input voltage plus output voltage divided by the turns ratio. Using the same 15-V to 5-V converter, the maximum switch voltage is $15 + 5/$*turns ratio,* or $15 + 5/0.25 = 35$.

### 8.4.9   Cuk

The Cuk converter in Fig. 8.19 (named after Slobodan Cuk, a professor at Cal Tech) is similar to a buck-boost or inverter in that input and output polarities are reversed. However, the Cuk configuration has the advantage of low ripple current at both the input and output. The need for two separate inductors can be eliminated by winding both on the same core with an exact 1:1 turns ratio.

From a troubleshooting standpoint, either input ripple current or output ripple current (but not both) can be forced to zero. This eases the requirements on the size and quality of input and output capacitors without requiring further filtering. Notice that the ripple current in C2 is equal to the output current, so C2 must be large. However, C2 can be an electrolytic, so physical size is not normally a problem.

## 8.5   Inductors and Transformers

This section is devoted to troubleshooting problems caused by inductors (or coils) and transformers used in switching-regulator circuits. As you will soon find out, the magnetic components are the greatest source of design problems and design failure in switching supplies. Most present-day switching circuits are of the single-inductor type, so the text is limited to such circuits.

### 8.5.1   Inductor basics

Figure 8.20 shows a classic PFM IC (the Maxim MAX641), connected in the boost or step-up configuration, using a single inductor (L). Figure 8.21 shows

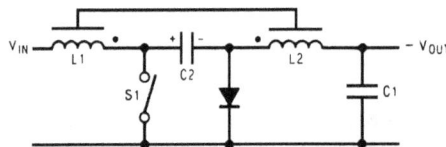

Figure 8.19.   Cuk converter (Linear Technology, Application Note 19, p. 15).

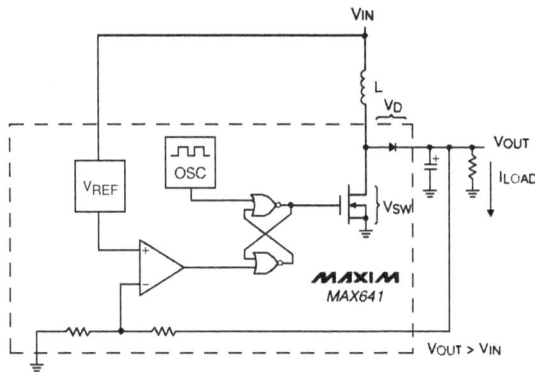

**Figure 8.20.** PFM IC connected in boost or step-up configuration (*Maxim Engineering Journal,* Vol. 4, p. 4).

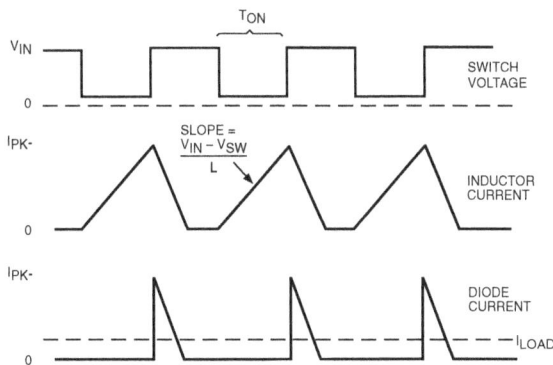

**Figure 8.21.** Boost or step-up waveforms (*Maxim Engineering Journal,* Vol. 4, p. 4).

the troubleshooting waveforms for the circuit of Fig. 8.20, including inductor current, diode current, and switch voltage.

From a troubleshooting standpoint, the inductor or coil in Fig. 8.20 must meet seven criteria:

- *Value*   The inductance value must be low enough to store adequate energy at the worst-case low input voltage, but high enough to avoid excessive current at the worst-case high on time.

- *Saturation*   The coil must show the correct value of inductance at the worst-case high peak operating current.

- *Dielectric strength*   Although this is not usually a problem, the winding insulation must be able to withstand the inductor flyback voltage.

- *Dc resistance*   The winding resistance must not cause excessive self-heating. In low-voltage, clocked-PFM circuits, the resistance should not degrade the load regulation.

- *Adequate Q*   *Q,* the ratio of coil resistance to reactance, must be such that core losses do not cause low efficiency or excessive self-heating.

■ *Electromagnetic interference*    EMI must not upset the regulator IC or other nearby circuits.

■ *Stray capacitance*    The coil's self-resonant frequency (SRF) must be 5 to 10 times greater than the switching frequency.

### 8.5.2    Correct inductance value

The correct inductance value depends on the switching frequency and the voltage applied across the inductor, both of which determine the peak inductor current in a given circuit. From a troubleshooting standpoint, wrong values can cause excessive current in the switching transistor or insufficient energy storage. In turn, insufficient storage causes slow start-up and slow transient-response time. In PFM regulators, insufficient storage can also cause poor load regulation. As a guide when troubleshooting an experimental circuit, lower the inductance to value just short of introducing problems related to excessive current (such as inefficiency, component stress, and high ripple).

As shown in Fig. 8.21, the supply voltage and inductance value determine the slope of the inductor current waveform. When operating in the discontinuous mode (current returns to zero on every cycle), the average load current is directly proportional to the peak inductor current. In turn, the peak current depends on slope and the slope depends on the inductance value. An inductor with an excessive value cannot transfer energy on each oscillator cycle.

### 8.5.3    High-inductance problems

The problem of excessive inductance value in a clocked PFM regulator often remains hidden until the design is in production. Then, random failures of the regulator include *low output current and poor load regulation.*

### 8.5.4    Low-inductance problems

Low inductance causes many odd symptoms, including low efficiency, rattling heatsinks, whining coils, and increased output ripple. Very low inductance can cause burned windings and shattered, smoking transistors and ICs. Coil inductance must never be so low that peak currents saturate the core or overstress the switching transistor. This rule applies to both PFM and PWM regulators.

### 8.5.5    Inductors for PWM regulators

Unlike PFM regulators, most PWM regulators show worst-case peak currents at a minimum supply voltage. Because PWM regulators generally operate in the continuous mode at high duty cycles (greater than 50%), the PWM inductance values are limited only by winding factors and the need for reasonable start-up times and transient response times.

In the continuous mode, the inductor current fluctuates, but never returns to zero. Because the current might increase in staircase fashion over a period of several cycles, the rate of increase (determined by the inductance value) does not limit the maximum inductor-current level or average load level. To sum up, the exact inductance value is not as crucial for PWM as for PFM. The minimum value is set primarily by the inductor's power losses (so-called "$I^2R$ losses") and the switch-transistor current capability. Because of these factors, PWM is better suited to high-power applications than PFM. As a guideline, use PFM where the power is less than 10 W and use PWM for all applications greater than 10 W.

## 8.5.6  Saturation effects

High-current spikes caused by saturation can lower efficiency, increase noise, and endanger power transistors in switching supplies. When the core of an inductor saturates, the apparent inductance values fall off and current begins to rise exponentially. Power losses cause drop in circuit efficiency and the inductor stores no additional energy. For these reasons, the worst-case peak current of any switching supply must be less than the peak-current or incremental-current rating of the inductor. When selecting or replacing an inductor for an experimental circuit, and the inductor literature does not have a "dc-current" rating, or shows only an "ac amps" rating, beware. Such inductors are often prone to saturation!

From a troubleshooting standpoint, the effects or presence of saturation can best be observed with a scope and current probe (such as a Tektronix AM503 clamp-on type). Figure 8.22 shows waveforms produced by inductor current in a switching supply using a MAX743 regulator IC. A nonlinear increase in inductor current occurs near the peaks when the inductor is saturated (or approaches saturation).

If a current probe is not available, a less direct, but still effective, method of measuring inductor current is to measure the voltage across a small *sense resistor* in series with the inductor. The resistor value should be 1 Ω (or less if the inductor current is more than a few hundred mA) and the resistor must be capable of dissipating the related power. Typically, a 10-W resistor will do the job for most PFW regulators. The waveforms produced by measuring the voltage across a series resistor will not be identical to the current waveforms in Fig. 8.22; the voltage reading must be converted to current, using $I=E/R$. (A 1-Ω or 0.1-Ω resistance is most convenient.)

## 8.5.7  Air gaps and core materials

All other factors being equal, the inductor core material determines both saturation and the amount of energy that can be stored. One way to increase the energy is to provide the core with an air gap. This extends the effective length of the core's magnetic path and amplifies the inductor's ability to store energy. The air gaps can be built in or created by grinding/machining operations. Although air gaps can be a benefit, they can also cause problems. The following section summarizes the effect of air gaps in various core materials from a troubleshooting standpoint.

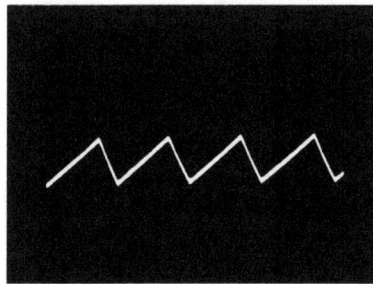

a   Normal Operation
    Linear charge and discharge slopes

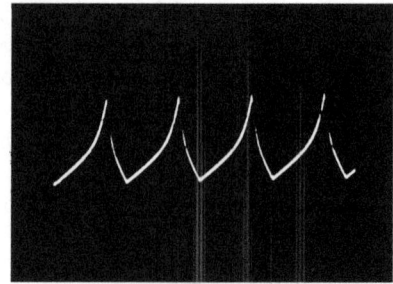

b   Saturation
    Non-linear increase in inductor
    current near peaks

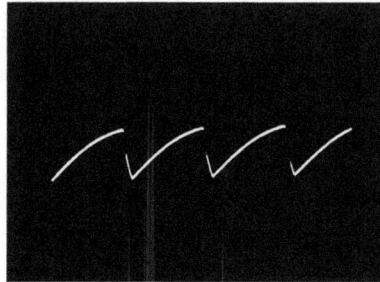

c   Excessive Resistance
    1.  High Winding Resistance
    2.  High Transistor $R_{ON}$
    3.  High Source Resistance

**Figure 8.22.**   Inductor current slope patterns in a switching supply (*Maxim Engineering Journal,* Vol. 4, p. 5).

Ferrite and other high-permeability core materials are very susceptible to saturation and must be treated accordingly. For example, the closed magnetic path of a *ferrite toroid* is good for containing EMI, but the short magnetic path, combined with the high permeability of a ferrite, makes toroids prone to saturation. Ferrite toroids make good transformer cores, but are less suitable for simple boost regulators (with large dc-offset currents).

When the power is about 5 W or less, the *powder-type cores* with distributed air gaps are often a better choice than ferrite toroids or pot cores. After cutting an air-gap slot in the core of a ferrite toroid, the machining costs and EMI level will be so high that a *ferrite bobbin* (with the large inherent air gap) is a better choice. The cylindrical core, although noisy, is best for low-power ferrite applications because the shape is easy to wind and thus inexpensive to manufacture.

When the power is above 5 W, *ferrite pot cores* are good because of low EMI emissions—especially where high-level currents and magnetic fields are involved. Pot cores, although self-shielding, are not cost-effective in applications that use less than 5 W because of the additional manufacturing steps.

For many low-power applications, *iron-powder* and *molypermalloy* (*MPP*) toroid tubes offer the best combination of cost, size, and EMI level. These materials have built-in air gaps that allow the core to saturate gradually as the magnetizing force increases. This is because of many tiny air gaps created by the binder material. Each gap saturates at a slightly different level of magnetizing force.

## 8.5.8  Core material trade-offs

Figure 8.23 summarizes the trade-offs for various inductor types used in low-power switching supplies, from the standpoint of troubleshooting experimental circuits.

| TYPE | EMI | COMMENT |
|---|---|---|
| Ferrite bobbin | High | Makes compact, low cost, axial-lead (cylindrical) inductors. Low core losses support high efficiency. |
| Ferrite bobbin with ferrite shield | Low | Efficient but prone to saturation |
| Ferrite pot core | Low | Efficient. Easily gapped to the correct value. Best for high-current or high-frequency applications. |
| Molded (low cost) | High | OK for light loads. Prone to saturation and often inefficient. Observe current ratings carefully. |
| Silicon steel toroid | Low | Tape wound; similar to iron powder. Use thinner tape for higher frequencies. |
| Ferrite toroid | Low | Prone to saturation |
| Molypermalloy powder (MPP) toroid | Low | Best available for frequencies less than 400kHz. Low EMI, low losses, compact, and expensive. Use high-flux type. |
| Iron-powder toroid | Low | Specify core material carefully to achieve low losses. Best is "Kool-mu" (Magnetics, Inc.). |

**Figure 8.23.**  Trade-offs for various inductor types (*Maxim Engineering Journal*, Vol. 4, p. 7).

Ferrites are attractive because of the combined low cost and high resistivity, which minimizes eddy-current losses. (Ferrites are the only choice for switching frequencies of 500 kHz and higher.) On the other hand, the high permeability of ferrites usually requires an air gap with all of the associated complications (high EMI for bobbin types and extra assembly steps for pot-core types).

Powder MPP combines good saturation characteristics with low hysteresis losses. However, MPP is expensive (it contains nickel) and requires many processing steps. Iron powder and silicon-steel tape, in spite of a tendency to sustain eddy-current and hysteresis losses, are inexpensive and well suited to general-purpose applications.

### 8.5.9   High-flux MPP cores

High-flux MPP cores combine good EMI performance with small size. Ordinary MPP cores for radio-frequency (RF) applications contain 80% nickel, plus iron and molybdenum. High-flux MPP contains 50% nickel, which is not good for RF but is good for switching supplies. Although high-flux MPP is expensive, it is sometimes more cost-effective than ferrite because of the precision air-gap requirements.

### 8.5.10   Dc winding resistance and $I^2R$ losses

The core material affects the power level available for a given inductor size. However, dc winding resistance can also limit available output current and circuit efficiency. As shown in Fig. 8.22, a high dc winding resistance alters the inductor-current waveform (causes a slope, rather than the desired linear rise). The resulting $I^2R$ losses affect the overall efficiency and cause the core temperature to rise. Dc winding resistance is significant for battery-powered and low-voltage applications of about 3 V or less, in which the inductance values must be low to get an acceptable slope for the inductor charging current.

### 8.5.11   Temperature rise

Inductor specifications often include two current ratings: *continuous or RMS* and *dc saturation* (sometimes called *peak* or *incremental current*). The continuous rating accounts for the temperature rise caused by winding resistance, as well as the operating-temperature range and properties of insulation or potting material. The continuous rating is usually higher than the dc-saturation rating (but not always, especially for high-value inductors).

When troubleshooting experimental circuits, be sure that the average current of the inductor is less than the continuous rating. If high frequencies are involved, include a safety margin (at least 10%) for additional temperature rise because of core losses.

### 8.5.12   High-frequency losses and inductor Q

High-frequency losses in a switching-supply inductor include three major components: those from hysteresis, those from eddy currents in the core, and those from eddy currents in the wire. (Residual loss, a fourth component, is usually insignificant.)

*Magnetic hysteresis,* which occurs as the flux density nears the saturation point, becomes a problem in iron-powder cores at switching frequencies of 100 kHz or less. One cure is to reduce the peak flux density at high currents by enlarging the core. However, larger cores make the eddy-current problem worse by providing more low-resistance paths for the eddy currents. *Core eddy current* is a function of frequency squared; it quickly becomes excessive when the frequency approaches 300 to 400 kHz. When troubleshooting experimental circuits, use another core material, rather than change the core size to offset the core eddy-current problem.

The table in Fig. 8.24 shows the frequency limits for some common core materials. High switching frequencies (about 100 kHz and higher) applied to iron-powder and steel-tape cores can cause significant eddy currents that contribute to a rise of core temperature and efficiency loss. Because the regulator might appear to be operating properly, the eddy-current problem can be difficult to detect, except as an unexplained efficiency loss or core-heating effect.

*Winding eddy currents* (circulating currents within the wire) can also be a problem at frequencies of 500 kHz and higher. The practical solution is to use a minimum wire thickness. Litz wire (ultra-thin multi-stranded wire) or windings made from PC traces are common approaches to the winding eddy-current problem. It also helps to position the windings within the core as far as possible from the air gap.

One practical troubleshooting approach to the inductor-loss problem is to measure the inductor $Q$ with a sine-wave inductance bridge (Q meter), at the switching frequency. As a guideline, if the $Q$ value is 25 (or greater), the efficiency loss produced by the inductor is 5% (or less). If you are not familiar with inductor $Q$ measurements, read the author's *McGraw-Hill Electronic Testing Handbook,* 1994.

## 8.5.13  EMI problems

The solutions to EMI problems (interference to adjacent circuits or to nearby equipment) usually depend on application trade-offs. For example, if EMI must be kept to a minimum, the inductors are shielded. However, shielded inductors tend to be larger, more expensive, and more difficult to mount than unshielded types. If moderate amounts of EMI can be tolerated, unshielded

| to 100kHz | Standard iron powders and steel tape |
|-----------|--------------------------------------|
| to 200kHz | Low-permeability, high-frequency iron powders |
| to 400kHz | High-flux MPP |
| to 500kHz | Standard MPP |
| to 1MHz   | Manganese-zinc ferrite |
| to 10MHz  | Nickel-zinc ferrite |

Figure 8.24.  Frequency limits for some common core materials (*Maxim Engineering Journal,* Vol. 4, p. 8).

bobbin-type inductors are used. Usually, such cores are about half the price and size of an electrically equivalent pot core or toroid. When troubleshooting an experimental circuit with bobbin inductors, mount the inductors at 90° angles to other magnetic components and point them away from sensitive circuits. This is necessary because bobbin inductors generate the highest magnetic fields near the ends along the axis.

EMI-producing *fringe fields* are set at the cut, whenever material must be cut to provide an air gap (such as with ferrite cores). Fringe fields also occur in cores with large inherent gaps, such as between the ends of a bobbin inductor wound on a cylindrical core. Pot cores or similar designs allow air gaps in the ferrite material, but keep the EMI from radiating. The gap spacings in powdered-material cores are so small that EMI is seldom a problem, provided that the core is a toroid or similar design with a closed magnetic path.

### 8.5.14  Self-resonant frequency (SRF)

All inductors have some *distributed capacitance* that combines with the inductance to form a resonant circuit. The frequency of this self-resonance should be between 5 and 10 times the switching frequency (but never at an exact multiple of the switching frequency!). Because the inductance value is set by circuit requirements, the SRF is determined by the distributed capacitance (a higher capacitance produces a lower SRF).

When the SRF is low, the normal linear ramp of the inductor current (Fig. 8.22) is preceded by a sudden jump of current when the switching transistor turns on. This results in so-called switching losses, which lower the overall efficiency. When troubleshooting experimental circuits, keep distributed capacitance at a minimum so that the SRF will be high and will not seriously affect the inductor current. Distributed capacitance can be lowered when the toroid is wound, either by overlapping the ends of the winding somewhat or by leaving a gap between the winding ends (rather than ending the winding as one full layer).

### 8.5.15  Inductor kits

When troubleshooting experimental circuits, it is often convenient to try various inductors of different values and peak-current ratings. This usually means going through inductors that you might have in stock, or through catalogs and data sheets. Inductor kits offer a shortcut to this process. A typical example is the Model 845 Inductor Selection Kit, available from Pulse Engineering, Inc., P.O. Box 12235, San Diego, CA 92112, telephone (619) 268-2400. This kit contains 18 inductors of various ratings and sizes. These can be used in the experimental circuit and the results can be compared with the desired performance.

## 8.6  Switching-Supply Tests

The test procedures described for linear supplies in Chapter 7 can generally be applied to switching supplies. One major exception is the ripple test. Because

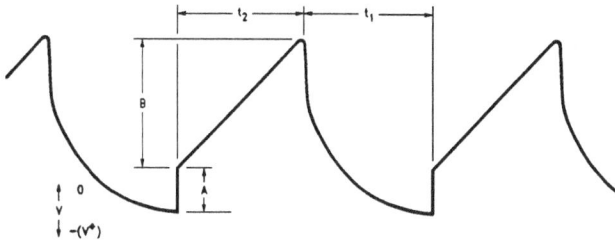

**Figure 8.25.**   Theoretical ripple waveforms (Harris Semiconductor, *Linear & Telecom ICs,* 1991, p. 2-101).

switching supplies have a battery input, any ripple voltage is the result of switching action, rather than the ac-line input.

### 8.6.1   Measuring ripple in switching supplies

Figure 8.25 shows the theoretical ripple waveform for a typical voltage converter (the Harris Semiconductor ICL7662). As in the case of linear supplies, the scope must be set to the ac mode when measuring ripple in switching supplies. This blocks the dc output of the supply and prevents the dc voltage from driving the display off the scope screen. When measuring the amplitude of the ripple on the scope's voltage-calibrated vertical scale, both the A and B components shown in Fig. 8.25 are included.

When measuring the frequency of the ripple on the scope's horizontal scale, use *frequency* = 1/*period* of a complete cycle. For example, $t_1$ and $t_2$ in Fig. 8.25 represent one complete cycle. If the total period of $t_1$ and $t_2$ is 40 μs, then the frequency is: $1/(40^{-6}) = 25$ kHz.

## 8.7   Switching-Supply Troubleshooting

The following notes describe some of the most common troubleshooting problems for switching-supply circuits—especially experimental circuits. In general, most switching-supply problems are the result of wiring mistakes (which you never make), defective (or inadequate) components, and possibly test errors, all of which can be located by basic voltage checks, resistance checks, and point-to-point wiring checks. However, switching supplies and switching regulator/converters present particular problems.

### 8.7.1   Test-measurement techniques

The following notes apply specifically when connecting test instruments to an experimental switching-supply circuit.

*Ground loops*   Figure 8.26 shows a typical ground-loop condition. A generator is driving a 5-V signal into 50 Ω on the experimental circuit, which results in a 100-mA current. The return path for this current divides between the ground from the generator (typically, the shield on a BNC

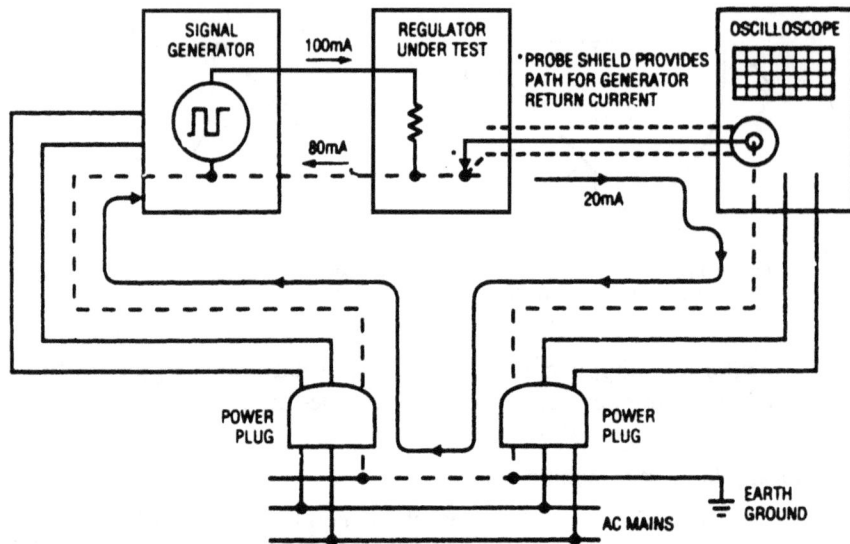

**Figure 8.26.** Ground-loop errors (Linear Technology, Application Note 44, p. 44).

cable), the secondary ground "loop" that is created by the scope-probe ground clip (shield), and the two "third-wire" connections on the generators and the scope.

In this example, assume that 20 mA flows in the parasitic ground loop. If the scope ground lead has a resistance of 0.2 $\Omega$, the scope will show a 4-mV "bogus" signal. The problem gets much worse for higher currents and for fast-signal edges, where the inductance of the scope-probe shield is important.

In practical troubleshooting, use an isolation transformer for the scope. For a quick check, touch the scope probe tip to the probe ground clip, with the clip connected to the experimental-circuit ground. The scope should show a flat line. Any signal displayed on the scope is a ground-loop problem or pickup problem.

*Scope probe compensation*   Always check that the scope probe is properly compensated when troubleshooting switching supplies. It is especially important for the ac attenuation (on a 10× probe, for example) to match the dc attenuation exactly. If not, low-frequency signals are distorted and high-frequency signals have the wrong amplitude. At typical switching frequencies, the waveshape might appear to be good because the probe acts purely capacitive, so the wrong amplitude might not be immediately obvious.

*Ground-clip pickup*   Do not make any test measurements on a switching regulator with a standard (alligator) ground-clip lead. Replace the alligator clip with a special soldered-in probe terminator, which can be obtained from many probe manufacturers. The standard alligator ground-clip lead can act as an "antenna" and can pick up magnetic and other radiated signals. Make the test described for ground loops if you suspect that the scope probe is receiving other signals.

*Measure at the component*    Make all measurements (output voltage, ripple, etc.) at the component, not at a wire that is connected to a component, because wires are not shorts. For example, switching regulators that deliver square waves to an output capacitor can generate about 2-V-per-inch "spikes" in the lead inductance of the capacitor. The further you measure from the capacitor, the greater the spike voltage.

### 8.7.2  EMI suppression

EMI is present with all switching regulator circuits. The EMI can be *conducted,* which travels along input and output wiring, or *radiated,* which takes the form of electric and magnetic fields. Although these fields do not usually cause regulator problems, they can create problems for surrounding circuits. The following guidelines are helpful when troubleshooting EMI problems.

*Avoid long high-current grounds and feedback nodes*    Figure 8.27 shows the correct and incorrect ways to make ground and feedback connections to switching regulators. Even though low-power switching-regulator ICs are generally easy to use, pay some attention to PC layout and routing—especially at power levels greater than 1 W or when high-speed PWM ICs are used.

Trace out the high-current paths and minimize their length—especially at the ground trace. Use a *star ground* in which all grounds are brought to one point. Place input filter capacitors physically close to the IC. Minimize stray capacitance at the feedback (FB) pin. Return all compensation capacitors and bypass capacitors to quiet, well-filtered points (such as an analog ground pin).

*Use inductors with good EMI characteristics, such as toroids*    Avoid using rod inductors. If you must use rod inductors, keep them in the output filter, where ripple current is low (one hopes). Figure 8.22 shows some typical current waveforms that are produced by good and bad inductors.

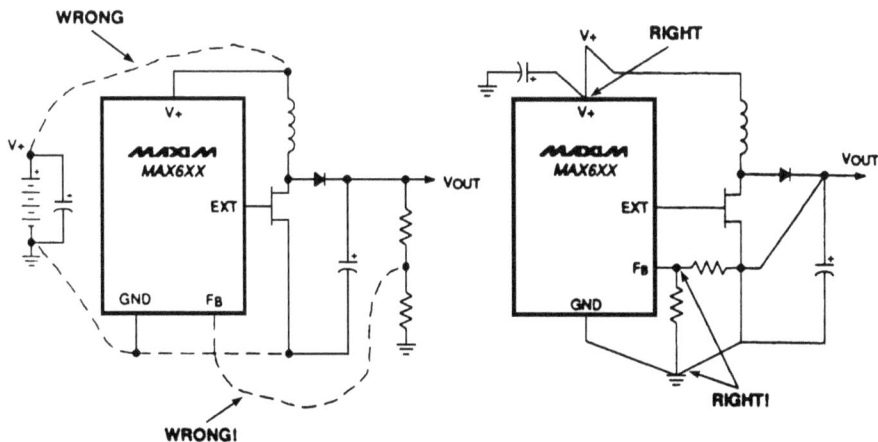

**Figure 8.27.**   Correct and incorrect ground and feedback connections for switching regulators (*Maxim Applications and Product Highlights,* 1992, p. 4-13).

When troubleshooting experimental circuits, most inductor problems (other than using the incorrect inductance) can be traced to inadequate saturation (peak current) ratings or excessive dc resistance. If an inductor saturates, the current rises exponentially with time (rapidly at the peak). With excessive resistance, a distinct LR characteristic (inductance-resistance slope, Fig. 8.22c) can be seen in the waveform. If the waveform takes small (but strange) bends, the inductor might be producing both effects.

*Route all traces carrying high ripple current over a ground plane*   This will minimize radiated fields. Be sure to include diode leads, input and output capacitor leads, snubber leads, inductor leads, IC input and switch-pin leads, and input-power leads. Keep these leads short, and keep the components close to the ground plane.

*Keep sensitive low-level circuits as far away as possible*   If this is not practical, use field-canceling tricks, such as twisted-pair differential lines.

*In very crucial applications, add a "spike killer" bead*   Such beads are sometimes added to the diode to suppress high harmonics. Unfortunately, this can create higher transient switch voltages at switch turn-off, so check the switch waveforms carefully.

*Add an input filter choke*   Such chokes are generally used only when radiation from input lines is a problem. A low-value (a few microhenries) choke in the input will allow the regulator input capacitor to swallow most of the ripple current that is created at the regulator input.

## 8.8   Troubleshooting Summary

The following notes apply specifically when troubleshooting an inoperative or poorly performing switching supply or switching regulator/converter, particularly one that is in experimental form. If none of the following conditions apply to a particular switching circuit, refer to the table in Fig. 8.28 that summarizes some classic IC switching-regulator problems.

### 8.8.1   Circuit totally inoperative

Look for such things as transformers wired backward (always check the polarity dots on transformers), electrolytic capacitors wired backward (usually, you will find this out shortly after power is applied), and IC pins reversed (compare the data sheet to the wiring that appears on the schematic).

### 8.8.2   Circuit works but results are poor

If there is no smoke, fire, or explosion, but the results are poor (low efficiency, low output voltage or current, line or load regulation out of tolerance, etc.), one of the first troubleshooting steps is to display the inductor-current waveform on a scope (as covered in Section 8.5.6). Then compare the actual scope waveforms with those shown in Fig. 8.22.

| Symptom | Possible Problems |
|---|---|
| Draws excessive supply current on start-up. | Battery not "stiff" — inadequate supply bypass capacitor. Inductance value too low. Operating frequency too low. |
| Output voltage is low. | Inductance value too high for $F_O$ or core saturating. |
| Inductor "sings" with audible hum. | Not potted well or bolted loosely. |
| Lx pin appears noisy — scope will not synchronize. | Normal operating condition. |
| 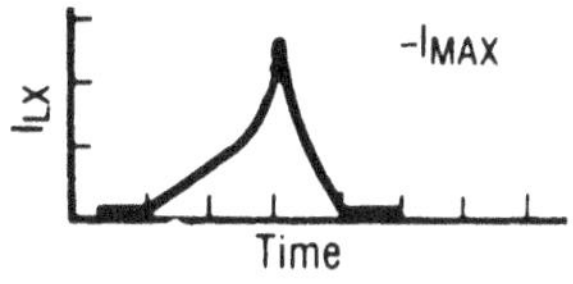  Inductor current shows nonlinear waveform. | Inductor is saturating: 1. Core too small. 2. Core too hot. 3. Operating frequency too low. |
| 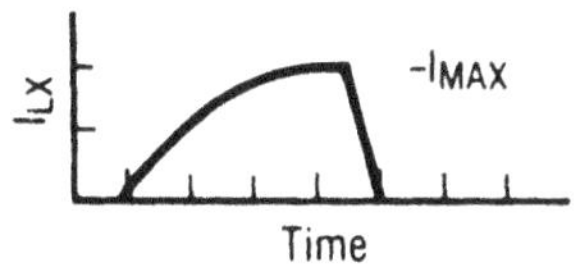  Inductor current shows nonlinear waveform. | Waveform has resistive component: 1. Wire size too small. 2. Power transistor lacks base drive. 3. Components not rated high enough. 4. Battery has high series resistance. |
| 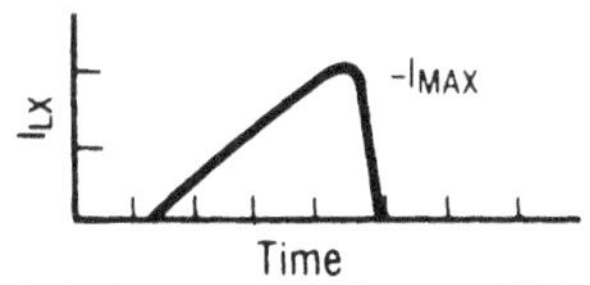  Inductor current is linear until high current is reached. | External transistor lacks base drive or beta is too low. |
| Poor efficiency. | Core saturating. Diode or transistor: 1. Not fast enough. 2. Not rated for current level (high SAT). High series resistance. Operating frequency too high. |
| Motorboating (erratic current pulses). | Loop stability problem — needs feedback capacitor from $V_{OUT}$ to pin 7 (100 to 1000pF). |

**Figure 8.28.** Classic IC switching-regulator problems (*Raytheon Linear Integrated Circuits,* 1989, p. 9-20).

Both the charge and discharge slopes of the inductor-current waveforms should be linear (Fig. 8.22*a*). If so, but the circuit does not perform as desired, try changing the inductor value, oscillator frequency, output capacitance value, and diode, in that order. If any (or all) of these changes produce the desired results, you have solved the problem. However, be certain that changes do not produce a nonlinear inductor-current waveform or some other problem.

For example, if the inductor value is changed to modify output voltage (for example, the inductance is decreased to increase output), the inductor might saturate and produce a nonlinear current near the peaks (Fig. 8.22*b*). On the other hand, if the inductor resistance value is changed (resistance increased), the waveform might be nonlinear during the entire charge time (Fig. 8.22c).

Instead of changing the inductor, it might be simpler to change the oscillator frequency. However, if the oscillator frequency is too high, efficiency will be poor. A low oscillator frequency can result in excessive supply current being drawn during the start up.

### 8.8.3  Input voltage appears to dip

If the input voltage dips, it is possible that the input leads (from the battery to the switching IC, for example) are too long when connected in experimental form. (The problem might clear up when the circuit is in PC form.) Switching regulators draw current in pulses. Long input wires can cause dips in the input voltage—especially at the switching frequency. (That is, the input-voltage pulses or dips occur at the same frequency as the switching oscillator.) In practical troubleshooting, even though the schematic does not show an input capacitor, add a 100-$\mu$F (or larger) input capacitor close to the regulator during the experimental stage. If this cures the problem, it might be necessary to add some input capacitance when the circuit is in final form.

### 8.8.4  Irregular start up

If a nonbattery input supply does not start switching or if the start up is irregular, it is possible that the input supply cannot deliver the necessary start-up current. Switching regulators have negative input resistance at start up and draw high current. This can latch some input supplies into a low or off condition. In the case of a battery-input supply, it is possible that the batteries are not "stiff" (not capable of delivering a large momentary current at the start up).

### 8.8.5  Low efficiency

If much more power goes into the supply than is coming out, suspect a problem with the inductors. Core or copper (winding) losses might be the problem. Of course, the problem could be an accumulation of all losses (inductor, capacitor, diode, etc.), which results in an inefficient supply/regulator circuit.

### 8.8.6  Faulty inductor

If you suspect that the inductor is faulty, try operating the inductor in a standard test circuit, such as shown in Fig. 8.29. Compare the waveforms obtained with the inductor in the test circuit against the waveforms shown in Figs. 8.30 through 8.33. In each case, trace A is the voltage at the $V_{\text{SWITCH}}$ pin of the LT1074, and trace B is the current (measured with a clip probe at the $V_{\text{SWITCH}}$ pin). Inductor current flows when the $V_{\text{SWITCH}}$ pin voltage is low.

With a high inductance, as shown in Fig. 8.30, current rises slowly with no saturation. When inductance decreases, the current rise is steeper (Figs. 8.31 and 8.32), but still no saturation occurs. However, the highest inductance

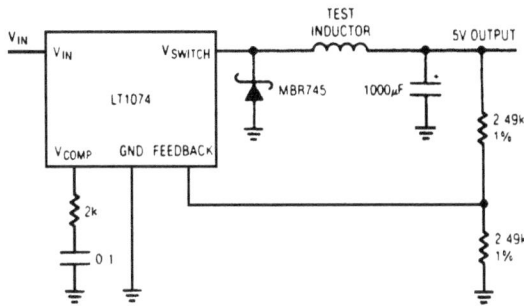

**Figure 8.29.**   Basic LT1074 test circuit (Linear Technology, Application Note 35, p. 21).

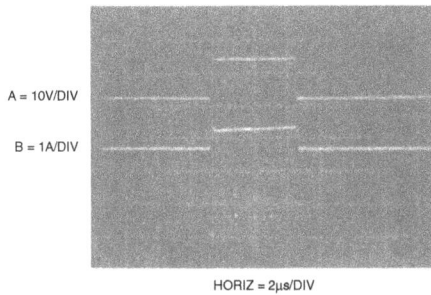

A = 10V/DIV

B = 1A/DIV

HORIZ = 2µs/DIV

**Figure 8.30.**   Waveforms for 450-µH high-capacity core (Linear Technology, Application Note 35, p. 21).

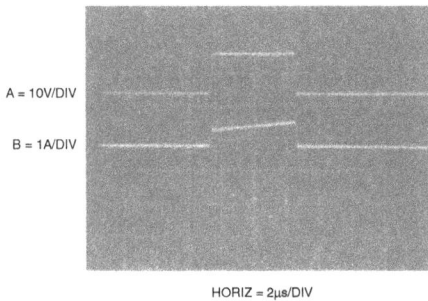

A = 10V/DIV

B = 1A/DIV

HORIZ = 2µs/DIV

**Figure 8.31.**   Waveforms for 170-µH high-capacity core (Linear Technology, Application Note 35, p. 21).

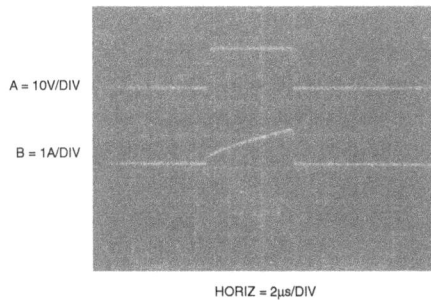

A = 10V/DIV

B = 1A/DIV

HORIZ = 2µs/DIV

**Figure 8.32.**   Waveforms for 55-µH high-capacity core (Linear Technology, Application Note 35, p. 21).

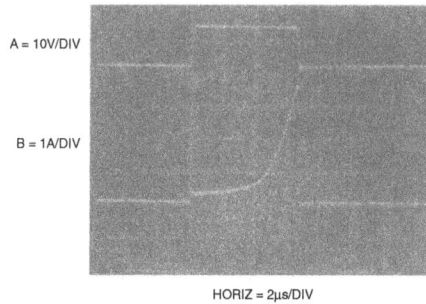

A = 10V/DIV

B = 1A/DIV

HORIZ = 2μs/DIV

**Figure 8.33.** Waveforms for 500-μH low-capacity core (Linear Technology, Application Note 35, p. 22).

(Fig. 8.33), but wound on a low-capacity core, produces extreme saturation and is totally unsuitable for switching-regulator use. Using the classic Linear Technology test circuit and procedure, you can narrow the selection of an inductor down to the best unit for your particular circuit (with regard to cost, size, etc.). The inductor kit described in Section 8.5.15 will also help in the inductor selection process.

### 8.8.7   Switch timing varies

If you monitor the circuit waveforms as shown in Figs. 8.30 through 8.33, and find that the switch-on time varies from cycle to cycle, the problem might be one of excessive ripple. Check for excessive ripple at the output, and at any feedback or compensations pins of the IC (such as at the $V_{COMP}$ and FEEDBACK pins of the LT1074 in Fig. 8.29). In practical troubleshooting, try connecting a capacitor (about 1000 to 3000 pF) across the output capacitor (and/or at the feedback/compensation pins) to ground. If any of these capacitors eliminate the variation in switch timing, you have located the problem area.

### 8.8.8   High output ripple or noise spikes

The output capacitor is the most likely suspect when you have high output ripple or noise spikes. Capacitors have a capacitance value (in μF or pF) and an equivalent series resistance (ESR). In the case of ripple, an increase in capacitance value will decrease ripple voltage. However, an increase in ESR will increase ripple. This can be understood by reference to Fig. 8.25.

The total ripple is determined by two voltages (A and B). Segment A is the voltage drop across the ESR of the output capacitor at the instant that the capacitor goes from being charged to being discharged through the load. Segment B is the voltage drop across the output capacitor during time $t_2$ (the half of the cycle when the capacitor supplies current to the load). The peak-to-peak ripple voltage is the sum of these voltage drops and can be expressed by:

$$Ripple\ voltage = \left( \frac{1}{2 \times frequency \times capacitance} + 2ESR \right) I_{OUTPUT}$$

where *frequency* = Switching frequency
   *capacitance* = Output capacitance
   *ESR*  = Equivalent series resistance of the output capacitor
   $I_{\text{OUTPUT}}$  = Output current

From a troubleshooting standpoint, an increase in switch frequency or output capacitance will decrease ripple voltage. An increase in capacitor ESR or output current will increase ripple. An output capacitor with low ESR will reduce ripple, all other factors being equal.

### 8.8.9    IC blows up

If the IC blows up (with the right components, all properly connected, no shorted outputs, no reversed electrolytics or battery polarities), it is possible that start-up surges are causing momentary large switch voltages.

### 8.8.10    IC runs hot

If the IC runs hot, it is possible that you need a heatsink. For example, a TO-220 package has a thermal resistance of about 50 to 55°C/W with no heatsink. A 5-V, 3-A output (15 W) with 10% switch loss will dissipate over 1.5 W in the IC. This means a 75°C temperature rise, or 100°C case temperature at 25°C room temperature (which is hot!). Simply soldering the TO-220 tab to an enlarged copper pad on the PC board will reduce thermal resistance to about 25°C/W.

### 8.8.11    Poor load or line regulation

If you have poor load or line regulation, check in the following order: (1) output capacitor with high ESR (especially if the capacitor is outside the feedback loop); (2) a ground-loop error in the scope (see Fig. 8.26); (3) improper connections of the output divider resistors to current-carrying lines (Fig. 8.28); (4) excess output ripple; (5) too high switching frequency. (It is assumed that you have checked the inductor for saturation!)

### 8.8.12    Low efficiency

If the circuit is working, but efficiency is low, it is possible that the diodes could be at fault. If the diode is not fast enough or is not rated for the current level, efficiency will drop—even though the circuit might work (and show good operating waveforms). If the diode is hopelessly underrated, it will be destroyed.

Most diodes used in switching regulators are Schottky or ultra-fast diodes. The major advantage of Schottky versus ultra-fast diodes is efficiency. As a guideline when troubleshooting experimental circuits, if the output voltage is below 12 V, a Schottky diode will provide about 5% improvement in efficiency over an ultra-fast diode. Above 12-V output, the advantage is less significant. Of course, the breakdown rating of any diode must exceed the input voltage.

# Charge-Pump Circuit Troubleshooting

This chapter is devoted to troubleshooting for charge-pump circuits. It starts with a review of charge-pump circuits used as voltage converters. (Such circuits are similar to the voltage-doubling circuits described in Section 7.1.5.) It concentrates on voltage-converter ICs that use the charge-pump principle, and describes circuit operation from a troubleshooting standpoint. In addition to the basic operating principles and theoretical power-efficiency considerations, it covers a number of practical circuits using the ICs, concentrating on how circuit elements relate to troubleshooting.

## 9.1 Charge-Pump Voltage-Converter Basics

Figures 9.1, 9.2, and 9.3 show the functional diagram, pin configurations, and electrical characteristics for a Harris Semiconductor ICL7660S super voltage converter. The IC operates on the charge-pump principle, and performs supply-voltage conversion from positive to negative for an input range of 1.5 to 12 V, resulting in complementary output voltages of $-1.5$ V to $-12$ V. (A CMOS voltage converter, ICL7662, provides a similar function for input voltages from 4.5 to 20 V.)

Only two noncrucial external capacitors are needed for charge-pump and charge-reservoir functions. Although not a true switching dc-dc converter, the ICL7660S can be connected to function as a voltage doubler (as well as an inverter) and can generate up to 22.8 V with a 12-V input. The IC can also be used as a voltage amplifier or voltage doubler.

As shown in Fig. 9.1, the chip contains a series dc supply regulator, RC oscillator, voltage-level translator, and four output-power MOS switches. The oscillator, when unloaded, oscillates at a nominal frequency of 10 kHz, for an input supply voltage of 5 V. This frequency can be lowered by the addition of an external capacitor to the OSC terminal or the oscillator can be driven by an external clock.

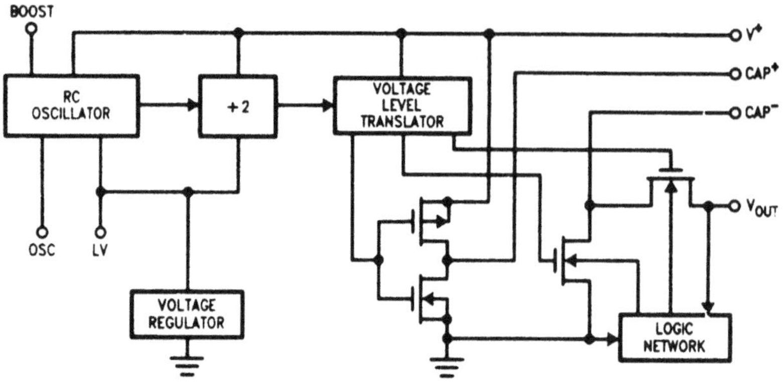

**Figure 9.1.**  Functional diagram for super voltage converter (Harris Semiconductor, *Linear and Telecom ICs,* 1991, p. 2-106).

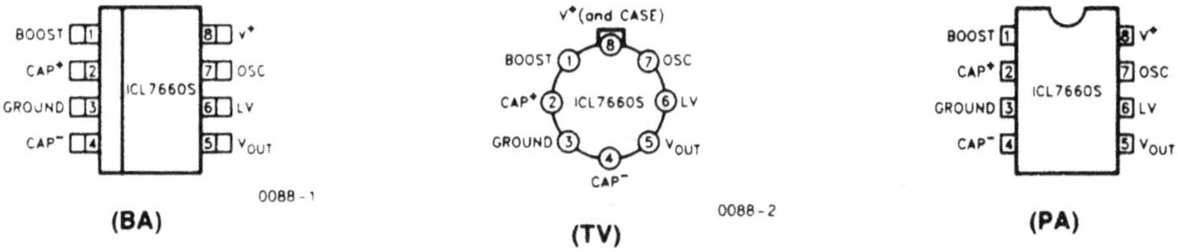

**Figure 9.2.**  Pin configurations for super voltage converter (Harris Semiconductor, *Linear and Telecom ICs,* 1991, p. 2-105).

## 9.2   Basic Operating Principle

Figure 9.4 shows the circuits of an idealized negative voltage converter. Capacitor C1 is charged to voltage $V+$ for the half cycle when switches S1 and S3 are closed. (Switches S2 and S4 are open during this half cycle.) During the second half cycle of operation, switches S2 and S4 are closed, with S1 and S3 open. This shifts capacitor C1 in the negative direction by $V+$ volts. The charge is then transferred from C1 to C2 so that the voltage on C2 is exactly $V+$, assuming ideal switches and no load on C2.

The four switches shown in Fig. 9.4 are MOS power switches. S1 is a P-channel device and the remaining switches are N-channel. The main difficulty with this approach is that in integrating the switches, the substrates (S3 and S4) must always remain reverse-biased with respect to their sources, but so much as to degrade the on resistance. Also, at the circuit start up and under output short-circuit condition (where $V_{OUT} = V+$), the output voltage must be sensed and the substrate bias adjusted accordingly. Failure to do so would result in high power losses and probable device latchup. This problem is eliminated in the ICL7660S by a logic network that senses the output voltage ($V_{OUT}$) together with level translators that switch the substrates of S3/S4 to the correct level (and thus maintain the necessary reverse bias).

The voltage-regulator portion of the IC is an integral part of the anti-latchup circuits. However, the inherent voltage drop of the regulator can

| Symbol | Parameter | Test Conditions | Limits | | | Units |
|---|---|---|---|---|---|---|
| | | | Min | Typ | Max | |
| $I^+$ | *Supply Current*<br>(Note 3) | $R_L = \infty$, 25°C | | *80* | *160* | μA |
| | | 0°C < $T_A$ < +70°C | | | *180* | |
| | | −25°C < $T_A$ < +85°C | | | *180* | |
| | | −55°C < $T_A$ < +125°C | | | *200* | |
| $V_H^+$ | *Supply Voltage Range—Hi*<br>(Note 4) | $R_L$ = 10K, LV Open<br>$T_{min}$ < $T_A$ < $T_{max}$ | 3.0 | | *12* | V |
| $V_L^+$ | Supply Voltage Range—Lo | $R_L$ = 10K, LV to GROUND<br>$T_{min}$ < $T_A$ < $T_{max}$ | 1.5 | | 3.5 | V |
| $R_{OUT}$ | *Output Source Resistance* | $I_{OUT}$ = 20 mA, $T_A$ = 25°C | | 60 | 100 | Ω |
| | | $I_{OUT}$ = 20 mA, 0°C < $T_A$ < +70°C | | | 120 | |
| | | $I_{OUT}$ = 20 mA, −25°C < $T_A$ < +85°C | | | 120 | |
| | | $I_{OUT}$ = 20 mA, −55°C < $T_A$ < +125°C | | | 150 | |
| | | $I_{OUT}$ = 3 mA, $V^+$ = 2V, LV = GND,<br>0°C < $T_A$ < +70°C | | | *250* | |
| | | $I_{OUT}$ = 3 mA, $V^+$ = 2V, LV = GND,<br>−25°C < $T_A$ < +85°C | | | *300* | |
| | | $I_{OUT}$ = 3 mA, $V^+$ = 2V, LV = GND,<br>−55°C < $T_A$ < +125°C | | | *400* | |
| $f_{OSC}$ | *Oscillator Frequency* | $C_{OSC}$ = 0, Pin 1 Open or GND | *5* | 10 | | kHz |
| | | Pin 1 = $V^+$ | | *35* | | |
| PEff | *Power Efficiency* | $R_L$ = 5 kΩ | *96* | 98 | | % |
| | | $T_{min}$ < $T_A$ < $T_{max}$ | *95* | *97* | | |
| $V_{OUT}$ Eff | *Voltage Conversion Efficiency* | $R_L = \infty$ | *99* | 99.9 | | % |
| $Z_{OSC}$ | Oscillator Impedance | $V^+$ = 2V | | 1 | | MΩ |
| | | $V^+$ = 5V | | 100 | | kΩ |

NOTE 1: Connecting any terminal to voltages greater than $V^+$ or less than GROUND mat cause destructive latchup. It is recommended that no inputs from sources operating from external supplies be applied prior to "power up" of ICL7660s.

2: Derate linearly above 50°C by 5.5 mW/°C.

3: In the test circuit, there is no external capacitor applied to pin 7. However, when the device is plugged into a test socket, there is usually a very small but finite stray capacitance present, of the order of 5 pF.

4: The Harris ICL7660S can operate without an external diode over the full temperature and voltage range. This device will function in existing designs which incorporate an external diode with no degradation in overall circuit performance.

5: All significant improvements over the industry-standard ICL7660 are highlighted in *bold italics*.

**Figure 9.3.**  Electrical characteristics for super voltage converter (Harris Semiconductor, *Linear and Telecom ICs,* 1991, p. 2-107).

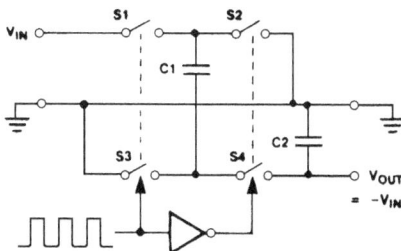

**Figure 9.4.**  Idealized negative voltage converter (Harris Semiconductor, *Linear and Telecom ICs,* 1991, p. 2-110).

degrade operation at low voltages. To improve low-voltage operation, the LV pin should be connected to ground, disabling the regulator. For supply voltages greater than 3.5 V, the LV terminal must be left open to ensure latchup-proof operation and to prevent device damage.

## 9.3    Theoretical Power-Efficiency Considerations

In theory, this type of converter can approach 100% efficiency—if certain conditions are met:

- The drive circuitry consumes minimum power.
- The output switches have extremely low on resistance and virtually no off-set.
- The impedances of the pump and reservoir capacitors are negligible at the pump frequency.

In operation, the 100% efficiency is not reached, but efficiency is much greater than for the typical switching regulator operated in the inverting mode. The charge-pump efficiency is a typical 98%. The table in Fig. 9.5 shows a comparison of charge-pump, switching-regulator, and linear-regulator circuits.

The ICL7660S approaches the 100% efficiency condition if large values of $C_1$ and $C_2$ are used. Energy is lost only in the transfer of charge between

| Feature | Linear regulation | Charge pump | Switcher |
|---|---|---|---|
| Output voltage regulation | Excellent | Poor | Excellent |
| Voltage step-up | No | Yes, but only in discrete multiples of input voltages | Yes |
| Voltage step-down | Yes | Yes, but only in discrete fractions of input voltage | Yes |
| Current step-up | No | Yes, but only in discrete fractions of input current | Yes |
| Power conversion efficiency | Poor | Excellent—can approach 100% | Good 80% |
| High power control | Yes | Not suitable | Excellent |
| Inductor required | No | No | Yes |
| Output noise | Low | Includes clock frequency plus harmonics | Includes clock frequency plus harmonics, hf hash |
| Minimum of external components | 0 | 2 capacitors | 1 inductor 1 capacitor |

**Figure 9.5.**  Comparison of dc-dc converters (*Maxim Seminar Applications Book,* 1989, p. 132).

capacitors if a change in voltage occurs. The energy lost is defined by: $\frac{1}{2}C_1(V_1{}^2 - V_2{}^2)$, where $V_1$ and $V_2$ are the voltages on C1 during the pump and transfer cycles.

If the impedances of C1 and C2 are relatively high at the pump frequency, compared to the value of the load resistance $R_L$, there will be a substantial difference in voltages $V_1$ and $V_2$. Thus it is desirable, not only to make C2 as large as possible to eliminate output-voltage ripple, but also to use a correspondingly large value for $C_1$ to get maximum efficiency.

## 9.4  Troubleshooting Precautions

The following manufacturer's recommendations apply when troubleshooting any of the circuits described in this chapter:

- Do not exceed maximum supply voltages.

- Do not connect the LV terminal to ground for supply voltages greater than 3.5 V.

- Do not short-circuit the output to the $V+$ supply for supply voltages greater than 5.5 V for extended periods. However, transient conditions (including the start up) should be no problem.

- When using polarized capacitors, the positive terminal of C1 must be connected to pin 2, and the positive terminal of C2 must be connected to ground.

- If the voltage supply driving the IC has a large source impedance (25 to 30 $\Omega$), then 2.2 $\mu$F from pin 8 to ground might be required (to limit the rate of input voltage rise to less than 2 V/$\mu$s).

- Be certain that the output (pin 5) does not go more positive than GND (pin 3). Device latchup will occur under these conditions.

- A 1N914 or similar diode in parallel with C2 will prevent the IC from latching up. The diode anode should be at pin 5, with the cathode at pin 6.

## 9.5  Basic Negative Voltage Converter

Figure 9.6 shows the ICL7660S connected in the basic negative voltage converter configuration. The output at pin 4 is approximately equal to the input at pin 8, except that the voltage polarity is inverted. A positive supply of $+1.5$ V to $+12$ V is converted to a negative output of $-1.5$ V to $-12$ V. Remember that pin 6 (LV) is tied to the supply negative (GND) for supply voltages less than 3.5 V.

The output impedance $(R_o)$ at pin 5 is a function of the on resistance of the internal MOS switches (Fig. 9.4), the switching frequency, the capacitances $(C_1$ and $C_2)$, and the ESR $(C_1/C_2)$. In troubleshooting an experimental circuit, $R_o$ should be at a minimum value so that the available output is maximum with a given output voltage. $R_o$ decreases with an increase in $C_1/C_2$ and switching frequency, but increases with an increase in ESR.

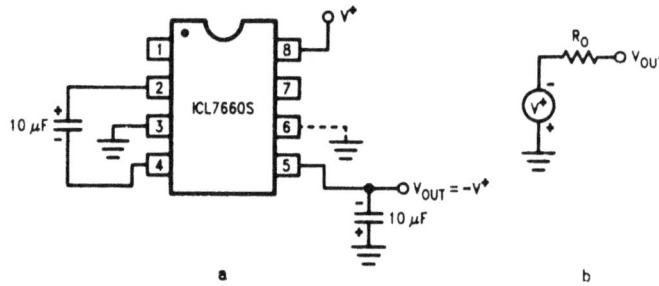

**Figure 9.6.**   Basic negative voltage converter configuration (Harris Semiconductor, *Linear and Telecom ICs,* 1991, p. 2-111).

# TYPICAL PERFORMANCE CHARACTERISTICS

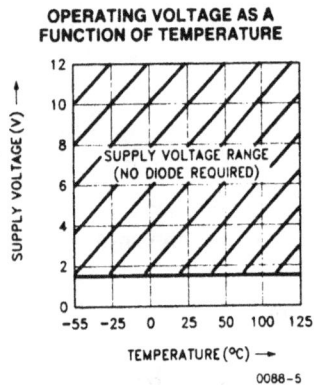

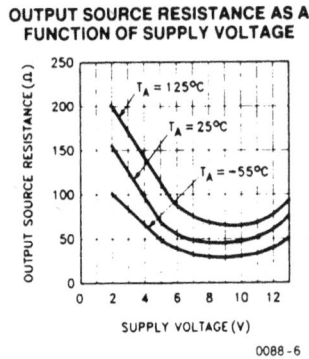

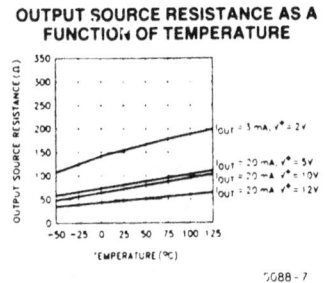

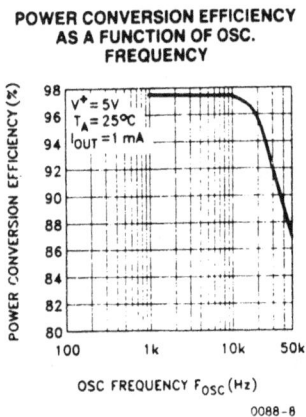

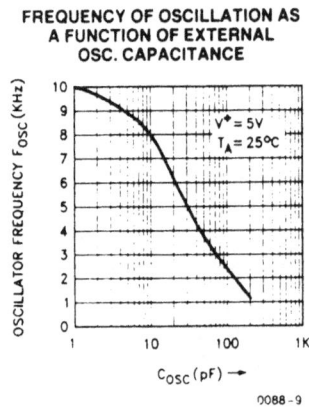

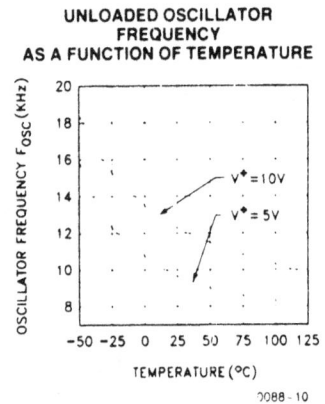

**Figure 9.7.**   Negative voltage converter characteristics (Harris Semiconductor, *Linear and Telecom ICs,* 1991, pp. 2-108, 2-109).

In theory, the output resistance can be decreased to any value by increasing the switching frequency and $C_1/C_2$. However, a practical limit is set by the ESR of $C_1/C_2$. (A small increase in ESR will offset large increases in frequency and capacitance values.)

Figure 9.7 shows the effect of oscillator frequency on output resistance, together with many other performance characteristics of the basic negative voltage-converter function. The circuit of Figure 3 (referred to in Fig. 9.7) is the test circuit shown in Fig. 9.8.

The information in Fig. 9.7 can be used in troubleshooting experimental circuits. For example, the oscillator or switching frequency can be altered as described in Section 9.9. The graph of Fig. 9.7 shows that the frequency is about 6 kHz when $C_{OSC}$ is 20 pF. The frequency drops to about 1 kHz when $C_{OSC}$ is increased to 200 pF.

## TYPICAL PERFORMANCE CHARACTERISTICS

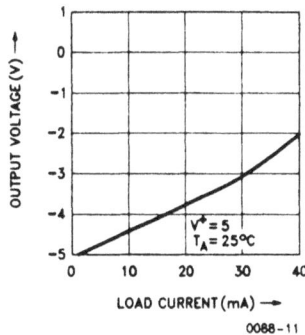

**OUTPUT VOLTAGE AS A FUNCTION OF OUTPUT CURRENT**

0088-11

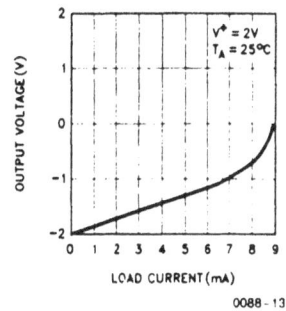

**SUPPLY CURRENT & POWER CONVERSION EFFICIENCY AS A FUNCTION OF LOAD CURRENT**

0088-12

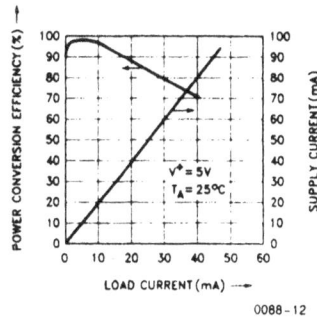

**OUTPUT VOLTAGE AS A FUNCTION OF OUTPUT CURRENT**

0088-13

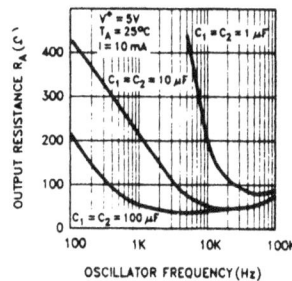

**SUPPLY CURRENT & POWER CONVERSION EFFICIENCY AS A FUNCTION OF LOAD CURRENT**

0088-14

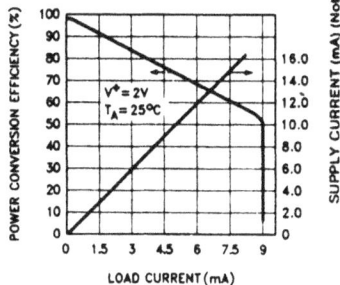

**OUTPUT SOURCE RESISTANCE AS A FUNCTION OF OSCILLATOR FREQUENCY**

0088-15

NOTE 6: These curves include in the supply current that current fed directly into the load $R_L$ from V+ (see Figure 3). Thus approximately half the supply current goes directly to the positive side of the load, and the other half, through the ICL7660S, to the negative side of the load. Ideally, $V_{OUT} \approx 2 V_{IN}$, $I_S \approx 2 I_L$, so $V_{IN} \times I_S \approx V_{OUT} \times I_L$.

**Figure 9.7.**  (Continued)

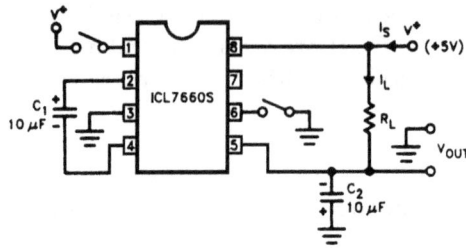

**Figure 9.8.** Voltage converter test circuit (Harris Semiconductor, *Linear and Telecom ICs,* 1991, p. 2-110).

0088-16

**NOTE 1:** For large values of $C_{OSC}$ ( > 1000pF) the values of $C_1$ and $C_2$ should be increased to 100μF.

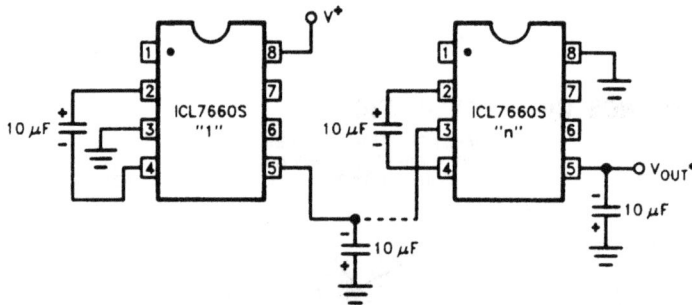

*NOTE 1: $V_{OUT} = -nV^+$ for 1.5V ≤ V$^+$ ≤ 12V.

**Figure 9.9.** Circuit for cascading super voltage converters (Harris Semiconductor, *Linear and Telecom ICs,* 1991, p. 2-112).

## 9.6  Ripple Voltage and ESR

As in the case of switching supplies, the ESR of capacitors C1 and C2 affect the ripple voltage. From a practical troubleshooting standpoint, a capacitor with low ESR produces a lower output ripple, all other factors being equal.

## 9.7  Parallel Charge-Pump Circuits

Any number of ICL7660S converters can be connected in parallel to reduce output resistance. In troubleshooting, remember that the reservoir capacitor (C2) serves all devices, but each device requires a separate pump capacitor (C1). The resultant output resistance will be approximately:

$$R_{OUT} = \frac{R_{OUT}\ (\text{of ICL7660S})}{n\ (\text{number of devices})}$$

## 9.8  Cascade Charge-Pump Circuits

Figure 9.9 shows the circuit for cascading devices to increase output voltage. The ICL7660S can be cascaded (as shown) to produce a larger negative multiplication of the initial supply voltage. When troubleshooting an experimental circuit, the practical limit is 10 devices for light loads. This is because of the finite efficiency of each device.

The output voltage is defined by:

$$V_{OUT} = -N(V_{IN})$$

where $N$ is an integer that represents the number of devices cascaded. The resulting output resistance is approximately the weighted sum of the individual ICL7660S $R_{OUT}$ values.

## 9.9    Changing Oscillator Frequency

Because of noise or other considerations, it might be desirable to alter the oscillator frequency. This can be done by one of several methods described in the following paragraphs.

The oscillator frequency can be increased by about 3.5 times when the boost pin (1) is connected to $V+$. From a troubleshooting standpoint, the result is a decrease in output impedance and ripple. This is of major importance for surface-mount applications where capacitor size and cost are crucial. With the increased frequency, smaller capacitors (such as 0.1 μF) can be used to get output currents comparable to those when the IC is operated at the free-running frequency of about 10 kHz and with $C_1/C_2$ at 10 μF or 100 μF. This is shown in Fig. 9.7 (output source resistance as a function of oscillator frequency).

The oscillator frequency can also be increased by overdriving from an external clock as shown in Fig. 9.10. To prevent latchup when troubleshooting an experimental circuit, use a 100-kΩ resistance in series with the clock. If the clock source is TTL, rather than the CMOS shown, connect a 10-kΩ pullup resistor to the $V+$ supply. Notice that the pump frequency is one half of the clock or free-running oscillator frequency. Output transitions occur on the positive-going edge of the clock.

It is also possible to increase the conversion efficiency of the ICL7660S (at low load levels) by lowering the oscillator frequency. This is shown in Fig. 9.7 (power conversion efficiency as a function of oscillator frequency). In troubleshooting, the oscillator frequency can be lowered to about 1 kHz by connecting a 100-μF capacitor between pins 7 and 8, as shown in Fig. 9.11. Unfortunately, lowering the oscillator frequency increases the impedance of both C1 and C2. This can be overcome by increasing $C_1/C_2$ by the same factor

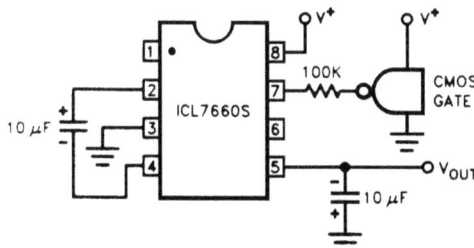

Figure 9.10.    Circuit for external clocking (Harris Semiconductor, *Linear and Telecom ICs*, 1991, p. 2-113).

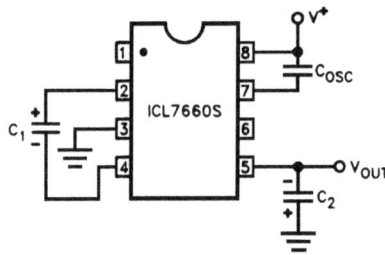

**Figure 9.11.** Circuit for lowering oscillator frequency (Harris Semiconductor, *Linear and Telecom ICs,* 1991, p. 2-113).

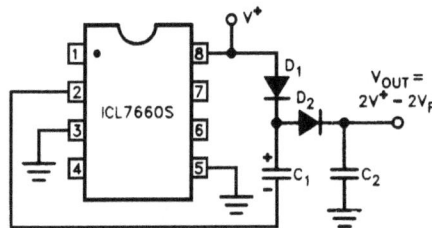

**Figure 9.12.** Circuit for positive voltage doubling (Harris Semiconductor, *Linear and Telecom ICs,* 1991, p. 2-113).

**NOTE:** $D_1$ & $D_2$ can be any suitable diode.

that the frequency is reduced. For example, if the frequency is decreased by a factor of 10 (1 kHz to 10 kHz), both $C_1$ and $C_2$ must be increased by the same factor (from 10 µF to 100 µF).

## 9.10    Positive Voltage Doubling

Figure 9.12 shows the ICL7660S connected to produce positive voltage doubling. In this application, the pump inverter switches are used to charge C1 to a voltage level of $(V+)-V_F$, where $V+$ is the supply voltage and $V_F$ is the forward voltage drop of diode D1. On the transfer cycle, the voltage on C1, plus the supply voltage $(V+)$ is applied through diode D2 to capacitor C2.

In troubleshooting the positive voltage-doubling circuit, remember that the voltage thus created on C2 becomes $(2\,V+)-(2\,V_F)$, twice the supply voltage minus the combined forward voltage drops of diodes D1 and D2. The source impedance of $V_{OUT}$ depends on the output current. For a $V+$ of 5 V and an output current of 10 mA, the source impedance is about 60 Ω.

## 9.11    Combined Positive and Negative Outputs

Figure 9.13 shows a circuit that combined the functions shown in Figs. 9.6 and 9.12 (negative voltage conversion and positive supply doubling). As an example, this approach would be suitable for generating +9 V and −5 V from an existing +5-V supply. In this configuration, capacitors C1 and C3 perform the pump and reservoir functions, respectively, for the generation of negative voltage. Capacitors C2 and C4 are pump and reservoir, respectively, for the doubled positive voltage. In troubleshooting, remember that the source impedances of the generated supplies are somewhat higher because of the finite impedance of the common charge-pump driver at pin 2. In turn, this reduces the available output current for a given output voltage.

## 9.12 Voltage Splitting

Figure 9.14 shows how the bidirectional characteristics of the IC can be used to split a higher supply voltage in half. The combined load ($R_{L1}$, $R_{L2}$) will be evenly shared between the two sides. In troubleshooting, remember that the output impedance is much lower than in the standard circuit, and higher currents can be drawn. This is because the switches share the load in parallel.

## 9.13 Regulated Negative Supply Voltage

In some cases, the output impedance of the IC can be a problem, particularly if the load current varies substantially. The circuit of Fig. 9.15 overcomes this problem. A low-power op amp is used to maintain a constant output voltage. Variations in output voltage (caused by load changes) are fed back to the ICL7660S through the op amp to offset the variations. (Direct feedback, without an op amp, is not recommended because the ICL7660S output does not respond instantly to changes in input, but only after the switching delay.)

In troubleshooting the regulated circuit, remember that the circuit supplies enough delay to accommodate the IC, and maintains adequate feedback. As a result, an increase in pump and storage or reservoir capacitor values is recommended. The values shown provide an output impedance of less than 5 $\Omega$ to a load of 10 mA.

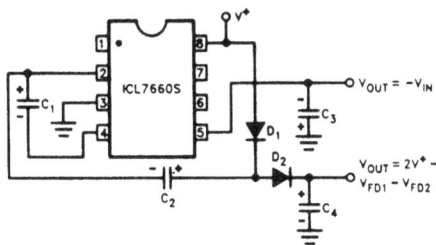

**Figure 9.13.** Combined negative voltage converter and positive doubler (Harris Semiconductor, *Linear and Telecom ICs*, 1991, p. 2-114).

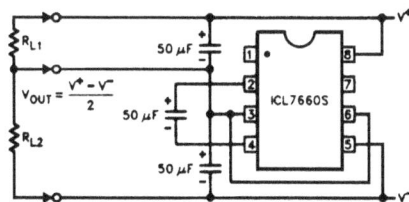

**Figure 9.14.** Circuit for splitting a power supply (Harris Semiconductor, *Linear and Telecom ICs*, 1991, p. 2-114).

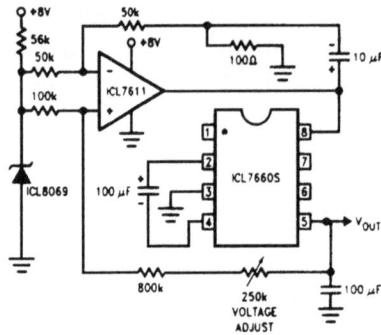

**Figure 9.15.**  Circuit for regulating output with variable load (Harris Semiconductor, *Linear and Telecom ICs,* 1991, p. 2-114).

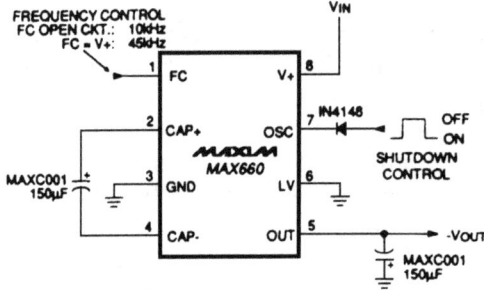

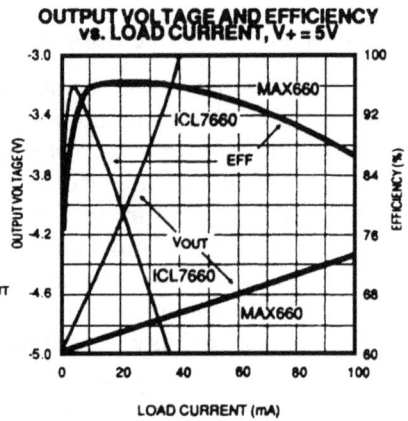

**Figure 9.16.**  Charge pump with shutdown (*Maxim Applications and Product Highlights,* 1992, p. 4-17).

## 9.14   Charge Pump with Shutdown

Figure 9.16 shows a Maxim MAX660 charge pump connected as a voltage inverter, similar to those described in this chapter. This circuit creates a negative voltage of approximately equal magnitude to that of the input voltage. The important characteristic to remember in troubleshooting is that the MAX660 has no feedback mechanism, and so is unregulated. However, the output is a stiff, accurate supply when the charge pump is operated from a regulated input.

An output current of 100 mA results in a typical voltage loss of 0.65 V (the drop at 10 mA is less than 100 mV). The FC pin selects a 10-kHz or 45-kHz oscillator frequency. The circuit provides an optional shutdown that disables the internal oscillator and reduces the supply current to less than 1 μA.

# Data-Converter Circuit Troubleshooting

This chapter is devoted to troubleshooting for data-converter circuits. It starts with a review of basic data-converter techniques, including operation of typical DAC (digital-to-analog converter) and ADC (analog-to-digital converter) ICs. This is followed by basic ADC/DAC testing and troubleshooting information. An entire section is devoted to data-converter terms. Another section describes how design characteristics relate to troubleshooting problems.

Before starting, let's resolve certain differences in terms. Some manufacturers refer to analog-to-digital converters as ADCs. Other manufacturers use the term *A/D converter*. The same is true of digital-to-analog converters, which are referred to as *DACs* by some and *D/A converters* by others. I prefer the terms *ADC* and *DAC*, but do not be surprised to find both terms in this book.

## 10.1 Basic Data-Conversion Technique

This section describes the various common ADC and DAC techniques. Here we concentrate on explanations of the basic principles of data conversion. By studying this information, you should be able to understand operation of the converter ICs described throughout the chapter. It is assumed that you are familiar with basic digital electronics. If not, read *Lenk's Digital Handbook* (McGraw-Hill, 1993).

### 10.1.1 Typical BCD signal formats

Figure 10.1A shows the relationship of the three most common BCD (binary coded decimal) signal formats used in ADC/DAC circuits. These include NRZL (nonreturn-to-zero level). NRZM (nonreturn-to-zero-mark), and RZ (return-to-zero).

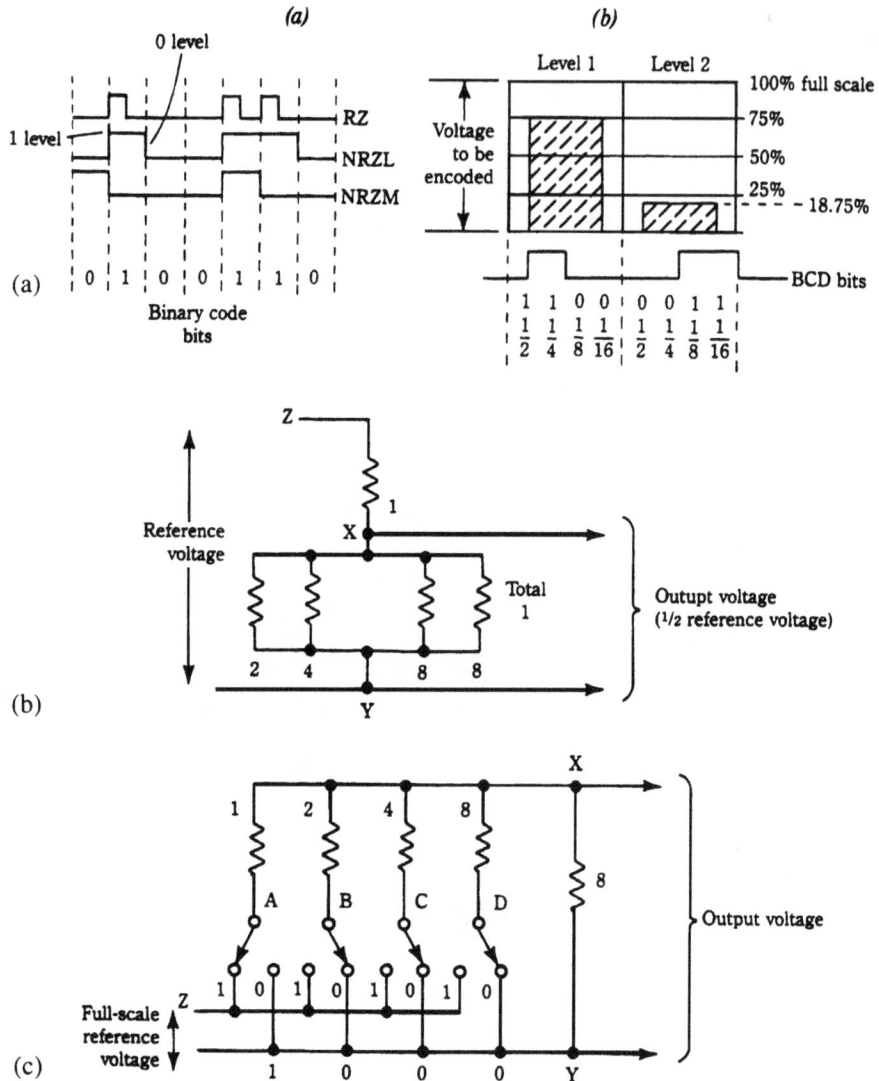

**Figure 10.1**   ADC and DAC conversion basics.

In NRZL, a 1 is one signal level, and a 0 is another signal level. These levels can be 5 V, 10 V, 0 V, −5 V, or any other selected values, provided that the 1 and 0 levels are entirely different.

In RZ, a 1-bit is represented by a pulse of definite width (usually a ½-bit width) that returns to a zero-signal level and the 0-bit is represented by a zero-level signal.

In NRZM, the level of the pulse has no meaning. A 1 is represented by a change in level, and a 0 is represented by no change.

## 10.1.2  Four-bit system

Figure 10.1B shows the relationship between two voltage levels to be convert-ed, and the corresponding binary code (in NRZL form), in a basic ADC. In a practical circuit, the four-bit ADC (sometimes called a *binary encoder*) samples the voltage level to be converted and compares the voltage to $1/2$ scale, $1/4$ scale, $1/8$ scale, and $1/16$ scale (in that order) of a given full-scale voltage. The ADC then produces four data bits, in sequence, with the comparison made on the most significant ($1/2$ scale) first.

As shown in Fig. 10.1B, each of the two voltage levels is divided into four equal time increments. The first time increment is used to represent the $1/2$-scale bit, the second increment the $1/4$-scale bit, etc.

In voltage level 1, the first two time increments are at binary 1 and the sec-ond two increments are at 0. This produces 1100, decimal 12. Twelve is $3/4$ of 16. Thus, level 1 is 75% of full scale. For example, if full scale is 10 V, level 1 is 7.5 V.

In level 2, the first two increments are at 0 and the second two increments are at 1. This is represented as 0011 or 3. Thus, level 2 is $3/16$ of full scale (1.875 V). This can be expressed in another way. In the first ($1/2$ scale) increment, the converter produces a 0 because the voltage (1.875 V) is less than $1/2$ scale (5 V). The same is true of the second ($1/4$ scale) increment (1.875 V is less than 2.5 V).

In the third ($1/8$ scale) increment of level 2, the converter produces a 1, as it does in the fourth ($1/16$ scale) increment because the voltage being compared is greater than $1/8$ of full scale (1.875 is greater than 0.625 V). Thus, the $1/2$-scale and $1/4$-scale increments are at 0, and the $1/8$-scale and $1/16$-scale increments are at 1 (also, $1/8 + 1/16 = 3/16$, 18.75%).

## 10.1.3  ADC conversion ladder

Figure 10-1C shows a conversion ladder which is the heart of many ADC cir-cuits. The ladder provides a means of implementing a four-bit binary-coding system and produces an output that is equivalent to switch positions. The switches can be moved to either a 1 or a 0 position, which corresponds to a four-place binary number. The output voltage describes a percentage of full-scale reference voltage, depending on the switch positions. For example, if all switches are at the 0 position, there is no output voltage. This produces a binary 0000, represented by 0 V.

If switch A is at 1 and the remaining switches are at 0, this produces a binary 1000 (decimal 8). Because the total in a four-bit system is 16 (0 to 15), 8 rep-resents $1/2$ full scale. Thus, the output voltage is half of the full-scale reference voltage. This conversion is done as follows.

The 2-, 4-, and 8-$\Omega$ switch resistors and the 8-$\Omega$ output resistor are con-nected in parallel. This produces a value of 1 $\Omega$ across points X and Y. The ref-erence voltage is applied across the 1-$\Omega$ switch resistor (across points Z and Y). In effect, this is the same as two 1-$\Omega$ resistors in series. Because the full-scale reference voltage is applied across both resistors in series and the output is measured across only one of the resistors, the output voltage is half of the reference voltage.

In a practical converter, the same basic ladder is used to supply a comparison voltage to a comparison circuit, which compares the voltage to be converted against the binary-coded voltage from the ladder. The resultant output of the comparison circuit is a binary code representing the voltages to be converted.

The mechanical switches shown in Fig. 10.1C are replaced by electronic switches, usually flip-flops (FFs). When the switch is on, the corresponding ladder resistor is connected to the reference voltage. The switches are triggered by four pulses (representing each of the four binary bits) from the digital system clock. An enable pulse is used to turn the comparison circuit on and off, so that as each switch is operated, a comparison can be made among the four bits.

### 10.1.4  Typical ADC operating sequence

Figure 10.2A is a simplified diagram of a typical ADC. Here the reference voltage is applied to the ladder through the electronic switches. The ladder output (comparison voltage) is controlled by switch positions, which are controlled by pulses from the clock.

The following paragraphs outline the sequence of events necessary to produce a series of four binary bits that describe the input voltage as a percentage of full scale (in $\frac{1}{16}$ increments). Assume that the input voltage is 75% of full scale.

When pulse 1 arrives, switch 1 is turned on and the remaining switches are off. The ladder output is a 50% voltage that is applied to the differential amplifier. The balance of this amplifier is set so that the output is sufficient to turn on one AND gate and turn off the other AND gate if the ladder voltage is greater than the input voltage. Similarly, the differential amplifier reverses the AND gates if the ladder voltage is not greater than the input voltage. Both AND gates are enabled by the clock pulse.

In this example (75% of full scale), the ladder output is less than the input voltage when pulse 1 is applied to the ladder. As a result, the nongreater AND gate turns on and the output FF is set to the 1 position. Thus, for the first of the four bits, the FF output is 1.

When pulse 2 arrives, switch 2 is turned on, and switch 1 remains on. Both switches 3 and 4 remain off. The ladder output is now 75% of the full-scale voltage. (The ladder voltage equals the input voltage.) However, the ladder output is still not greater than the input voltage. Consequently, when the AND gates are enabled, the AND gates remain in the same condition. Thus, the output FF remains at 1.

When pulse 3 arrives, switch 3 is turned on. Switches 1 and 2 remain on with switch 4 off. The ladder output is now 87.5% of full-scale voltage, and thus is greater than the input voltage. As a result, when the AND gates are

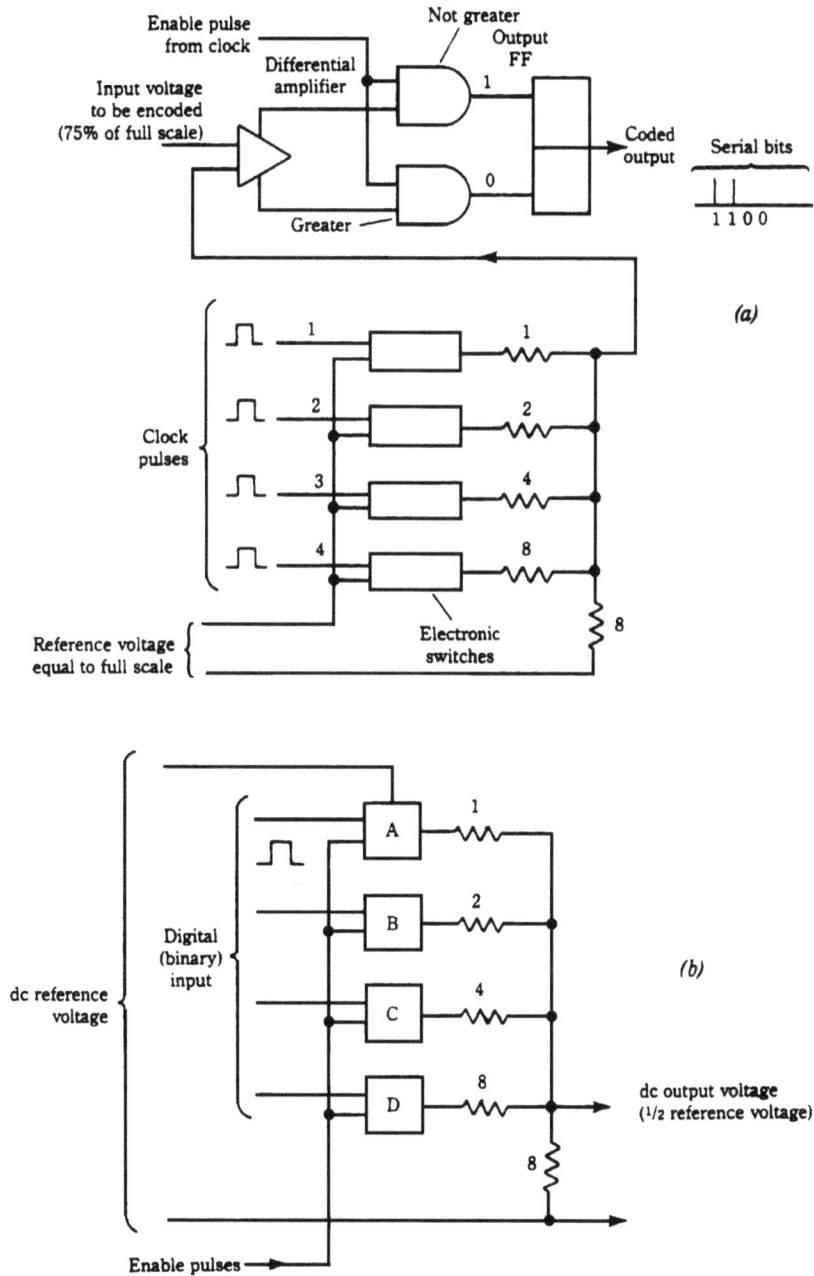

**Figure 10.2**  ADC and DAC conversion circuits.

enabled, they reverse. The not-greater AND gate turns off and the greater AND gate turns on. The output FF then sets to 0.

When pulse 4 arrives, switch 4 is turned on. All switches are now on. The ladder is at maximum (full scale) and is greater than the input voltage. As a result, when the AND gates are enabled, they remain in the same condition. The output FF remains at 0.

The four binary bits from the output are 1, 1, 0, and 0, or 1100. This is a binary 12, which is 75% of 16. In a practical ADC, when the fourth pulse has passed, all switches are reset to the OFF position. This places them in a condition to spell out the next four-bit binary word.

### 10.1.5  Typical DAC operating sequence

Figure 10.2B is a simplified diagram of a typical DAC. This circuit performs the opposite function of the ADC just described (the DAC produces an output voltage that corresponds to the binary code). A conversion ladder is also used in the DAC. The conversion-ladder output is a voltage that represents a percentage of the full-scale reference voltage.

The output voltage from the DAC also depends on switch positions. In turn, the switches are set to ON or OFF by corresponding binary pulses. If the information is applied to the switch in four-line (parallel) form, each line can be connected to the corresponding switch. If the information is in serial form, the data must be converted to parallel by a register (shift or storage register). The switches in a DAC are essentially a form of AND gate. Each gate completes the circuit from the reference voltage to the corresponding ladder resistor when both the enable pulse and binary pulse coincide.

Assume that the digital number to be converted is 1000 (decimal 8). When the first pulse is applied, switch A is enabled and the reference voltage is applied to the 1-$\Omega$ resistor. When switches B, C, and D receive their enable pulses, there are no binary pulses (or the pulses are in the 0 condition). Thus, switches B, C, and D do not complete the circuits to the 2-, 4-, and 8-$\Omega$ ladder resistors. These resistors combine with the 8-$\Omega$ output resistor to produce a 1-$\Omega$ resistance in series with the 1-$\Omega$ ladder resistance. This divides the reference voltage in half to produce 50% of full-scale output. Because 8 is half of 16, the 50% output voltage represents 8.

### 10.1.6  High-speed ADCs

Although there are a number of ADC schemes, the three basic types are: parallel, serial, and combination. In parallel, all bits are converted simultaneously by many circuits. In series, each bit is converted in sequence, one at a time. The combination ADC includes features of both types. Generally, parallel is faster, but more complex, than serial. The combination types are a compromise between speed and complexity. The remaining paragraphs in this section describe some classic high-speed ADC circuits.

### 10.1.7  Parallel (flash) ADC

Figure 10.3A shows the basic parallel (or flash) ADC circuit, where all bits of the digital representation are converted simultaneously by a bank of voltage comparators. For $N$ bits of binary information, the system requires $2^{N-1}$ comparators, and each comparator determines one LSB (least-significant bit) level. This requires a number of circuits. Another disadvantage of parallel ADC is that comparator output is not directly usable information. The output must be converted to binary information with a decoder.

### 10.1.8  Tracking ADC

Figure 10.3B shows the basic tracking ADC circuit. A tracking ADC continuously tracks the analog input voltage and is often used in communication systems or similar applications in which the input is a continuously varying signal. The accuracy of this system is no better than the DAC used in the system. (An 8-bit DAC is shown in Fig. 10.3B.)

### 10.1.9  Successive approximation ADC

Figure 10.3C shows the basic successive approximation ADC circuit. Notice that this circuit is essentially the same as the ADC of Fig. 10.2A. The D/A block of Fig. 10.3C represents the electronic switches and ladder of Fig. 10.2A. The successive approximation (S/A) storage register (often called an *SAR*) of Fig. 10.3C represents the AND gates and FF of Fig. 10.2A. However, four bits are shown in Fig. 10.2A, whereas eight bits are used in Fig. 10.3C.

The successive-approximation type of ADC is relatively slow compared with other types of high-speed ADCs, but the low cost, ease of construction, and system features make up for the lack of speed. From a troubleshooting standpoint, the important point to remember about SARs is that the bits of the D/A are enabled, one at a time, starting with the MSB (most significant bit). As each bit is enabled, the comparator produces an output indicating that the input signal is greater, or not greater, in amplitude than the output of the D/A. If the D/A output is greater than the input signal, the bit is reset or turned off. The system does this with the MSB first, then the next most significant bit, then the next, etc. After all bits of the D/A are tried, the conversion cycle is complete, and another cycle is started. Notice that the S/A type of ADC provides a serial output during conversion and a parallel output between conversion cycles.

## 10.2  Typical DAC IC

Figure 10.4 shows the functional block diagram of a typical DAC IC (the classic DAC-08). The following paragraphs summarize the DAC-08 characteristics as they relate to troubleshooting.

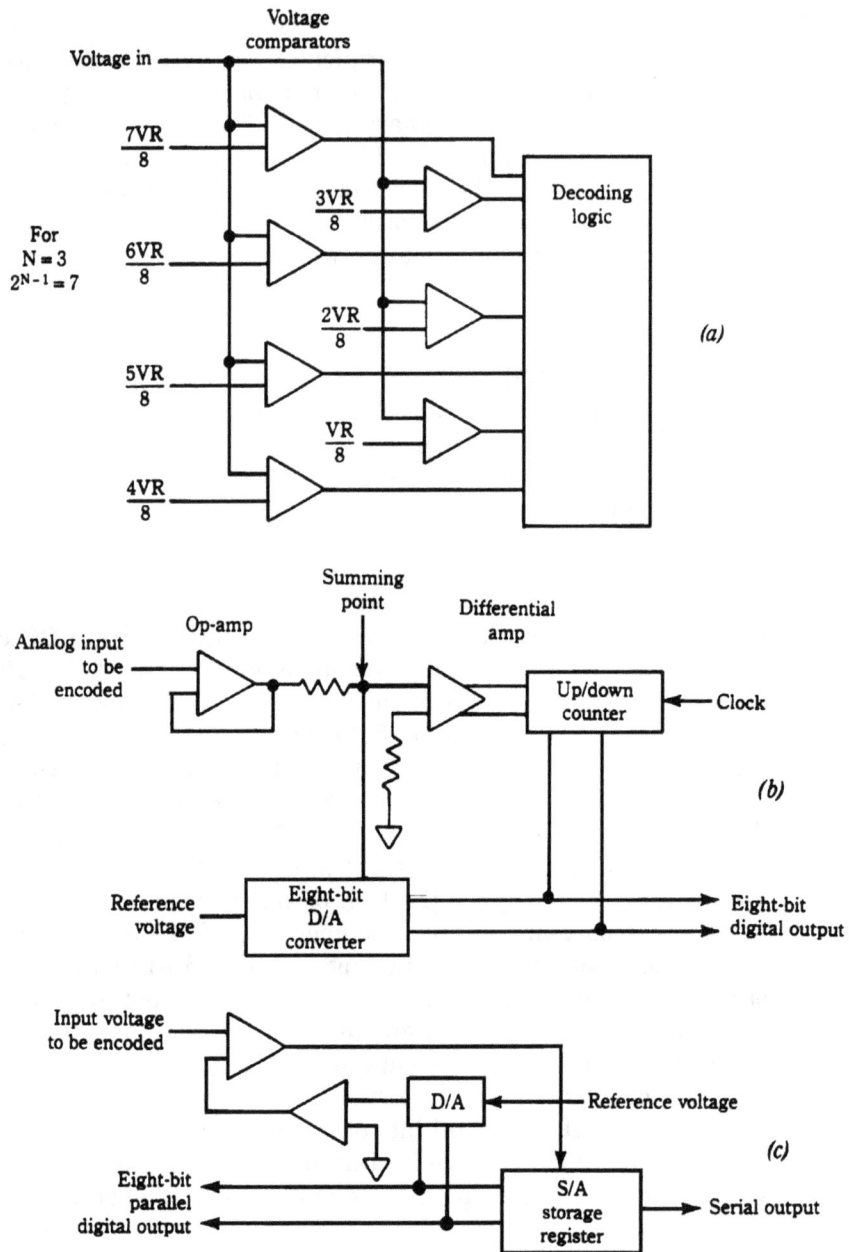

Figure 10.3  High-speed ADC converters.

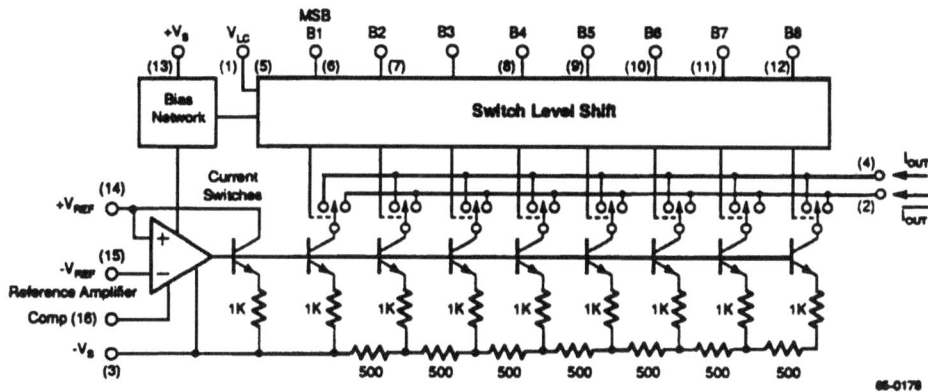

**Figure 10.4**  Functional block diagram of typical DAC (*Raytheon Semiconductor Data Book*, 1994, p. 3-141).

### 10.2.1  DAC circuit operation

The DAC-08 is a multiplying DAC in which the output current is the product of a digital number and the input reference current. This is somewhat different from the theoretical DAC of Fig. 10.2B, but the net result is the same. A digital or binary word is converted to an output current (which can be further converted to voltage by passing the current through a resistor or load).

In the circuit of Fig. 10.5, the reference current can be fixed or it might vary from 100 μA to 4 mA. The full-scale output current ($I_{FS}$) is a linear function of the reference current ($I_{REF}$) and is given by $I_{FS} = (255/256)$ times $I_{REF}$. In the positive-reference application shown, the external reference forces current through R14 into the $+V_{REF}$ terminal (pin 14) of the reference amplifier (Fig. 10.4).

In some circuits, a negative reference is applied to the $-V_{REF}$ terminal (pin 15). This negative-reference connection has the advantage of a very high impedance presented at pin 15. The voltage at pin 14 is equal to and tracks the voltage at pin 15. In other circuits, R15 is eliminated with only minor increases in tracking error. (R15 is used to cancel any input bias current errors in the reference amplifier.)

Figure 10.6 shows how bipolar references can be accommodated. Either $V_{REF}$ or pin 15 can be offset. The negative common-mode range of the reference amplifier is given by $V_{CM} = -V_S + (I_{REF} \times 1 \text{ k}\Omega) + 2.5$ V. The positive common-mode range of the reference amplifier is $V_S$ less 1.5 V. When the reference is dc, a reference bypass capacitor is recommended. If a regulated power supply is used as a reference, R14 should be split into two resistors with the junction bypassed to ground with a 0.1-μF capacitor. (A 5-V TTL-logic supply is not recommended as a reference.)

Figure 10.7 shows how the full-scale circuit can be adjusted. In troubleshooting a typical experimental circuit, this is not necessary because of the close relationship between $I_{REF}$ and $I_{FS}$. It is also possible to adjust full-scale by substituting a pot for R14. However, this is not recommended because of the temperature-coefficient ($T_C$) effects of a pot.

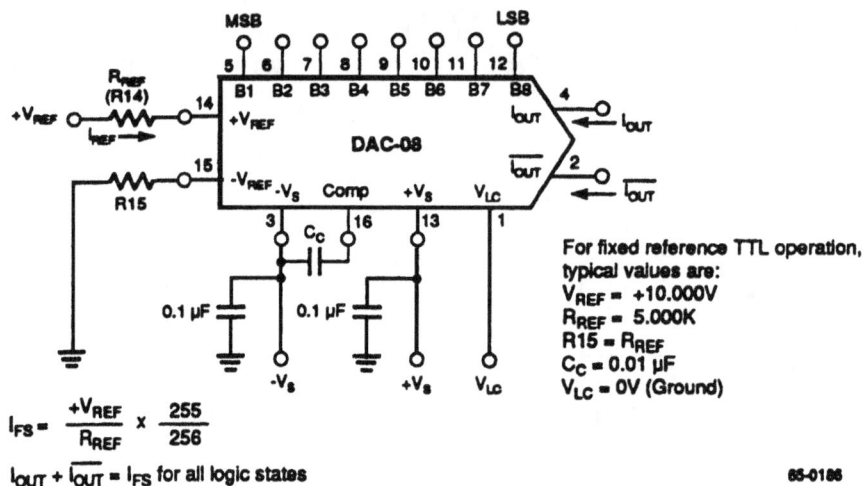

$$I_{FS} = \frac{+V_{REF}}{R_{REF}} \times \frac{255}{256}$$

$I_{OUT} + \overline{I_{OUT}} = I_{FS}$ for all logic states

Figure 10.5    DAC-08 connected for basic positive-reference operation (*Raytheon Semiconductor Data Book*, 1994, p. 3-148).

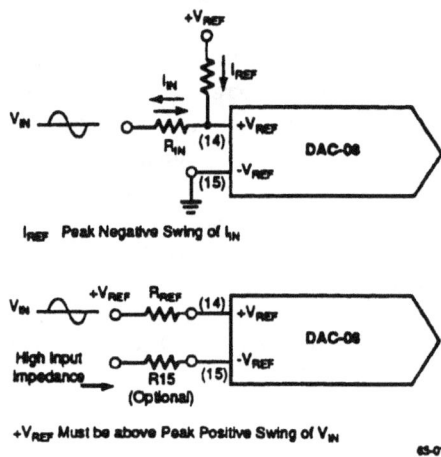

Figure 10.6    Accommodating bipolar references (*Raytheon Semiconductor Data Book*, 1994, p. 3-148).

Figure 10.8 shows the basic negative-reference operation. Using lower values of reference current reduces negative power-supply current and increases reference-amplifier negative common-mode range. The recommended range for operation with a dc reference current is +0.2 mA to +4.0 mA. With either positive or negative reference operation, the reference amplifier must be compensated by a capacitor $(C_c)$ connected from pin 16 to $-V_S$, as shown in Fig. 10.5. If the reference is a fixed dc voltage, a value of 0.01 μF is recommended for $C_c$. If the reference is ac or pulse, the value of $C_c$ must be selected as described next in Section 10.2.2.

## 10.2.2  Reference amplifier compensation techniques

Compensating capacitor $C_c$ is often a problem when troubleshooting experimental DAC circuits. The following section summarizes the problems. The value of $C_c$ depends on the impedance presented to pin 14. For example, for R14 values of 1.0, 2.5, and 5.0 k$\Omega$, the minimum values of $C_c$ are 15, 37, and 75 pF (for this particular DAC), when the reference is ac. Larger values of R14 require a proportionately increased value of $C_c$ to ensure proper phase margin. When the reference input is a pulse, use a low value for R14. This makes it possible to use smaller values for $C_c$.

If pin 14 is driven by a high impedance, such as a transistor current source, the recommended values for compensation do not apply. This is because the reference amplifier must be heavily compensated. Unfortunately, such compensation decreases overall bandwidth and slew rate (as is the case when an op amp is heavily compensated, Chapter 1).

As a point of reference (for this DAC), for an R14 of 1 k$\Omega$ and a $C_c$ of 15 pF, the reference amplifier slews 4.0 mA/$\mu$s and results in a transition from $I_{REF} = 0$ to $I_{REF} = 2.0$ mA in 500 ms. When a low full-scale transition time is crucial, use 200 $\Omega$ for R14 and make $C_c = 0$. Under these conditions, full-scale transition (0 to 2.0 A) occurs in 120 ns. This produces a slew rate of 16 mA/$\mu$s.

## 10.2.3  Input circuits

The DAC-08 can be interfaced directly to all popular logic families with a maximum of noise immunity. This is because of the large input-swing capability, a

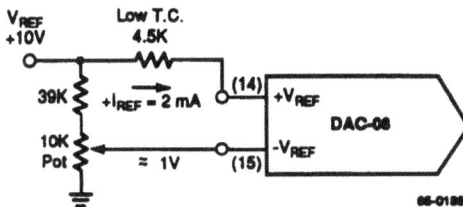

**Figure 10.7**  Typical full-scale adjustment circuit for DAC-08 (*Raytheon Semiconductor Data Book*, 1994, p. 3-148).

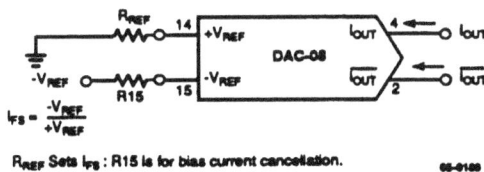

**Figure 10.8**  DAC-08 connected for basic negative-reference operation (*Raytheon Semiconductor Data Book*, 1994, p. 3-149).

2.0-μA logic-input current, and adjustable logic-threshold voltage. For example, when the supply is −15 V, the logic inputs can swing between −10 V and +18 V. This allows direct interface with +5-V CMOS logic—even when the supply is +5 V. The minimum input-logic swing and minimum logic threshold are given by: supply plus ($I_{REF}$ times 1 kΩ) plus 2.5 V.

The logic threshold can be adjusted over a wide range by placing an appropriate voltage at the logic threshold control, $V_{LC}$ (pin 1). The logic-threshold voltage $V_{TH}$ is nominally 1.4 V above $V_{LC}$. For TTL and DTL interfaces, simply ground pin 1 as shown in Fig. 10.5. For interfacing ECL, an $I_{REF}$ of 1 mA is recommended. Figure 10.9 shows typical interface circuits for ECL, CMOS, and PMOS/NMOS.

When troubleshooting an experimental circuit, notice that pin 1 will source or sink 100 μA (typical), so external circuits are designed to accommodate this current. When a fast settling time (Section 10.2.7) is essential, keep in mind that the fastest times are obtained when pin 1 "sees" a low impedance. For example, pin 1 can be connected to a 1-kΩ divider and bypassed to ground with a 0.01-μF capacitor.

### 10.2.4 Output circuits

Figures 10.10 through 10.12 show typical unipolar, bipolar, and offset binary output circuits and the relationships between digital inputs and voltage/current outputs. As shown, both true and complemented output sink currents are provided where $I_{OUT} + I_{OUT} = I_{FS}$.

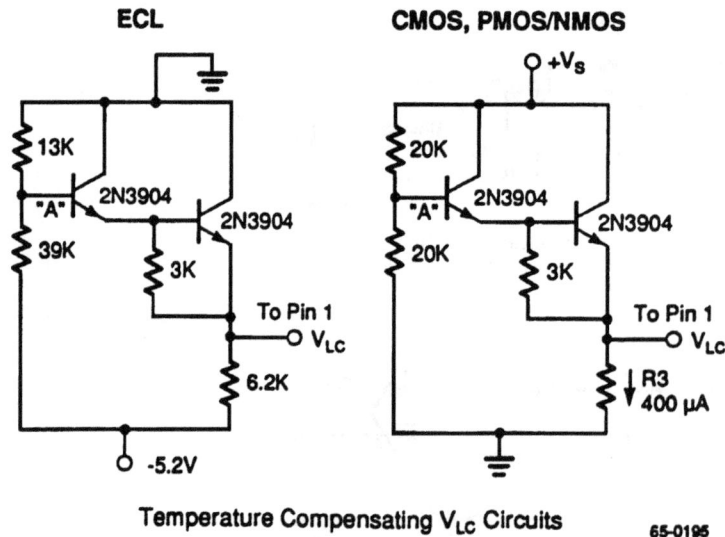

**Figure 10.9** Typical interface circuits for ECL, CMOS, and PMOS/NMOS (Raytheon Semiconductor Data Book, 1994, p. 3-150).

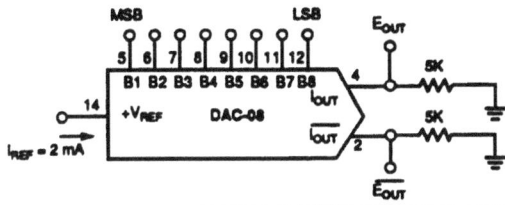

| Scale | B1 | B2 | B3 | B4 | B5 | B6 | B7 | B8 | $I_{OUT}$mA | $\overline{I_{OUT}}$mA | $E_{OUT}$ | $\overline{E_{OUT}}$ |
|---|---|---|---|---|---|---|---|---|---|---|---|---|
| Full Scale | 1 | 1 | 1 | 1 | 1 | 1 | 1 | 1 | 1.992 | 0.008 | -9.960 | -0.000 |
| Half Scale +LSB | 1 | 0 | 0 | 0 | 0 | 0 | 0 | 1 | 1.008 | 0.984 | -5.040 | -4.920 |
| Half Scale | 1 | 0 | 0 | 0 | 0 | 0 | 0 | 0 | 1.000 | 0.992 | -5.000 | -4.960 |
| Half Scale -LSB | 0 | 1 | 1 | 1 | 1 | 1 | 1 | 1 | 0.992 | 1.000 | -4.960 | -5.000 |
| Zero Scale +LSB | 0 | 0 | 0 | 0 | 0 | 0 | 0 | 1 | 0.008 | 1.984 | -0.040 | -9.920 |
| Zero Scale | 0 | 0 | 0 | 0 | 0 | 0 | 0 | 0 | 0.000 | 1.992 | 0.000 | -9.960 |

**Figure 10.10**   DAC-08 connected for basic unipolar-negative operation (*Raytheon Semiconductor Data Book*, 1994, p. 3-149).

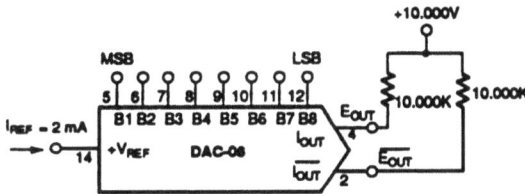

| Scale | B1 | B2 | B3 | B4 | B5 | B6 | B7 | B8 | $E_{OUT}$ | $\overline{E_{OUT}}$ |
|---|---|---|---|---|---|---|---|---|---|---|
| Pos Full Scale | 1 | 1 | 1 | 1 | 1 | 1 | 1 | 1 | -9.920 | +10.000 |
| Pos Full Scale - LSB | 1 | 1 | 1 | 1 | 1 | 1 | 1 | 0 | -9.840 | +9.920 |
| Zero Scale + LSB | 1 | 0 | 0 | 0 | 0 | 0 | 0 | 1 | -0.080 | +0.160 |
| Zero Scale | 1 | 0 | 0 | 0 | 0 | 0 | 0 | 0 | 0.000 | +0.080 |
| Zero Scale - LSB | 0 | 1 | 1 | 1 | 1 | 1 | 1 | 1 | +0.080 | 0.000 |
| Neg Full Scale + LSB | 0 | 0 | 0 | 0 | 0 | 0 | 0 | 1 | +9.920 | -9.840 |
| Neg Full Scale | 0 | 0 | 0 | 0 | 0 | 0 | 0 | 0 | +10.000 | -9.920 |

**Figure 10.11**   DAC-08 connected for basic bipolar-output operation (*Raytheon Semiconductor Data Book*, 1994, p. 3-149).

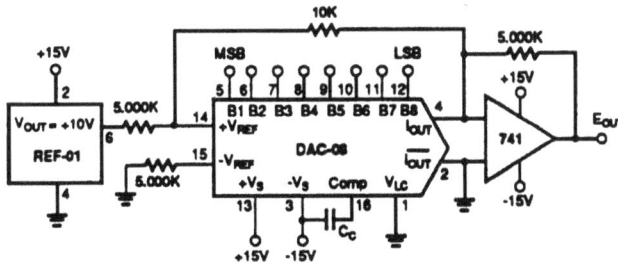

| Scale | B1 | B2 | B3 | B4 | B5 | B6 | B7 | B8 | $E_{OUT}$ |
|---|---|---|---|---|---|---|---|---|---|
| Pos Full Scale | 1 | 1 | 1 | 1 | 1 | 1 | 1 | 1 | +4.960 |
| Zero Scale | 1 | 0 | 0 | 0 | 0 | 0 | 0 | 0 | 0.000 |
| Neg Full Scale + 1 LSB | 0 | 0 | 0 | 0 | 0 | 0 | 0 | 1 | -4.960 |
| Neg Full Scale | 0 | 0 | 0 | 0 | 0 | 0 | 0 | 0 | -5.000 |

**Figure 10.12**   DAC-08 connected for basic offset-binary operation (*Raytheon Semiconductor Data Book*, 1994, p. 3-149).

Current appears at the true output when a 1 is applied to each logic input. As the binary count increases, the sink current at pin 4 increases proportionally (a positive-logic DAC). When a 0 is applied to any input bit, that current is turned off at pin 4 and turned on at pin 2. A decreasing logic count increases $I_{\text{OUT}}$ proportionally (a negative-logic or inverted-logic DAC).

Both outputs can be used simultaneously and both have a wide voltage compliance. This allows fast current-to-voltage conversion through a resistor tied to ground or other voltage source (Figs. 10.10 and 10.11). Positive compliance is 36 V above the negative supply and is independent of the positive supply. Negative compliance is given by: *supply voltage* $+ (I_{\text{REF}} \times 1 \text{ k}\Omega) + 2.5$ V.

The dual outputs allow double the usual peak-to-peak logic swing when driving loads in a quasidifferential manner. This feature is especially useful in cable-drive or CRT-deflection circuits and other balanced applications, such as driving center-tapped coils and transformers. In troubleshooting an experimental circuit, if one of the outputs is not required, it must be connected to ground or to a point that is capable of sourcing $I_{\text{FS}}$. Do not leave an unused output pin open. This is shown in Figs. 10.13 and 10.14, where the pin 4 output is connected to an op amp. With these circuits, the output impedance is lowered to a value determined by the op amp, not the DAC. In both cases, the

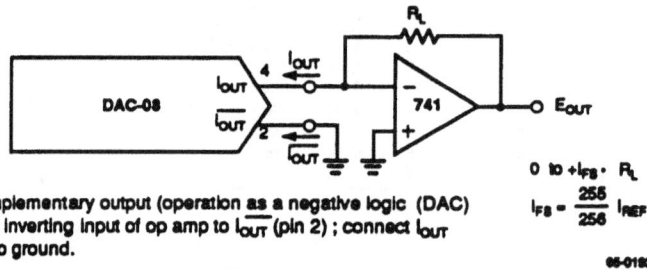

For complementary output (operation as a negative logic (DAC) connect inverting input of op amp to $\overline{I_{\text{OUT}}}$ (pin 2) ; connect $I_{\text{OUT}}$ (pin 4) to ground.

$$0 \text{ to } +I_{\text{FS}} \cdot R_L$$
$$I_{\text{FS}} = \frac{255}{256} I_{\text{REF}}$$

05-0193

**Figure 10.13**  DAC-08 connected for positive low-impedance output operation (*Raytheon Semiconductor Data Book,* 1994, p. 3-150).

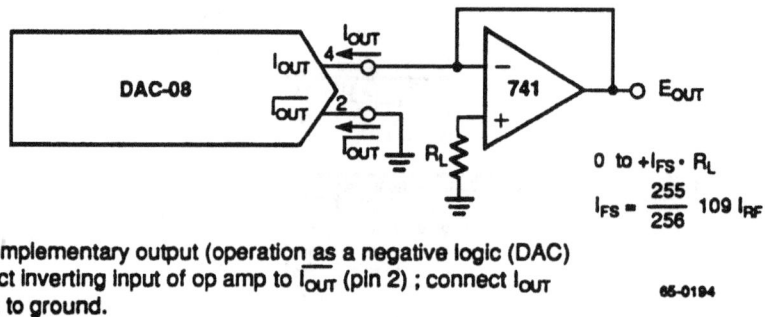

For complementary output (operation as a negative logic (DAC) connect inverting input of op amp to $\overline{I_{\text{OUT}}}$ (pin 2) ; connect $I_{\text{OUT}}$ (pin 4) to ground.

$$0 \text{ to } +I_{\text{FS}} \cdot R_L$$
$$I_{\text{FS}} = \frac{255}{256} \, 109 \, I_{\text{RF}}$$

05-0194

**Figure 10.14**  DAC-08 connected for negative low-impedance output operation (*Raytheon Semiconductor Data Book,* 1994, p. 3-150).

unused DAC output is connected to ground and the converted output is taken from the op amp.

### 10.2.5  Power supply problems

Symmetrical supplies are not required. The DAC will operate satisfactorily with standard ±5.0-V supplies, or any combination in which the total is 9 V to 36 V. If ±5.0 V is used, the $I_{REF}$ should not exceed 1 mA for this DAC. In troubleshooting, notice that low reference-current operation decreases power consumption and increases negative compliance, reference-amplifier negative common-mode range, negative-logic input range, and negative-logic threshold range. This is shown in the various graphs of Fig. 10.15.

### 10.2.6  Temperature characteristics in troubleshooting

The nonlinearity and monotonicity specifications of the DAC-08 are guaranteed to apply over the entire rated operating temperature range. (See Section 10.5 for more information concerning monotonicity.) Full-scale output-current drift is typically ±10 ppm/°C. Zero-scale output current and drift are essentially negligible to $\frac{1}{2}$ LSB.

When troubleshooting an experimental circuit to get minimum overall full-scale drift, the temperature coefficient of the reference resistor (Fig. 10.5) should match and track that of any output resistor. Settling times (Section 10.2.7) decrease about 10% at −55°C. At +125°C, an increase of about 15% is typical (for this DAC).

### 10.2.7  Settling time tests

Figure 10.16 shows a settling-time test circuit for the DAC-08. With slight modification, this circuit can also be used for many other DACs. When testing a DAC as part of the troubleshooting process, the settling time is usually the single most crucial factor in operation because it determines overall speed of digital-to-voltage conversion. The DAC-08 requires 35 ns for each of the eight data bits. Settling time to with $\frac{1}{2}$ LSB is therefore 35 ns, with each progressively larger bit taking successively longer. The MSB settles in 85 ns, thus setting the overall settling time of 85 ns.

Figure 10.15 shows a typical scope response for measurement of full-scale settling time, indicating about 85 ns from the start of the logic input to the point where the output settles to a constant level. Settling to 6-bit accuracy requires about 65 to 70 ns.

As is the case when troubleshooting any digital IC in the experimental stage, the fastest operation is obtained using short leads, minimum output capacitance, and adequate bypassing for the supply. The bypass need not be electrolytic because the supply current drain is independent of the input logic state. The 0.1-μF bypass capacitors shown in Fig. 10.16 provide full transient

## Typical Performance Characteristics

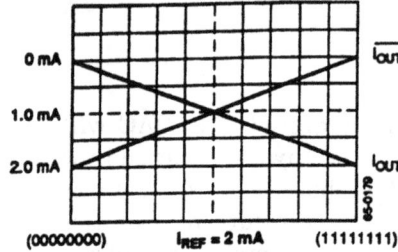

**True and Complementary Output Operation**

$I_{REF} = 2$ mA

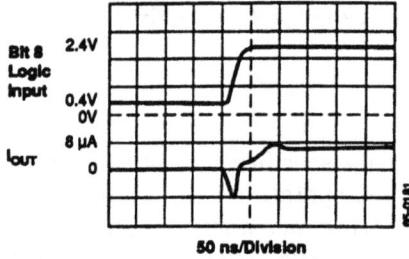

**LSB Switching**

50 ns/Division

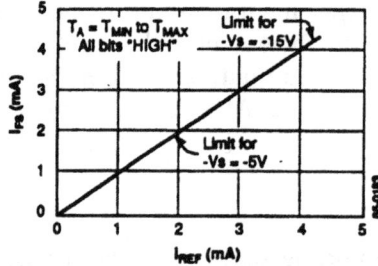

**Full Scale Output Current vs. Reference Current**

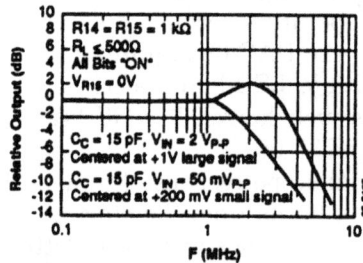

**Reference Input Frequency Response**

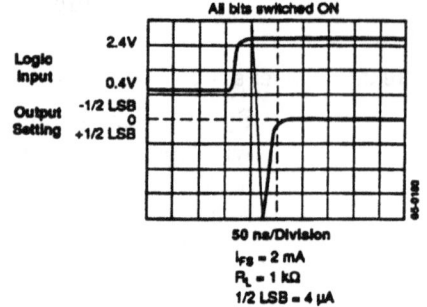

**Full Scale Settling Time**

$I_{FS} = 2$ mA
$R_L = 1$ kΩ
1/2 LSB = 4 μA

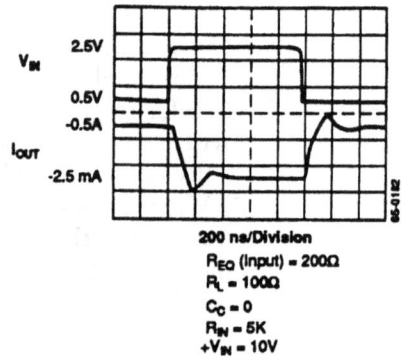

**Fast Pulsed Reference Operation**

$R_{EQ}$ (Input) = 200Ω
$R_L = 100Ω$
$C_C = 0$
$R_{IN} = 5K$
$+V_{IN} = 10V$

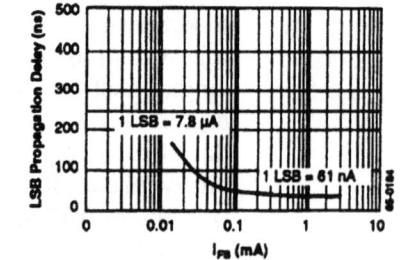

**LSB Propagation Delay vs. Full Scale Output Current**

1 LSB = 7.8 μA
1 LSB = 61 nA

**Figure 10.15**  Typical performance characteristics of DAC-08 (*Raytheon Semiconductor Data Book,* 1994, p. 3-145).

protection. The output capacitance of the DAC-08 (including the package) is about 15 pF. Therefore, the output RC time constant determines settling time when the load is greater than 500 $\Omega$.

Measurement of settling time requires the ability to accurately resolve $\pm 4.0$ $\mu A$, so a 1-k$\Omega$ load is needed to provide adequate drive ($\pm 0.4$ V) for most scopes. The circuit of Fig. 10.16 uses a cascade design to allow driving the 1-k$\Omega$ load with less than 5 pF of parasitic capacitance at the measurement mode. At $I_{REF}$ values less than 1 mA, excessive RC damping of the output is difficult to prevent if adequate sensitivity is maintained. However, the major carry from 01111111 to 10000000 provides an accurate indicator of settling time. This code change does not require the normal 6.2 time constants to settle to within $\pm 0.2\%$ of the final value, so settling times can be observed at lower values of $I_{REF}$.

The settling time remains essentially constant for $I_{REF}$ values down to 1 mA, with gradual increases for lower $I_{REF}$ values. The main advantage of higher $I_{REF}$ values is the ability to obtain a given output level with lower load resistors, reducing the output RC time constant. Switching transients are generally low and can be further reduced by small capacitive loads at the output. Of course, this increase in RC time constant does increase the settling time.

## 10.3   Typical ADC IC

Figure 10.17 shows the functional block diagram of typical ADC ICs (the industry standard MAX174, MX574A, and MX674A). Figure 10.18 shows the

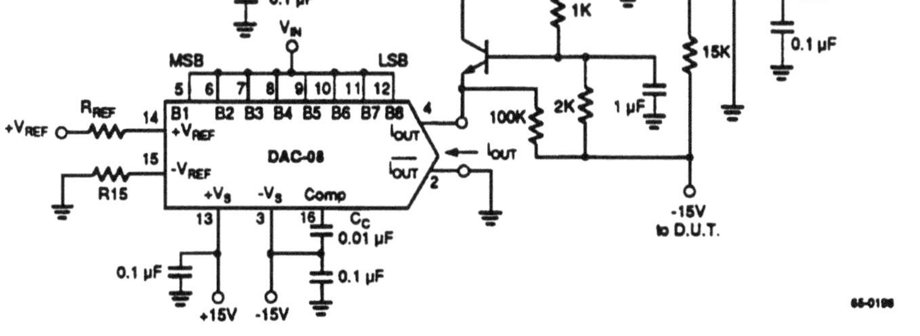

**Figure 10.16**  Settling-time test circuit for DAC-08 (*Raytheon Semiconductor Data Book, 1994,* p. 3-151).

pin functions. The following paragraphs summarize the ADC characteristics as they relate to troubleshooting.

### 10.3.1  ADC circuit operation

These ICs use the successive-approximation technique described in Section 10.1.9 to convert an unknown analog input to a 12-bit digital output code. Compare the circuit of Fig. 10.17 with that of Fig. 10.3C. The control logic

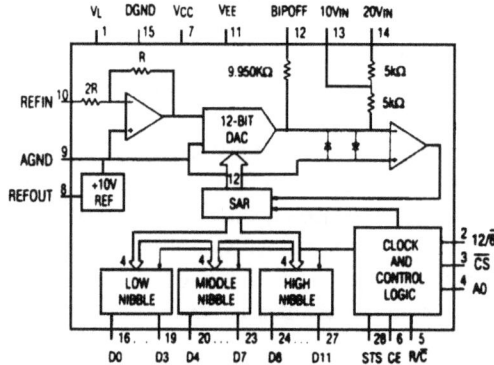

**Figure 10.17** Functional block diagram of typical ADC IC (*Maxim New Releases Data Book,* 1992, p. 7-81).

| PIN # | NAME | FUNCTION |
|---|---|---|
| 1 | $V_L$ | Logic Supply, +5V |
| 2 | 12/$\overline{8}$ | Data Mode Select Input |
| 3 | $\overline{CS}$ | Chip-Select Input.  Must be low to select device. |
| 4 | A0 | Byte Address/Short Cycle Input.  When starting a conversion, controls number of bits converted (low = 12 bits, high = 8 bits). When reading data, if 12/$\overline{8}$ = low, enables low byte (A0 = high) or high byte (A0 = low). |
| 5 | R/$\overline{C}$ | Read/Convert Input.  When high, the device will be in the data-read mode. When low, the device will be in the conversion start mode. |
| 6 | CE | Chip-Enable Input.  Must be high to select device. |
| 7 | $V_{CC}$ | +12V or +15V Supply |
| 8 | REFOUT | +10V Reference Output |
| 9 | AGND | Analog Ground |
| 10 | REFIN | Reference Input |
| 11 | $V_{EE}$ | -12V or -15V Supply |
| 12 | BIPOFF | Bipolar Offset Input.  Connect to REFOUT for bipolar input range. |
| 13 | 10V$_{IN}$ | 10V Span Input |
| 14 | 20V$_{IN}$ | 20V Span Input |
| 15 | DGND | Digital Ground |
| 16-27 | D0-D11 | Three-State Data Outputs |
| 28 | STS | Status Output |

**Figure 10.18**  Pin descriptions for typical ADC IC (*Maxim New Releases Data Book,* 1992, p. 7-87).

function of Fig. 10.17 provides easy interface with most microprocessors. The internal voltage output DAC is controlled by an SAR (Section 10.1.9). The analog input is connected to the DAC output with a 5-k$\Omega$ resistor for the 10-V input and a 10-k$\Omega$ resistor for the 20-V input. The comparator is essentially a zero-crossing detector, with the output fed back to the SAR input. Figure 10.19 shows the equivalent of the analog input circuit.

In the IC of Fig. 10.17, the SAR is set to half scale as soon as a conversion starts. The analog input is compared to half of the full-scale voltage. The bit is kept if the analog input is greater than half scale or dropped if smaller. The next bit (bit 10) is then set with the DAC output either at quarter scale, if the MSB is dropped, or $^3/_4$ scale, if the MSB is kept. The conversion continues in this manner until the LSB is tried. At the end of the conversion, the SAR output is latched into the output buffers.

### 10.3.2 Digital control of ADCs

Operation of the ADCs is controlled by the CE (chip enable), $\overline{\text{CS}}$ (chip select), and R/$\overline{\text{C}}$ (read/convert) lines. Figure 10.20 shows the truth table for these lines and the A0 (pin 4) and 12/$\overline{8}$ (pin 2) lines. Although both CE and $\overline{\text{CS}}$ are asserted, the state of R/$\overline{\text{C}}$ selects whether a conversion (R/$\overline{\text{C}}$ = 0) or a data read (R/$\overline{\text{C}}$ = 1) is in progress. The register-control inputs, 12/$\overline{8}$ and A0, select the data

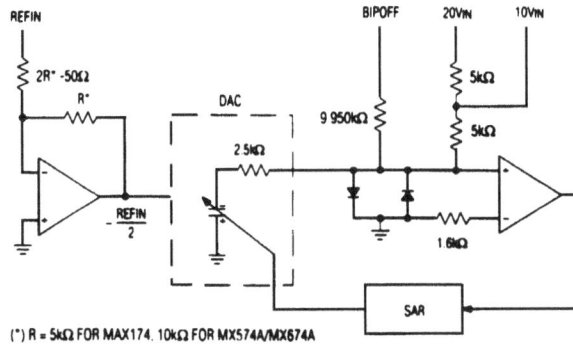

Figure 10.19  Analog equivalent for typical ADC IC (*Maxim New Releases Data Book,* 1992, p. 7-88).

| CE | $\overline{\text{CS}}$ | R/$\overline{\text{C}}$ | 12/$\overline{8}$ | A0 | OPERATION |
|---|---|---|---|---|---|
| 0 | X | X | X | X | None |
| X | 1 | X | X | X | None |
| 1 | 0 | 0 | X | 0 | Initiate 12-bit conversion |
| 1 | 0 | 0 | X | 1 | Initiate 8-bit conversion |
| 1 | 0 | 1 | 1 | X | Enable 12-bit parallel output |
| 1 | 0 | 1 | 0 | 0 | Enable 8MSBs |
| 1 | 0 | 1 | 0 | 1 | Enable 4LSBs + 4 trailing 0s |

Figure 10.20  Truth table for digital control of ADC IC (*Maxim New Releases Data Book,* 1992, p. 7-88).

format and conversion length. To perform full 12-bit conversion, set A0 low during a convert start. For a shorter 8-bit conversion, A0 must be high during a convert start.

### 10.3.3    Output data formats

Figure 10.21 shows the data format for an 8-bit bus, including typical hard wiring. Output data bits are formatted according to the control signal on the $12/\overline{8}$ input. If pin 2 is low, the output is a word broken into two 8-bit bytes. If pin 2 is high, the output is one 12-bit word. During a data read, A0 selects whether the three-state buffers contain the 8 MSBs (A0 = 0) or the 4 LSBs (A0 = 1) of the digital result. The 4 LSBs are followed by four trailing 0s (zeros). A0 can change state when a data-read operation is in effect.

To begin a conversion, the microprocessor must write to the ADC address. Then, because a conversion usually takes longer than a single clock cycle, the microprocessor must wait for the ADC to complete the conversion. Valid data bits are made available only at the end of the conversion, which is indicated by STS (pin 28, the output status). STS can be either polled or used to generate an interrupt upon completion. As an alternate, the microprocessor can be kept idle by inserting the appropriate number of NOP (no operation) instructions between the conversion-start and data-read commands.

After the conversion is completed, data can be obtained by the microprocessor. The ICs have the required logic for 8-, 12-, and 16-bit bus interfacing (as determined by the $12/\overline{8}$ input). If pin 2 is high, the ICs are configured for a 16-bit bus. Data lines D0 through D11 can be connected to the bus as either the 12 MSBs or the 12 LSBs. The other 4 bits must be masked out in software.

For 8-bit bus operation, pin 2 is low. The format is left-justified and the even address (A0 low) contains the 8 MSBs. The odd address (A0 high) contains the four LSBs, followed by four trailing 0s. There is no need to use a software

| | D7 | D6 | D5 | D4 | D3 | D2 | D1 | D0 |
|---|---|---|---|---|---|---|---|---|
| High Byte (A0 = 0) | MSB | D10 | D9 | D8 | D7 | D6 | D5 | D4 |
| Low Byte (A0 = 1) | D3 | D2 | D1 | D0 | 0 | 0 | 0 | 0 |

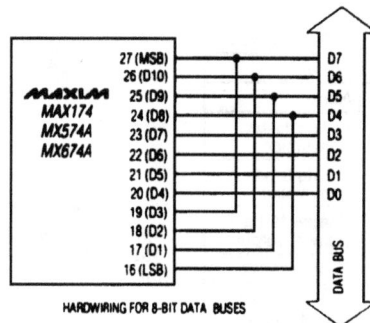

**Figure 10.21**  Data format for an 8-bit bus (*Maxim New Releases Data Book,* 1992, p. 7-89).

mask when the ICs are connected to an 8-bit bus. (Notice that the output cannot be forced to a right-justified format by rearranging the data lines on the 8-bit bus interface.)

### 10.3.4  Input ranges and digital output codes

Figure 10.22 shows the possible input ranges and "ideal" transition voltages. End-point errors can be adjusted in all ranges (Section 10.3.5 and 10.3.6).

### 10.3.5  Typical unipolar input

Figures 10.23 and 10.24 show the ideal transfer function and input connections, respectively, for unipolar input operation. In many cases, the gain (full scale) and offset need not be calibrated. This is because all internal resistors of the ADCs are trimmed for absolute calibration. (The absolute accuracy for each grade of ADC is given in the data-sheet specifications.)

For a 0-V to +10-V input range, the analog input is connected between AGND and 10 $V_{IN}$, as shown in Fig. 10.24. For a 0-V to +20-V input range,

| ANALOG INPUT VOLTAGE (Volts) | | | | DIGITAL OUTPUT | |
|---|---|---|---|---|---|
| 0 to +10V | 0 to +20V | ±5V | ±10V | MSB | LSB |
| +10.0000 | +20.0000 | +5.0000 | +10.0000 | 1111 1111 1111 | |
| +9.9963 | +19.9927 | +4.9963 | +9.9927 | 1111 1111 1110* | |
| +5.0012 | +10.0024 | +0.0012 | +0.0024 | 1000 0000 0000* | |
| +4.9988 | +9.9976 | -0.0012 | -0.0024 | 0111 1111 1111* | |
| +4.9963 | +9.9927 | -0.0037 | -0.0073 | 0111 1111 1110* | |
| +0.0012 | +0.0024 | -4.9988 | -9.9976 | 0000 0000 0000* | |
| 0.0000 | 0.0000 | -5.0000 | -10.0000 | 0000 0000 0000 | |

Note 1:  For unipolar input ranges, output coding is straight binary.
Note 2:  For bipolar input ranges, output coding is offset binary.
Note 3:  For 0V to +10V or ±5V ranges, 1LSB = 2.44mV.
Note 4:  For 0V to +20V or ±10V ranges, 1LSB = 4.88mV.

\* The digital outputs will be flickering between the indicated code and the indicated code plus one.

Figure 10.22  Input ranges and ideal digital output codes (*Maxim New Releases Data Book,* 1992, p. 7-9).

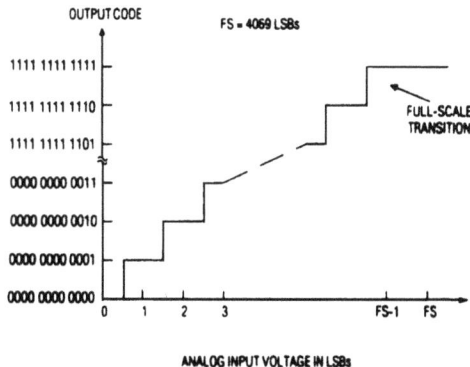

Figure 10.23  Ideal unipolar transfer function (*Maxim New Releases Data Book,* 1992, p. 7-93).

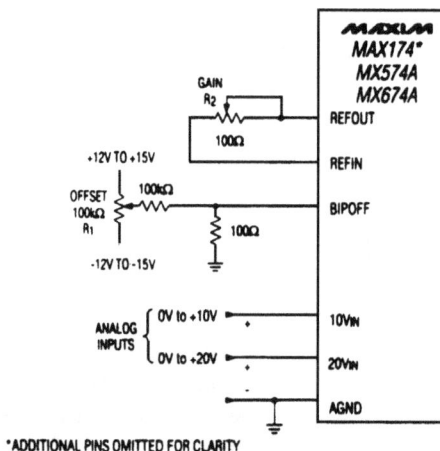

**Figure 10.24** Unipolar input connections (*Maxim New Releases Data Book,* 1992, p. 7-93).

*ADDITIONAL PINS OMITTED FOR CLARITY

connect the analog input between AGND and 20 $V_{IN}$. Notice that these ADCs can easily handle input signals beyond the supply voltage.

If a 10.24-V input range be required during troubleshooting, connect a 200-$\Omega$ trimmer in series with 10 $V_{IN}$. For a full-scale input range of 20.48 V, use a 500-$\Omega$ trimmer in series with 20 $V_{IN}$. The nominal input impedance into 10 $V_{IN}$ is 5 k$\Omega$, and into 20 $V_{IN}$ is 10 k$\Omega$. If the full-scale input and offset need not be trimmed, delete R1, R2, and the related wiring shown in Fig. 10.24. Connect BIPOFF directly to AGND. Connect a 50-$\Omega$ metal-film resistor ($\pm1\%$) between REFOUT and REFIN.

If the full-scale input and offset must be trimmed during troubleshooting, use the connections of Fig. 10.24 and proceed as follows. Adjust the offset first. Apply $\frac{1}{2}$ LSB (see Fig. 10.22 for voltages) at the analog input and adjust R1 until the digital output code flickers between 0000 0000 0000 and 0000 0000 0001. Then apply full-scale less $\frac{3}{2}$ LSB (Fig. 10.22 voltages) at the analog input adjust R2 until the output code changes between 1111 1111 1110 and 1111 1111 1111. (See Section 10.4 for more information concerning ADC trimming and adjustment.)

### 10.3.6  Typical bipolar input

Figures 10.25 and 10.26 show the ideal transfer function and input connections, respectively, for bipolar input operation. The full scale and offset need not be calibrated for all applications because of the internal-resistor accuracy and trimming. One or both of the trimmers (R1 and R2) can be replaced with a 50-$\Omega$ ($\pm1\%$) resistor if external trimming is not required. The analog input ranges can be either $\pm5$ V or $\pm10$ V, as needed.

If the full-scale and offset must be trimmed during troubleshooting, use the connections of Fig. 10.26 and proceed as follows. Adjust the offset first. Apply $\frac{1}{2}$ LSB above negative full-scale (see Fig. 10.22 for voltages) at the analog input and adjust R1 until the digital output code flickers between 0000 0000

0000 and 0000 0000 0001. Then apply a voltage $^3/_2$ LSB below the positive full-scale (Fig. 10.22) at the analog input and adjust R2 until the output code changes between 1111 1111 1110 and 1111 1111 1111.

## 10.4  Basic ADC/DAC Testing and Troubleshooting

This section is devoted to digital testing and troubleshooting basics. It is assumed that you are already familiar with digital troubleshooting at a level found in *Lenk's Digital Handbook* (McGraw-Hill, 1993). However, the following paragraphs summarize both testing and troubleshooting as they relate to ADC and DAC ICs.

Both testing and troubleshooting for data-converter circuits can be performed with conventional test equipment, such as meters, generators, and

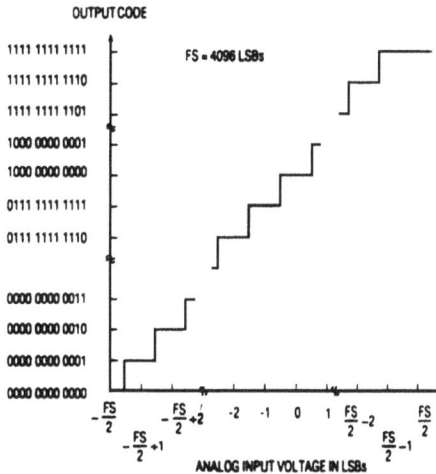

**Figure 10.25** Ideal bipolar transfer function (*Maxim New Releases Data Book,* 1992, p. 7-93).

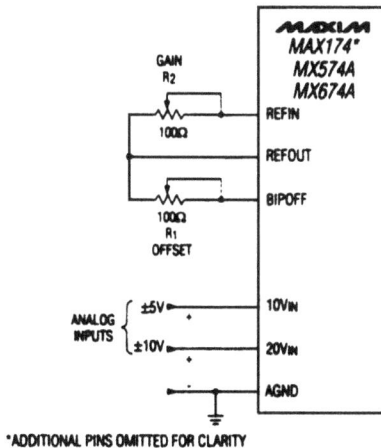

**Figure 10.26** Bipolar input connections (*Maxim New Releases Data Book,* 1992, p. 7-93).

scopes. However, a logic or digital probe and a digital pulser can make life much easier if you must regularly test and troubleshoot data converters (or any other digital device). So this section starts with brief descriptions of the probe and pulser.

### 10.4.1   Logic or digital probe

Logic probes are used to monitor the in-circuit pulse or logic activity. By means of a simple lamp indicator, a logic probe shows you the logic state of the digital signal and allows detection of brief pulses (the ones you might miss with a scope). Logic probes detect and indicate high and low (1 or 0) logic levels and intermediate or "bad" logic levels (indicating an open circuit) at the terminals of a logic element, such an ADC or DAC.

Not all logic probes have the same functions, and you must learn the operating characteristics for your particular probe. For example, on the more sophisticated probes, the indicator lamp can give any of four indications: off, dim (about half brilliance), bright (full brilliance), or flashing on and off.

The lamp is normally in the dim state and must be driven to one of the other three states by voltage levels at the probe tip. The lamp is dim for voltages between the 1 and 0 states and for open circuits. Pulsating inputs cause the lamp to flash at about 10 Hz (regardless of the input pulse rate). The probe is particularly effective when used with the logic pulser.

### 10.4.2   Logic pulser

The hand-held logic pulser (similar in appearance to the logic probe) is an in-circuit stimulus device that automatically outputs pulses of the required logic polarity, amplitude, current, and width to drive lines and other test points high or low. A typical pulser has several pulse burst and stream modes.

Logic pulsers are compatible with most digital devices. Pulse amplitude depends on the equipment supply voltage, which is also the supply voltage for the pulser. Pulse current and pulse width depend on the load being pulsed. A switch controls the frequency and number of pulses that are generated. A flashing LED indicator on the pulser tip indicates the output mode.

The logic pulser forces overriding pulses into lines or test points, and it can be programmed to output single pulses, steady pulse streams, or bursts. The pulser can be used to enable or clock ICs. The circuit inputs can also be pulsed while the effects of the circuit outputs are observed with a logic probe.

### 10.4.3   General digital troubleshooting tips

The following troubleshooting tips are not limited to ADCs and DACs. They also apply to all digital circuits in which most of the components are contained within ICs.

#### 10.4.4  Power and ground connections

The first step in tracing problems in a digital circuit with ICs is to check all power and ground connections to the ICs. Many ICs have more than one power and one ground connection. For example, the LTC1090 in Fig. 10.27 requires +5 V at the $V_{CC}$ pin and ground at the $V-$ pin. The IC also has both a digital ground (DGND) and an analog ground (AGND), as well as a common (COM) pin, and a minus-reference (REF−) pin that must be grounded. The REF+ pin must also be connected to 5-V power. If any of these power or ground connections is absent or abnormal, the IC cannot operate properly.

#### 10.4.5  Reset, chop-select, and start signals

With all power and ground connections confirmed, check that all the ICs receive reset, chip-select, and start signals, as required. For example, the DAC-4881 in Fig. 10.28 requires a chip-select at pin 1 and address-decode signals at pins 2 and 28. Likewise, the ADC0808/0809 in Fig. 10.29 requires start, address latch enable (ALE), end of conversion (EOC), and output enable signals (Fig. 10.29) from a microprocessor or control logic. If any of these signals is absent or abnormal (for example, incorrect amplitude, improper timing) circuit operation comes to an immediate halt.

In some cases, control signals to digital ICs are pulses (usually timed in a certain sequence), whereas other control signals are steady (high or low). If any of the lines carrying the signals to the IC are open, shorted to ground, or to power (typically +5 V or +12 V), the IC will not function. So if you find an IC pin that is always high, always low, or apparently is connected to nothing (floating), check the PC traces or other wiring to that pin carefully. This

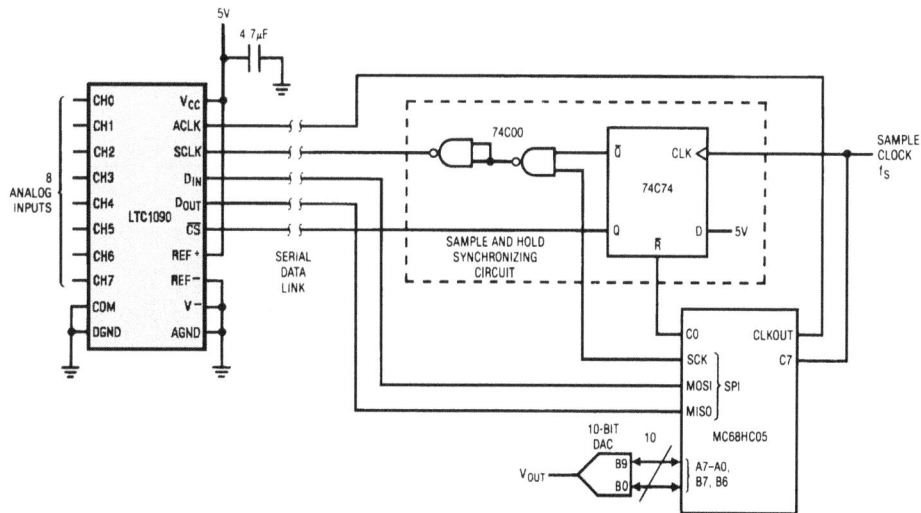

**Figure 10.27**  Typical power and ground connections for digital ICs (Linear Technology, *Linear Applications Handbook,* 1993, p. DN2-1).

+V_S  
10K to 100K  Gain Adjust  -V_S  
500K  
$R_{sum}$**  
DAC-4881  

18 Bipolar Offset   26 Gain Adjust   21 Sum Node  
20V R 20   10V R 19   $C_F$**  
+V_S   1M   Offset Adjust   10K to 100K   -V_S  
17 Ref In   1 μF*  
−  +   16   V_OUT   -10V to +10V  
27 Ref Out   +10V Reference  
Ref Amp Inputs   12-Bit DAC   I_O   $\bar{I}_O$   23   22  
24 Gnd  
+V_S 25   1 μF   -V_S 15   1 μF  
1 CS   2 ADH   Control Logic   4 Bit Latch   8 Bit Latch   28 ADL  
4 LSB   8 MSB   14 = LSB   3 = MSB  
14–11   10–3  
LSB Address Decode   MSB Address Decode  
4-Bit Data Bus  
WR   Address Bus  

Calibration Procedure:
1. Set inputs to all ones
2. Adjust offset until V_OUT equals − full scale
3. Set inputs to all zeros
4. Adjust gain until V_OUT equals − full scale — 1 LSB

*Optional — reduces reference noise  
**Optional — improves settling time (see table for values)

65-00573B

| Format | Output Scale | MSB B1 | B2 | B3 | B4 | B5 | B6 | B7 | B8 | B9 | B10 | B11 | LSB B12 | I_O(mA) | $\bar{I}_O$(mA) | V_OUT |
|---|---|---|---|---|---|---|---|---|---|---|---|---|---|---|---|---|
| Offset Binary, True Zero Output | Positive Full Scale | 1 | 1 | 1 | 1 | 1 | 1 | 1 | 1 | 1 | 1 | 1 | 1 | 3.999 | 0.000 | 9.9951 |
| | Positive Full Scale — LSB | 1 | 1 | 1 | 1 | 1 | 1 | 1 | 1 | 1 | 1 | 1 | 0 | 3.998 | 0.001 | 9.9902 |
| | + LSB | 1 | 0 | 0 | 0 | 0 | 0 | 0 | 0 | 0 | 0 | 0 | 1 | 2.001 | 1.998 | 0.0049 |
| | Zero Scale | 1 | 0 | 0 | 0 | 0 | 0 | 0 | 0 | 0 | 0 | 0 | 0 | 2.000 | 1.999 | 0.000 |
| | − LSB | 0 | 1 | 1 | 1 | 1 | 1 | 1 | 1 | 1 | 1 | 1 | 1 | 1.999 | 2.000 | -0.0049 |
| | Negative Full Scale +LSB | 0 | 0 | 0 | 0 | 0 | 0 | 0 | 0 | 0 | 0 | 0 | 1 | 0.001 | 3.998 | -9.9951 |
| | Negative Full Scale | 0 | 0 | 0 | 0 | 0 | 0 | 0 | 0 | 0 | 0 | 0 | 0 | 0.000 | 3.999 | -10.000 |
| 2's Complement, True Zero Output MSB Complemented (Need Inverter at B1) | Positive Full Scale | 0 | 1 | 1 | 1 | 1 | 1 | 1 | 1 | 1 | 1 | 1 | 1 | 3.999 | 0.000 | 9.9951 |
| | Positive Full Scale — LSB | 0 | 1 | 1 | 1 | 1 | 1 | 1 | 1 | 1 | 1 | 1 | 0 | 3.998 | 0.001 | 9.9902 |
| | + LSB | 0 | 0 | 0 | 0 | 0 | 0 | 0 | 0 | 0 | 0 | 0 | 1 | 2.001 | 1.998 | 0.0049 |
| | Zero Scale | 0 | 0 | 0 | 0 | 0 | 0 | 0 | 0 | 0 | 0 | 0 | 0 | 2.000 | 1.999 | 0.0000 |
| | − LSB | 1 | 1 | 1 | 1 | 1 | 1 | 1 | 1 | 1 | 1 | 1 | 1 | 1.999 | 2.000 | -0.0049 |
| | Negative Full Scale +LSB | 1 | 0 | 0 | 0 | 0 | 0 | 0 | 0 | 0 | 0 | 0 | 1 | 0.001 | 3.998 | -9.9951 |
| | Negative Full Scale | 1 | 0 | 0 | 0 | 0 | 0 | 0 | 0 | 0 | 0 | 0 | 0 | 0.000 | 3.999 | -10.000 |

Figure 10.28  Typical control signals for digital IC (*Raytheon Semiconductor Data Book,* 1994, p. 6-48).

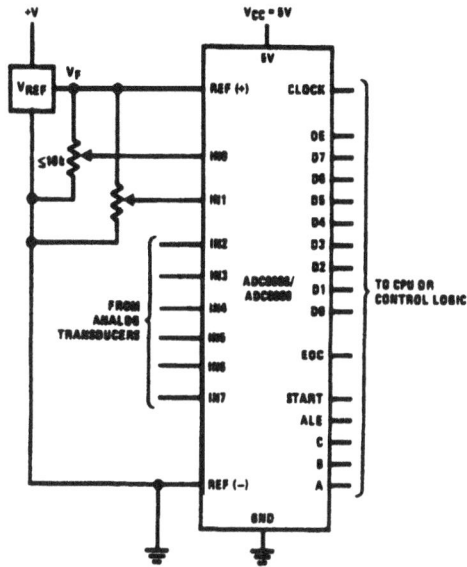

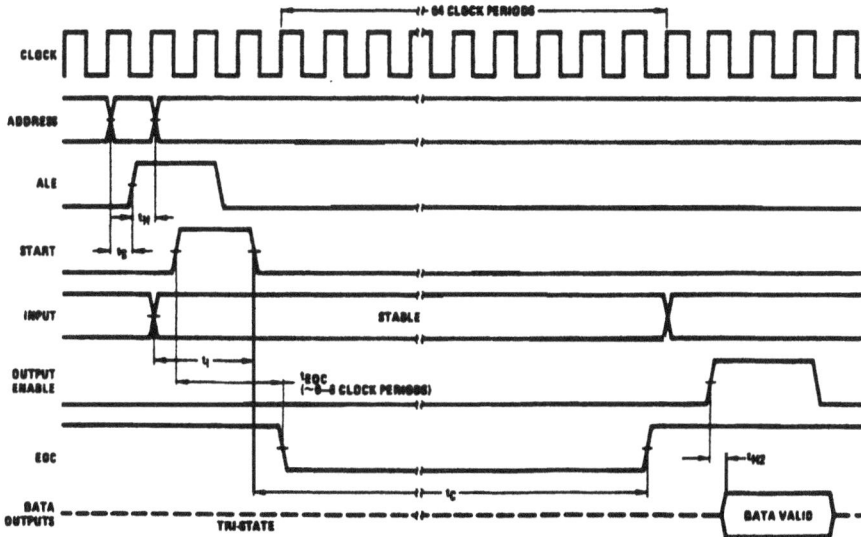

**Figure 10.29**   Typical control signals and timing diagram for digital ICs (Linear Technology, *Linear Applications Handbook,* 1994, p. 531/532).

applies to all control pins, unless the circuit calls for the control function to be steady (such as a steady +5 V on a pin to turn a circuit on). For example, if the DAC-4881 is connected as an 8-bit with complementary input DAC (as shown in Fig. 10.28), the chip-select (pin 1) must receive a write ($\overline{WR}$) signal, and the address-decode pins must receive address bits from the microprocessor.

### 10.4.6  Clock signals

Most digital ICs require clocks. For example, Fig. 10.29 shows the clock periods for the ADC. In this case, the clock comes from an external source. In other cases, the clock is part of the circuit. In general, the presence of pulse activity on any pin of a digital IC indicates the presence of a clock, but do not count on it. Check directly at the clock pins (all ICs that require a clock typically are connected to the same clock source or sources).

It is possible to measure the presence of a clock signal with a scope or logic probe. However, a frequency counter provides the most accurate measurement. If any ICs do not receive the required clock signals, the IC cannot function. On the other hand, if the clock is off frequency, all of the ICs might appear to have a clock signal, but the IC function can be impaired. Crystal-controlled clocks do not usually drift far off frequency, but can go into some overtone frequency (typically, a third overtone) beyond the capacity of the IC.

### 10.4.7  Input-output signals

When you are certain that all ICs are good and have proper power and ground connections and that all control signals (such as reset and chip-select) and clock signals are available, the next step is to monitor all input and output signals at each IC. This can be performed with either a scope or a probe. The following are some examples that apply to ADCs and DACs.

### 10.4.8  Basic ADC testing and troubleshooting

ADCs can be tested by applying precision voltages at the input and monitoring the output for corresponding digital values. For example, a fixed voltage between 0 and +10 V can be applied to the noninverting input of the 4805 in Fig. 10.30, and the corresponding value can be read out at the lines between the SAR-2504 and DAC-6012. The lines should go to +5 V for a digital 1 and to ground (or 0 V) for a digital 0. Also, a serial digital output is at the D0 pin of the 2504. This output must be monitored with a scope.

The rate at which the conversions are made is controlled by the clock at the CP pin of the 2504. Start (S) pin of the 2504 is connected to the conversion-complete (CC) pin to provide continuous digital output for the analog input. In the circuit of Fig. 10.30, the start pin must receive a conversion command from an external source (typically a microprocessor), whereas the conversion complete becomes an output to the microprocessor (indicating status and conversion complete or not complete).

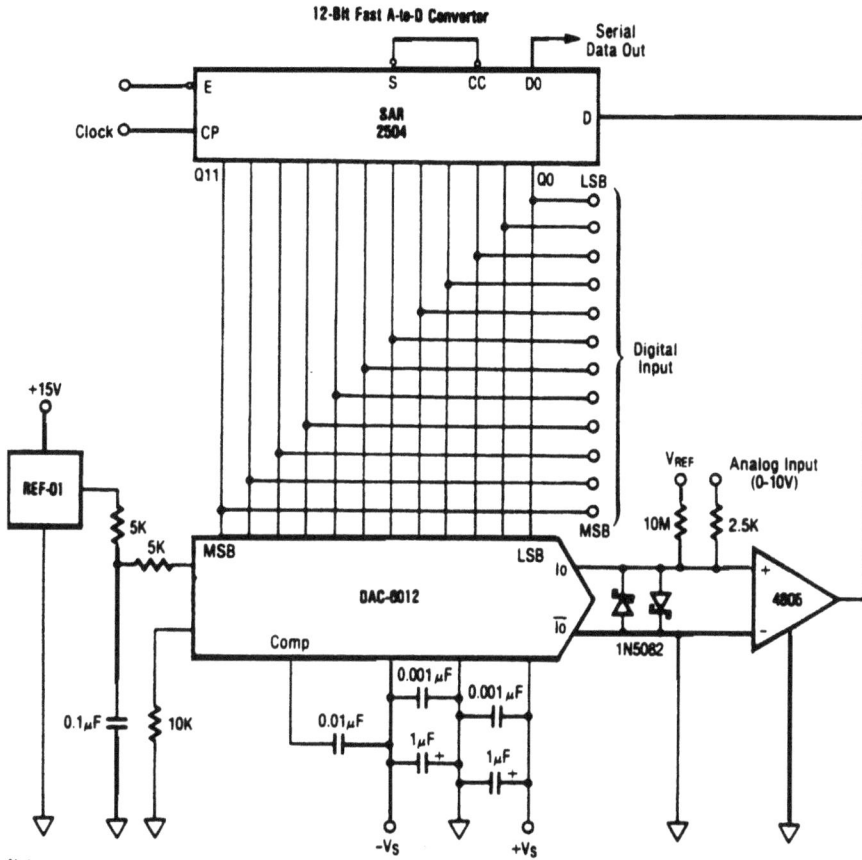

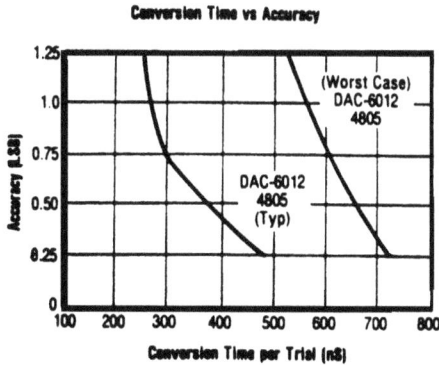

Figure 10.30   Typical test points for ADC circuits (*Raytheon Semiconductor Data Book,* 1994, p. 6-82).

If the output readings of the ADC are slightly off, try correcting the problem by adjustment. In the circuit of Fig. 10.30, the $V_{REF}$ voltage (at the noninverting input of the 4805) can be varied for zero (0 V at the analog input should make all digital outputs ground, 0 V). Also, REF-01 can be trimmed for full-scale output. This is done by connecting a 10-kΩ pot between the output (pin 6) of the REF-01 and ground (pin 4). The wiper of the pot is connected to trim (pin 5) of the REF-01. The accuracy of this circuit depends on the precision of the two 5-kΩ resistors between the REF-01 and pin 14 of the DAC-6012, and on the 2.5-kΩ resistor at the analog input.

### 10.4.9  Basic DAC testing and troubleshooting

DACs can be tested by applying digital inputs and monitoring the output for corresponding voltages. For example, the B1 through B10 inputs of the DAC-10 can be connected to ground (for a 0) or to +5 V (for a 1), and the output can be monitored with a precision voltmeter at pins 2 and 4 in the circuit of Fig. 10.31. If both voltages are slightly off, suspect a problem with the 2.00-mA references. If one of the output voltages is slightly off, suspect that the corresponding 1.25-kΩ resistors have failed. If the output voltages are absent or way off, suspect the DAC-10.

## 10.5  Data-Converter Test Parameters

Many specialized terms are used when testing ADCs and DACs against manufacturers' specifications as part of the troubleshooting process. The following is a summary of the most common terms.

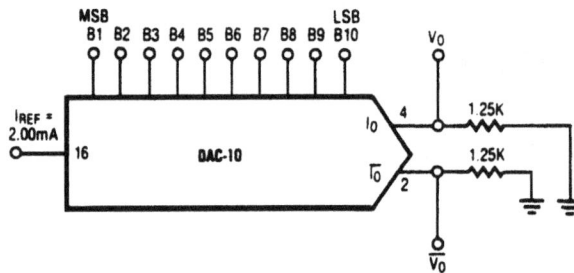

|  | B1 B2 B3 B4 B5 B6 B7 B8 B9 B10 | $I_O$ mA | $\bar{I_O}$ mA | $V_0$ | $\bar{V_0}$ |
|---|---|---|---|---|---|
| Full Scale | 1 1 1 1 1 1 1 1 1 1 | 3.996 | 0.000 | -4.995 | -0.000 |
| Half Scale +LSB | 1 0 0 0 0 0 0 0 0 1 | 2.004 | 1.992 | -2.505 | -2.490 |
| Half Scale | 1 0 0 0 0 0 0 0 0 0 | 2.000 | 1.996 | -2.500 | -2.495 |
| Half Scale -LSB | 0 1 1 1 1 1 1 1 1 1 | 1.996 | 2.000 | -2.495 | -2.500 |
| Zero Scale +LSB | 0 0 0 0 0 0 0 0 0 1 | 0.004 | 3.992 | -0.005 | -4.990 |
| Zero Scale | 0 0 0 0 0 0 0 0 0 0 | 0.000 | 3.996 | -0.000 | -4.995 |

**Figure 10.31**  Typical test points for DAC circuits (*Raytheon Semiconductor Data Book,* 1994, p. 6-34).

## 10.5.1   Resolution and accuracy

The terms *resolution* and *accuracy* are often interchanged (although incorrectly). In a DAC, *resolution* describes the smallest standard incremental change in output voltage. In an ADC, *resolution* is the amount of input-voltage change required to increment the output between one code change and the next adjacent code change. Both definitions of *resolution* differ from the definition of *accuracy,* which is the absolute error incurred in measurement of a signal. In many data converters, *accuracy* does not match *resolution.* Consider some examples.

A converter with $n$ switches can resolve one part in $2^n$. The least-significant increment is then $2^{-n}$, or 1 LSB. In contrast, the MSB carries a weight of $2^{-1}$. Resolution can be expressed in percentage of full scale or in binary bits. For example, an ADC with 12-bit resolution can resolve one part in $2^{12}$ (one part in 4096) = 0.0244% of full scale. A converter with 10-V full scale can resolve a 2.44-mV input change. Likewise, a 12-bit DAC shows an output-voltage change of 0.0244% of full scale when the binary input code is incremented one binary bit (1 LSB). Resolution is a design parameter, rather than a performance specification (because resolution says nothing about accuracy or linearity).

A *linearity* specification (see Section 10.5.2) is sometimes used in place of accuracy in data converters because linearity is more descriptive. When used, an accuracy specification describes the worst-case deviation of a DAC output voltage from a straight line drawn between zero and full scale.

For example, a 12-bit DAC cannot have a conversion accuracy better than $\pm^1/_2$ LSB or $\pm1$ part in $2^{12+1}$ ($\pm0.0122\%$ of full scale) because of the finite resolution. This is the case in Fig. 10.32, if there are no other errors. Note that *$\pm0.0122\%$ full scale* represents a deviation from 100% accuracy. Therefore, accuracy should be specified as 99.9878%. However, most data sheets will use 0.0122% as an accuracy specification, rather than an inaccuracy (tolerance or error) specification (to further confuse users).

In an ADC, *accuracy* describes the difference between the actual input voltage and the full-scale weighted equivalent of the binary output code, including all other errors (see Section 10.5.3). For example, if a 12-bit ADC is said to be $\pm1$ LSB accurate, this is equivalent to $\pm0.0245\%$, or twice the minimum possible quantizing error of 0.0122%. In effect, an accuracy specification describes the maximum sum of all errors.

## 10.5.2   Linearity

Linearity specifications describe the departure from a linear transfer curve for an ADC or DAC. Linearity error does not include quantizing, offset, zero, or scale errors (see Section 10.5.3). Thus, a specification of $\pm^1/_2$ LSB linearity implies an error, in addition to the inherent $\pm^1/_2$ LSB quantizing or resolution error. This is shown in Fig. 10.33, in which a linearity error allows one or more of the steps to be greater or less than the ideal shown.

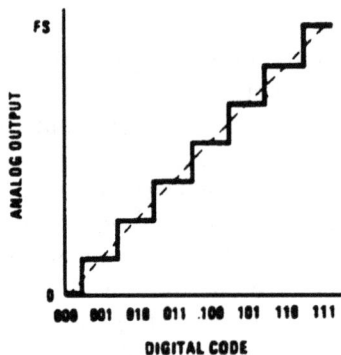

**Figure 10.32** Linear DAC transfer curve (National Semiconductor, *Linear Applications Handbook,* 1994, p. 333).

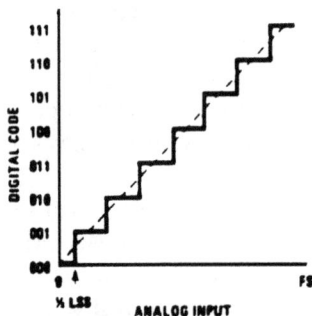

**Figure 10.33** ADC transfer curve with $^1/_2$ LSB offset at zero (National Semiconductor, *Linear Applications Handbook,* 1994, p. 334).

One of the problems with linearity specifications (or nonlinearity specifications, if you prefer) is that there are two testing approaches. Figure 10.34 shows how these two approaches (the *best-straight-line* and *endpoint-fit*) produce different specifications for an ADC.

The best straight-line approach makes no claim about zero error, full-scale error, or transfer-function slope, but simply quantifies (in LSBs or percentages) deviation from the straight line that best approaches the transfer function. In effect, the best-straight-line method provides the lowest (or "best looking") number. No points on the line are defined before the test. The result is a pure linearity specification that includes no other errors.

The endpoint-fit approach presets the ideal line between the measured endpoints of the data-converter transfer function. Deviations are measured without adjusting the position of the line for any optimum fit. As a result, the endpoint-fit linearity number is usually larger than that of the best-straight-line approach. However, both methods are valid ways of representing linearity (also called *integral nonlinearity, INLX,* on some data sheets).

Figure 10.35 shows a 3-bit DAC transfer curve with no more than $\pm^1/_2$ LSB nonlinearity, yet one step is of zero amplitude. This is within the specification, because the maximum deviation from the ideal straight line is $\pm 1$ LSB ($^1/_2$ LSB resolution error, plus $^1/_2$ LSB nonlinearity).

With any linearity error, there is *differential nonlinearity* (Section 10.5.9). A $\pm^1/_2$ LSB linearity specification guarantees monotonicity (Section 10.5.10) and an equal or better than $\pm 1$ LSB differential nonlinearity (DNL). In the example of Fig. 10-35, the code transition from 100 to 101 is the worst possible non-linearity (1 LSB high at code 100 to 1 LSB low at 110). Any fractional nonlinearity beyond $\pm^1\!/_2$ LSB allows for a nonmonotonic transfer curve. Figure 10.36 shows a typical nonlinear curve where the nonlinearity is $1^1\!/_4$ LSB, yet the curve is smooth and monotonic.

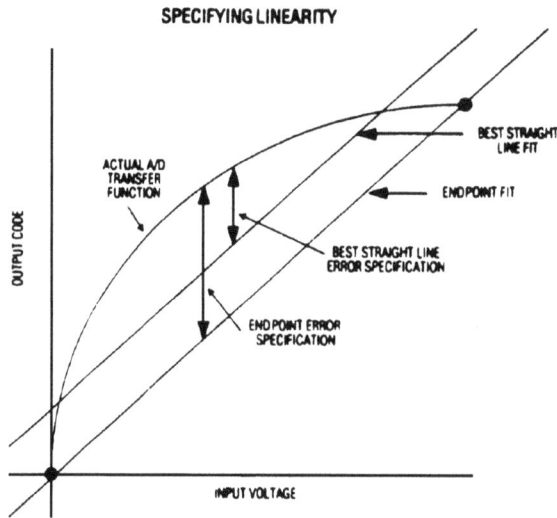

SPECIFYING LINEARITY

Figure 10.34  Best-straight-line and end-point linear curves (*Maxim New Releases Data Book,* 1992, p. 7-9).

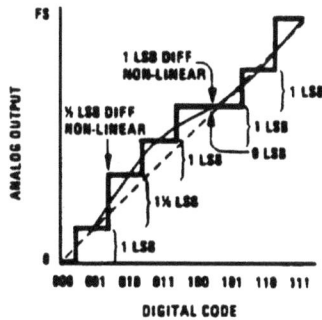

Figure 10.35  $\pm^1\!/_2$ LSB nonlinearity curve (National Semiconductor, *Linear Applications Handbook,* 1994, p. 335).

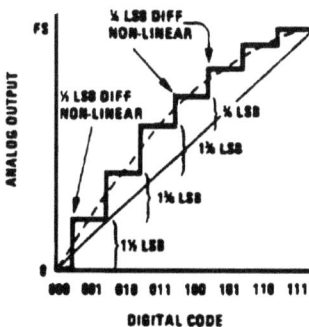

Figure 10.36  $\pm1^1\!/_4$ LSB nonlinearity curve (National Semiconductor, *Linear Applications Handbook,* 1994, p. 335).

TL/H/5612-5

### 10.5.3 Quantizing error

The term *quantizing error* is usually applied to an ADC. (The equivalent effect in a DAC is more properly called *resolution error.*) In any case, quantizing error is the maximum deviation from a straight-line transfer function of a perfect ADC. Because an ADC quantizes the analog input into a finite number of output codes, only an ADC with infinite resolution can show zero quantizing error. Figure 10.33 shows the transfer function of such an ADC, suitably offset $\frac{1}{2}$ LSB at zero scale. This transfer function shows $\pm\frac{1}{2}$ LSB maximum output error. If there were no offset, the error would be $-1/+0$ LSB (as shown in Fig. 10.37). For example, a perfect 12-bit ADC shows $\pm\frac{1}{2}$ LSB error of $\pm0.0122\%$, whereas the quantizing error of an 8-bit ADC is $\pm\frac{1}{2}$ part in $2^8=\pm0.195\%$ of full scale.

### 10.5.4 Scale error

Scale error (also known as *full-scale error*) is the departure from design output voltage of a DAC for a given input code (usually full-scale code). Figure 10.38 shows a transfer function with a linear 1 LSB scale error. In an ADC, scale error is the departure of actual input voltage from the design input voltage for a full-scale output code.

Scale errors can be caused by errors in reference voltage, ladder resistor values, or amplifier gain, and can be corrected by adjustments in output ampli-

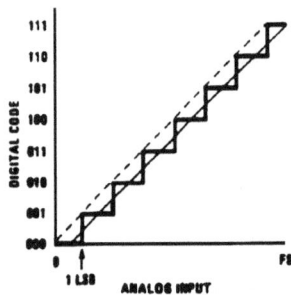

**Figure 10.37** ADC transfer curves with no offset (National Semiconductor, *Linear Applications Handbook*, 1994, p. 334).

TL/H/5612-2

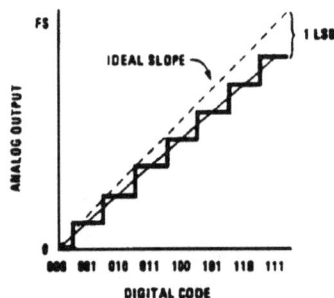

**Figure 10.38** Linear curve with 1 LSB scale error (National Semiconductor, *Linear Applications Handbook*, 1994, p. 334).

TL/H/5612-3

fier gain or reference voltage. For example, if the transfer curve resembles that of Fig. 10.36, a scale adjustment at $^3/_4$ scale could improve the overall ± accuracy, compared with an adjustment at full scale.

### 10.5.5  Gain error

In an ADC, gain error is essentially the same as scale error. In the case of a DAC with current and voltage-mode outputs, the current output can be to scale, but the voltage output might show some gain error. In any case, the amplifier-feedback resistors can be trimmed to correct gain-error problems.

### 10.5.6  Offset error

*Offset error* (also known as *zero error*) is the output voltage of a DAC, with zero-code input. In an ADC, offset error is the required mean value of input voltage to set a zero-code output. Figure 10.39 shows a DAC transfer curve with $^1/_2$ LSB offset at zero.

Offset error is usually caused by an input-offset voltage or current of the amplifier or comparator within the converter IC, and it can be trimmed to zero with an external offset-zero pot. Offset error can be expressed as a percentage of full scale or in a fraction of an LDB.

### 10.5.7  Hysteresis error

The term *hysteresis error* usually applies to an ADC. The error causes the voltage at which an ADC code transition occurs to depend on the direction from which the transition is approached. Hysteresis error in an ADC is usually caused by hysteresis in the internal comparator. Excessive hysteresis is reduced by design of the converter. However, some slight hysteresis is inevitable. For these purposes, hysteresis is objectionable if it approaches $^1/_2$ LSB.

### 10.5.8  Trimming data-converter errors

Some data-converter ICs provide pins for external trimming to offset any errors or to get a desired accuracy. In other cases, the accuracy is built in

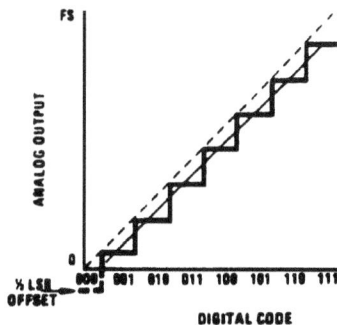

**Figure 10.39**  DAC transfer curve with $^1/_2$ LSB offset at zero (National Semiconductor, *Linear Applications Handbook,* 1994, p. 334).

TL/H/5612-4

(usually at a higher cost). Also, in some instances, greater accuracy or offset is simply not needed. Of course, if space is the ultimate consideration, you must use the converter without external components. Unfortunately, this might mean using a more expensive converter to get the required accuracy. The following points should be considered when pondering the trimmed (as opposed to untrimmed) accuracy tradeoff.

A typical low-cost 12-bit ADC (such as a MAX172) resolves a 5-V full-scale input range to 5 V/4096 = 1.2 mV (1 LSB), but is not accurate to this level. The untrimmed full-scale error limit of the B-grade IC is 15 LSB, with an offset limit of 6 LSB. These guarantees allow reduced cost without sacrificing linearity, which is guaranteed at $^1/_2$ LSB for the A-grade or 1 LSB for the B-grade device. Twelve-bit linearity is maintained (even on the lowest grade).

In practical troubleshooting, unadjusted full-scale and offset errors are often not crucial. This is because such errors are constant, and errors in other parts of the signal path are often trimmed. As a general guideline, ADC accuracy and offset specifications do not need full 12-bit precision when:

1. Only signal changes, not absolute voltage levels, are of interest.

2. Transducers or sensors that do not have precise accuracy specifications are being measured.

3. System calibration occurs elsewhere—either with manual trims or through a microprocessor or controller.

4. The ADC operates in a closed-loop control system in which gain and offset errors affect only loop dynamics and not accuracy.

### 10.5.9   Differential nonlinearity (DNL)

In an ADC, DNL indicates that the device is monotonic (see Section 10.5.10) or has no missing codes. DNL is the deviation of the analog span of each ADC output code from the ideal 1 LSB value. Figure 10.40 shows some typical DNL errors. For example, a DNL specification of $^1/_2$ LSB means that a code is at least $^1/_2$ LSB, but no more than 1.5 LSB wide. If DNL is less than 1 LSB, no missing codes is ensured. (Some data sheets list DNL and no missing codes as separate characteristics.)

In a DAC, DNL indicates the difference between actual analog voltage change and the ideal (1 LSB) voltage change at any code change. For example, a DAC with a 1.5 LSB step at a code change is said to show a DNL of $^1/_2$ LSB (see Figs. 10.35 and 10.36).

In troubleshooting, DNL specifications are as important as linearity specifications because the apparent quality of a data-converter curve can be markedly affected by DNL, even though linearity is good. For example, Fig. 10.35 shows a curve with $\pm^1/_2$ LSB linearity and $\pm1$ LSB DNL. Figure 10.36 shows a curve with $+1^1/_4$ LSB linearity and $\pm^1/_2$ LSB DNL. In many applications, the curve of Fig. 10.36 would be preferred over that of Fig. 10.35 because the Fig. 10.36

curve is smoother. (In simple terms, DNL describes the smoothness of the curve and is thus of great importance to the user.)

Figure 10.41 shows an exaggerated example of DNL in which DNL is ±2 LSB and the linearity specification is ±1 LSB. This results in a transfer curve with a grossly degraded resolution. For example, the normal 8-step curve is reduced to three steps and a 16-step curve (4-bit converter) with only 2 LSB DNL is reduced to six steps (a nonexistent 2.6-bit device!).

Unfortunately, DNL is not always listed on data sheets by all manufacturers. One reason for this is that DNL is difficult to measure on a production-line basis. Listed or not, DNL can be as much as twice nonlinearity, but never more.

## 10.5.10  Monotonicity

Figure 10.42 shows a nonmonotonic DAC transfer curve. For the curve to be nonmonotonic, the linearity error must exceed $\pm^1/_2$ LSB (by any amount, no matter how little). The greater the linearity error, the more significant the negative step might be. On the other hand, a monotonic curve has no change in the sign of the slope. All incremental elements of a monotonically increasing curve

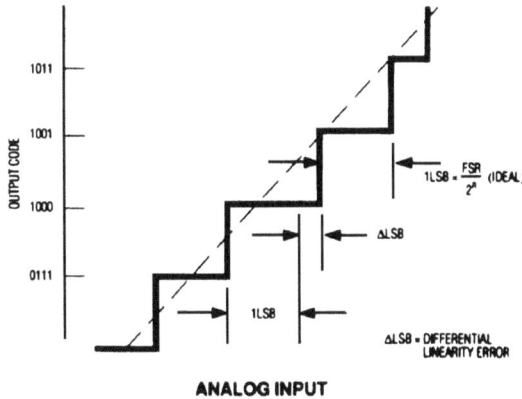

**Figure 10.40**  Typical DNL errors (*Maxim New Releases Data Book,* 1992, p. 7-10).

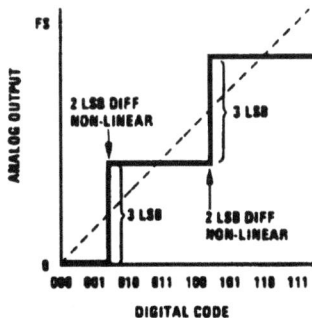

**Figure 10.41**  Exaggerated example of DNL (National Semiconductor, *Linear Applications Handbook,* 1994, p. 335).

TL/H/5612–6

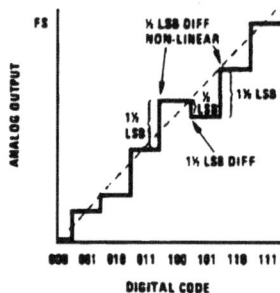

Figure 10.42 Nonmonotonic DAC transfer curve (National Semiconductor, *Linear Applications Handbook,* 1994, p. 336).

TL/H/5612-7

have positive or zero (but never negative) slope. The converse is true for decreasing curves.

A converter showing more than $\pm^1/_2$ LSB nonlinearity can still be monotonic up to a certain point, but not beyond that point. For example, a 12-bit DAC with $\pm^1/_2$-bit linearity to 10 bits (not a true $\pm^1/_2$ LSB) will be monotonic at 10 bits, but might not be monotonic at 12 bits (unless tested and guaranteed to be 12-bit monotonic). In troubleshooting, a nonmonotonic converter might be acceptable for some applications, but is disastrous in closed-loop servo systems, including DAC-controlled ADCs.

### 10.5.11  Settling time and slew rate

As covered in Section 10.2.7, settling time is a crucial factor in converter performance and is often listed along with slew rate. Settling time is the elapsed time after a code transition for DAC output to reach final value within specified limits, usually $\pm^1/_2$ LSB. Slew rate is an inherent limitation of the output amplifier in a DAC and it functions to limit the output voltage rate of change after code transitions.

Settling time is often summed with slew rate to get total elapsed time for the output to settle to the final value. This is shown in Fig. 10.43, which delineates the part of total elapsed time considered to be slew rate and the part this is settling time. As shown, the total time is greater for a major code change than a minor code change because of amplifier slew limitations. However, settling time can also be different, depending on amplifier overload-recovery characteristics.

### 10.5.12  Conversion rate

Both settling time and slew rate (as well as delay in counting circuits, ladder switches, and comparators) affect conversion rate (the speed at which an ADC or DAC can make repetitive data conversions). On some data sheets, conversion time is specified as a number of conversions per second. Other data sheets list conversion rate as the number of microseconds required to complete one conversion (including the effects of settling time). Some data sheets specify conversion rate for something less than full resolution, thus showing a misleading (high) rate.

### 10.5.13  Temperature coefficient and long-term drift

Temperature coefficient, or $T_C$, of the various components of a DAC or ADC can produce (or increase) any of several errors when the operating temperature varies. Zero-scale offset error can change because of the amplifier and comparator input-offset $T_C$. Scale error can occur because of shifts in the reference, changes in ladder resistance, change of beta in current switches, or drift in amplifier gain-set resistors. Linearity and monotonicity of the DAC can be affected by different temperature drifts of the ladder resistors and switches. Many other characteristics can be affected by $T_C$. In fact, with the possible exception of resolution and quantizing error, all data-converter characteristics can be affected by temperature changes.

Long-term drift, caused mainly by resistor and semiconductor aging, can affect all characteristics that temperature change can affect. The characteristics most commonly affected by long-term drift are linearity, monotonicity, scale, and offset. Scale changes because of reference-voltage changes are usually the most important long-term change or drift problem.

### 10.5.14  Overshoot and glitches

Both overshoot and glitches are essentially DAC problems. However, because most ADCs contain a DAC, ADCs can also be affected. Overshoot and glitches occur whenever a code transition occurs in a DAC. There are two causes. The current output of a DAC contains switching glitches because of possible asynchronous switching of the bit currents (expected to be worst at $1/2$-scale transitions when all bits are switched). Although such glitches are of extremely short duration, there could be $1/2$ scale in amplitude.

Although the glitches are generally attenuated at the DAC voltage output (because the amplifier is unable to slew at a very high rate), the glitches are coupled around the amplifier through the feedback network. In addition to the glitches, the output amplifier introduces some overshoot and some noncrucial damped ringing. These problems can be minimized, but not entirely eliminated

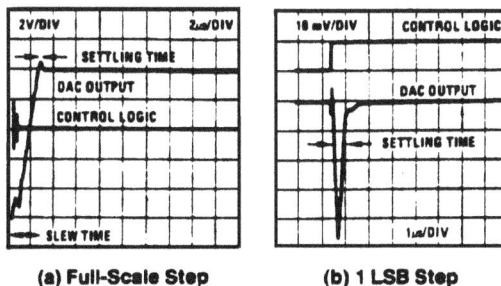

TL/H/5612-8

(a) Full-Scale Step          (b) 1 LSB Step

**Figure 10.43**   DAC slew and settling times (National Semiconductor, *Linear Applications Handbook,* 1994, p. 336).

(except at the expense of slew rate and settling time, which are much more important).

### 10.5.15    Power supply rejection

*Supply rejection* relates to the ability of a DAC or ADC to maintain scale, offset, $T_C$, slew rate, and linearity when the supply voltage is varied. The reference voltage must be constant (unless considering a multiplying DAC). Most affected are current sources (affecting linearity and scale), and amplifiers or comparators (affecting offset and slew rate). Supply rejection is usually specified as a percentage of full-scale change at or near full scale (at 25°C).

### 10.5.16    Input impedance and output drive

The *input impedance* of an ADC describes the load placed on the analog source. The *output drive* describes the digital load-driving capability of an ADC or the analog load-driving capability of an ADC or the analog load-driving capacity of a DAC. The output drive is usually given as a current level or a voltage output into a given load.

### 10.5.17    Clock rate

For an ADC, *clock rate* is the minimum or maximum pulse rate at which the counters can be driven. The fixed relationship between minimum conversion rate and clock rate depends on converter accuracy and type. All factors that affect the conversion rate of an ADC limit the clock rate.

### 10.5.18    Data-converter codes

Figure 10.44 shows some commonly used data-converter codes. Each code has advantages, depending on the system to be interfaced with the converter. The following is a summary of the most common data-converter codes.

*Natural (or simple) binary* is the usual $2^n$ code with 2, 4, 8, 16...$2^n$ progression. An input or output high (1) is considered a signal, whereas 0 is considered an absence of signal. This is a positive-true binary signal. Zero scale is all zeros and full scale is all ones.

*Complementary (or inverted) binary* is the negative-true binary system. It is identical to natural binary, except that all binary bits are inverted. Zero scale is all ones and full scale is all zeros.

*Binary coded decimal (BCD)* is the representation of decimal numbers in binary form. It is useful in ADC systems intended to drive decimal displays. The advantage of BCD over decimal is that only four lines are needed to represent 10 digits. The disadvantage of BCD is that a full 4 bits can represent 16 digits, whereas only 10 digits are represented in BCD. The full-scale resolution of a BCD-coded system is less than that of a binary-coded system. For example, a 12-bit BCD system has a resolution of only one part in 1000, com-

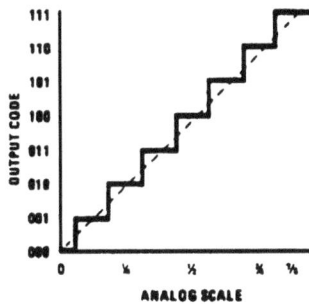

(a) Zero to + Full-Scale

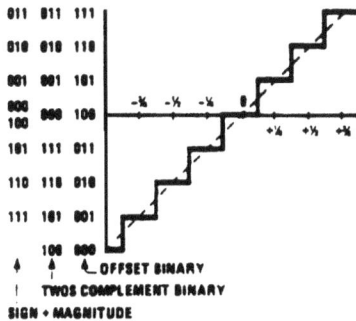

Figure 10.44  Typical ADC codes (National Semiconductor, *Linear Applications Handbook*, 1994, p. 337).

TL/H/5612-9

(b) ± Full-Scale

pared with one part in 4096 for a binary system. This represents a loss of resolution of more than 4:1.

*Offset binary* is a natural binary code, except that it is offset (usually $\frac{1}{2}$ scale) to represent negative and positive values. Maximum negative scale is represented as all zeros and with maximum positive scale all ones. Zero scale (actually center scale, as shown in Fig. 10.44) is then represented as a leading 1 and all remaining 0s.

*Two's complement binary* is widely used to represent negative values. With two's complement, zero and positive values are represented as in natural binary, with the all-negative values are represented in a two's complement form. That is, the two's complement of a number represents a negative value so that interface to a computer or microprocessor is simplified.

The two's complement is formed by complementing each bit and then adding a 1. Any overflow is neglected. For example, the decimal number −8 is represented in two's complement as follows: start with a binary code of decimal number 8 (off scale for ±representation in 4 bits, so not a value code in the ±scale of 4 bits), which is 1000. Complement this number to 0111 and add 0001 to get 1000.

The offset-binary representation of the ±scale differs from the two's complement representation only in that the MSB is complemented. The conversion from offset binary to two's complement requires that only the MSB be inverted.

*Sign plus magnitude* contains polarity information in the MSB. (When the MSB is 1, the number is negative.) All other bits represent magnitude only. One code is used up, providing a double code for zero (000 or 100). Sign plus magnitude code is used in certain instrument applications (such as digital voltmeters, DVMs) and in audio circuits. The advantage in both applications is that only one bit must be changed for small-scale changes when the value is near zero, and the plus-and-minus scales are symmetrical.

# Voltage-Frequency Converter Circuit Troubleshooting

This chapter is devoted to troubleshooting for voltage-frequency converter circuits. It starts with a review of basic voltage-frequency and frequency-voltage techniques. This is followed by basic voltage-frequency testing and troubleshooting. An entire section is devoted to voltage-frequency converter terms. Another section describes how design characteristics relate to troubleshooting.

Before getting started, let's resolve certain differences in terms. Some manufacturers refer to *voltage-to-frequency converters* as *VFCs*. Other manufacturers use the term *V/F converter*. The same is true for *frequency-to-voltage converters,* which are referred to as *FVCs* by some and as *F/V converters* by others. I prefer the terms *VFC* and *FVC,* but do not be surprised to find both terms in this book.

## 11.1   Basic Voltage-Frequency Conversion Techniques

This section describes the various VFC and FVC techniques in common use. This section concentrates on explaining the basic principles used in voltage-frequency conversion. By studying this information, you should be able to understand operation of the converter ICs that are described throughout the chapter. It is assumed that you are familiar with the principles of radio-frequency (RF) transmission and with basic digital electronics. If not, read *Lenk's RF Handbook* (McGraw-Hill, 1992) and *Lenk's Digital Handbook* (McGraw-Hill, 1993).

### 11.1.1   Basic VFC operation

Most present-day VFC and FVC circuits use a VFC IC as the basic converter element. Figure 11.1 shows such an IC connected as a VFC. The circuit is

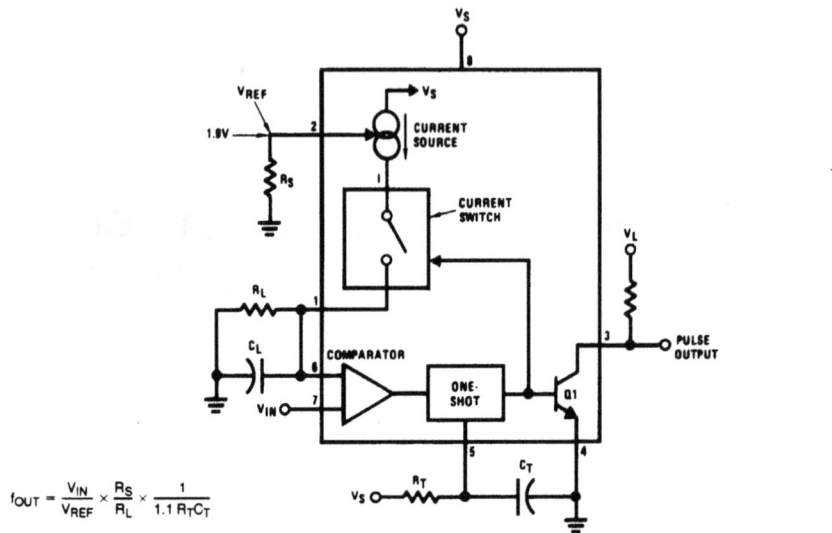

$$f_{OUT} = \frac{V_{IN}}{V_{REF}} \times \frac{R_S}{R_L} \times \frac{1}{1.1\,R_T C_T}$$

TL/H/8742-1

**Figure 11.1** VFC IC connected as a VFC (National Semiconductor, *Linear Applications Handbook,* 1994, p. 1247).

essentially a relaxation oscillator with an output frequency proportional to input voltage. The voltage to be converted is applied to a comparator within the IC at pin 7. The comparator output is applied to a one shot. Output pulses from the one-shot are applied to pin 3 (the circuit output) through Q1 and to a current switch.

The current switch interrupts currents to $R_L$ and it thus provides current pulses at pins 1 and 6. Except at zero, the current pulses keep the average voltage across $C_L$ (at pin 6 of the comparator) slightly greater than the input voltage, so the one-shot continues to produce pulses. Resistor $R_S$ is made adjustable so that the current source can be set to a given scale factor (frequency out for a given voltage in), with $R_T$ and $C_T$ selected for some given frequency range.

## 11.1.2    Basic FVC operation

The same VFC shown in Fig. 11.1 can also be used for frequency-to-voltage conversion. Figure 11.2 shows typical connections. Notice that for FVC, the frequency input (pulse or square wave) is applied to the comparator at pin 6. This connection produces pulses at the comparator output that correspond in frequency to the input.

The comparator pulses trigger the one shot. In turn, the one shot controls the amount of current at pin 1 (and thus the circuit output voltage) through the current switch. Resistor $R_S$ is made adjustable so that the current source can be set to produce a given scale factor (voltage out for a given frequency in), with $R_T$ and $C_T$ selected for some given frequency range. In the circuit of Fig. 11.2,

$R_S$ is adjusted so that the voltage output (pin 1) is 10 V when the input frequency is 10 kHz.

### 11.1.3 Ramp-comparator and charge-pump VFCs

In addition to the basic VFC of Fig. 11.2, there are other VFC circuits (usually found in discrete-component form). The most common VFC circuits are the ramp comparator and charge pump.

With the ramp-comparator VFC of Fig. 11.3, the input drives an integrator and the slope of the integrator ramp varies with the input-derived current. When the ramp crosses $V_{REF}$, the comparator turns on the switch, discharging the capacitor. This restarts the cycle. The frequency of this action directly relates to the input voltage. In some circuits, one op amp serves as both the integrator and the comparator.

With the charge-pump VFC of Fig. 11.4, the integrator is enclosed in a charge-dispensing loop. Capacitor C1 charges to $V_{REF}$ during the integrator-ramp time. When the comparator trips, C1 is discharged into the op-amp summing point, forcing the op amp high. After C1 discharges, the op amp begins to ramp and the cycle repeats (frequency is related to input voltage).

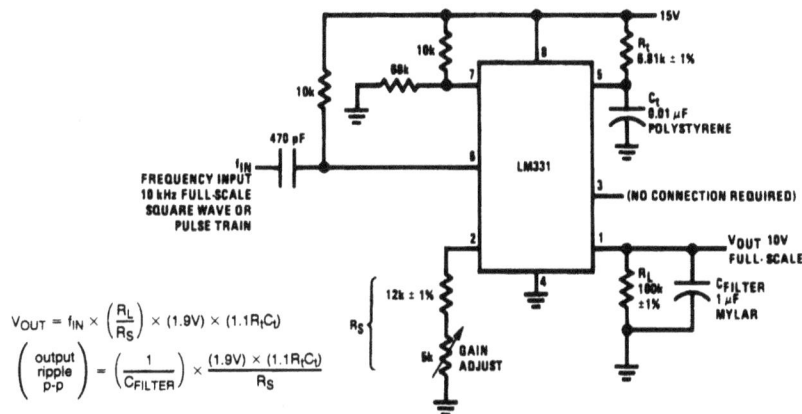

$$V_{OUT} = f_{IN} \times \left(\frac{R_L}{R_S}\right) \times (1.9V) \times (1.1R_tC_t)$$

$$\left(\begin{array}{c}\text{output}\\\text{ripple}\\\text{p-p}\end{array}\right) = \left(\frac{1}{C_{FILTER}}\right) \times \frac{(1.9V) \times (1.1R_tC_t)}{R_S}$$

**Figure 11.2** VFC IC connected as an FVC (National Semiconductor, *Linear Applications Handbook*, 1994, p. 1204).

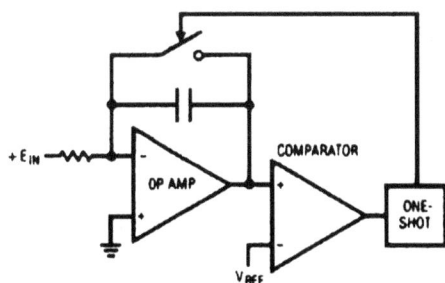

**Figure 11.3** Basic ramp-comparator VFC concept (Linear Technology, Application Note 3, p. 7).

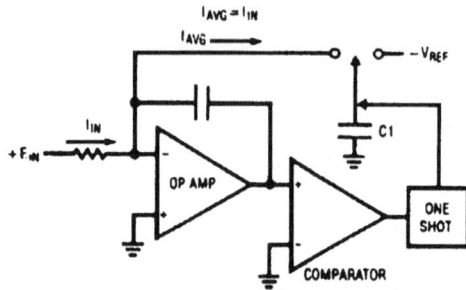

**Figure 11.4** Basic charge-pump VFC concept (Linear Technology, Application Note 3, p. 7).

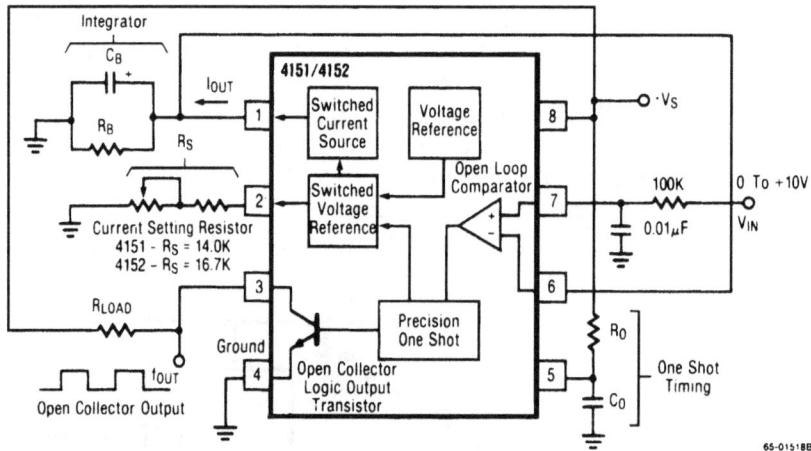

**Figure 11.5** Typical IC voltage-frequency converter (*Raytheon Semiconductor Data Book,* 1994, p. 7-6).

## 11.2   Voltage-Frequency Converter Terms and Data Sheets

Much troubleshooting information for a particular IC voltage-frequency converter can be obtained from the data sheet. Likewise, a typical data sheet describes a few specific circuits. However, converter data sheets often have two weak points. First, they assume that everyone understands all of the terms used. Of more importance, the data sheets do not show how the listed parameters relate to troubleshooting problems. To further complicate the situation, each manufacturer has a separate data sheet system. It is impractical to cover all data sheets here. Instead, this section includes typical information found on the converter data sheet and how this information affects circuit troubleshooting.

### 11.2.1   Typical data-sheet information

Figure 11.5 shows typical IC voltage-frequency converters (the Raytheon 4151/4152, which are covered in greater detail at the end of this chapter). Figures 11.6 through 11.8 show typical data-sheet information for the IC, including pinouts, maximum ratings, thermal characteristics, electrical characteristics, and typical performance characteristics.

## 11.2.2    IC VFC operation

Before covering the definition of terms, review how the IC operates and how this relates to practical troubleshooting. As shown in Fig. 11.5, the IC contains an open-loop comparator, a precision one-shot timer, a switched voltage reference, a switched current source, and an open-collector logic-output transistor.

**Connection Information**

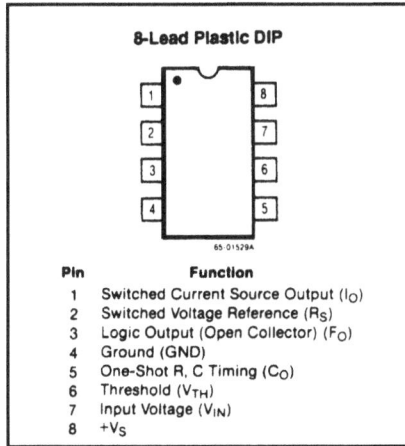

**Functional Block Diagram**

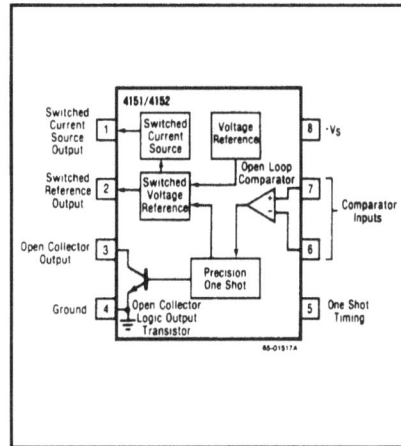

**Absolute Maximum Ratings**

Supply Voltage ..........................................+22V
Internal Power Dissipation ....................500 mW
Input Voltage ....................................-0.2V to +V$_s$
Output Sink Current
    (Frequency Output) ............................20 mA
Output Short Circuit to Ground ........Continuous
Storage Temperature
    Range ..................................-65˚C to +150˚C
Operating Temperature
    Range ......................................0˚C to +70˚C

**Ordering Information**

| Part Number | Package | Operating Temperature Range |
|---|---|---|
| RC4151N | N | 0˚C to +70˚C |
| RC4152N | N | 0˚C to +70˚C |

Notes:
N = 8- lead plastic DIP
Contact a Raytheon sales office or representative for ordering information on special package/temperature range combinations.

**Thermal Characteristics**

| | 8-Lead Plastic DIP |
|---|---|
| Max. Junction Temp. | 125˚C |
| Max. P$_D$ T$_A$ <50˚C | 468 mW |
| Therm. Res θ$_{JC}$ | — |
| Therm. Res. θ$_{JA}$ | 160˚C/W |
| For T$_A$ >50˚C Derate at | 6.25 mW/˚C |

**Figure 11.6**  Typical VFC data-sheet information (*Raytheon Semiconductor Data Book,* 1994, p. 7-3).

**Electrical Characteristics** ($V_S$ = +15V and $T_A$ = +25°C unless otherwise noted)

| Parameters | Test Conditions | 4151 Min | 4151 Typ | 4151 Max | 4152 Min | 4152 Typ | 4152 Max | Units |
|---|---|---|---|---|---|---|---|---|
| Power Supply Requirements (Pin 8) Supply Current | $V_S$ = +15V | | 4.5 | 7.5 | | 2.5 | 6.0 | mA |
| Supply Voltage | | +8.0 | +15 | +22 | +7.0 | +15 | +18 | V |
| Input Comparator (Pins 6 and 7) $V_{OS}$ | | | ±2.0 | ±10 | | ±2.0 | ±10 | mV |
| Input Bias Current | | | -100 | -300 | | -50 | -300 | nA |
| Input Offset Current | | | ±50 | ±100 | | ±30 | ±100 | nA |
| Input Voltage Range | | 0 | $V_S$-2 | $V_S$-3 | 0 | $V_S$-2 | $V_S$-3 | V |
| One Shot (Pin 5) Threshold Voltage | | 0.63 | 0.67 | 0.70 | 0.65 | 0.67 | 0.69 | $XV_S$ |
| Input Bias Current | | | -100 | -500 | | -50 | -500 | nA |
| Saturation Voltage | I = 2.2mA | | 0.15 | 0.5 | | 0.1 | 0.5 | V |
| Drift of Timing vs. Temperature[2] | T = 75μS 0°C to +70°C | | ±35 | | | ±30 | ±50 | ppm/°C |
| Drift of Timing vs. Supply | | | ±150 | | | ±100 | | ppm/V |
| Switched Current Source[1] (Pin 1) Output Current | 4151-$R_S$ = 14.0K/ 4152-$R_S$ = 16.7K | | +138 | | | +138 | | μA |
| Drift vs. Temperature[2] | 0°C to +70°C | | ±75 | | | ±50 | ±100 | ppm/°C |
| Drift vs. Supply Voltage | | | 0.15 | | | 0.10 | | %/V |
| Leakage Current | Off State | | 1.0 | 50 | | 1.0 | 50 | nA |
| Compliance | Pin 1 = 0V to +10V | 1.0 | 2.5 | | 1.0 | 2.5 | | μA |
| Reference Voltage (Pin 2) $V_{REF}$ | | 1.7 | 1.9 | 2.08 | 2.0 | 2.25 | 2.5 | V |
| Drift vs. Temperature[2] | 0°C to +70°C | | ±50 | | | ±50 | ±100 | ppm/°C |
| Logic Output (Pin 3) Saturation Voltage | $I_{SINK}$ = 3.0mA | | 0.1 | 0.5 | | 0.1 | 0.5 | V |
| Saturation Voltage | $I_{SINK}$ = 10mA | | 0.8 | | | 0.8 | | V |
| Leakage Current | Off State | | 0.2 | 1.0 | | 0.1 | 1.0 | μA |
| Nonlinearity % Error Voltage Sourced Circuit of Figure 3 | 1.0Hz to 10kHz | | 0.013 | | | 0.007 | 0.05 | % |
| Temperature Drift Voltage[2] Sourced Circuit of Figure 3 | 0°C to +70°C $F_0$ = 10kHz | | ±100 | | | ±75 | ±150 | ppm/°C |

Notes:
1. Temperature coefficient of output current source (pin 1 output) exclusive of reference voltage drift.
2. Guaranteed but not tested.

**Figure 11.7**  Typical VFC electrical characteristics (*Raytheon Semiconductor Data Book,* 1994, p. 7-4).

The basic IC is converted to a single-supply VFC by adding a few external resistors and capacitors.

The comparator output controls the one shot (which is actually a monostable timer). In turn, the one shot controls the switched current source, the switched reference, and the open-collector transistor. In troubleshooting, if the voltage on pin 7 is greater than the voltage at pin 6, the comparator switches and triggers the one shot. When the one shot is triggered, the timing period begins, and the switched current source, switching voltage reference, and the transistor are turned on.

The one shot creates its timing period much like the classic 555 timer, by charging a capacitor from a resistor tied to $+V_S$. For troubleshooting, the one-shot senses the voltage on the capacitor (pin 5) and ends the timing period when the voltage reaches the two-thirds of the supply voltage. The capacitor is discharged by the transistor at the end of the timing period.

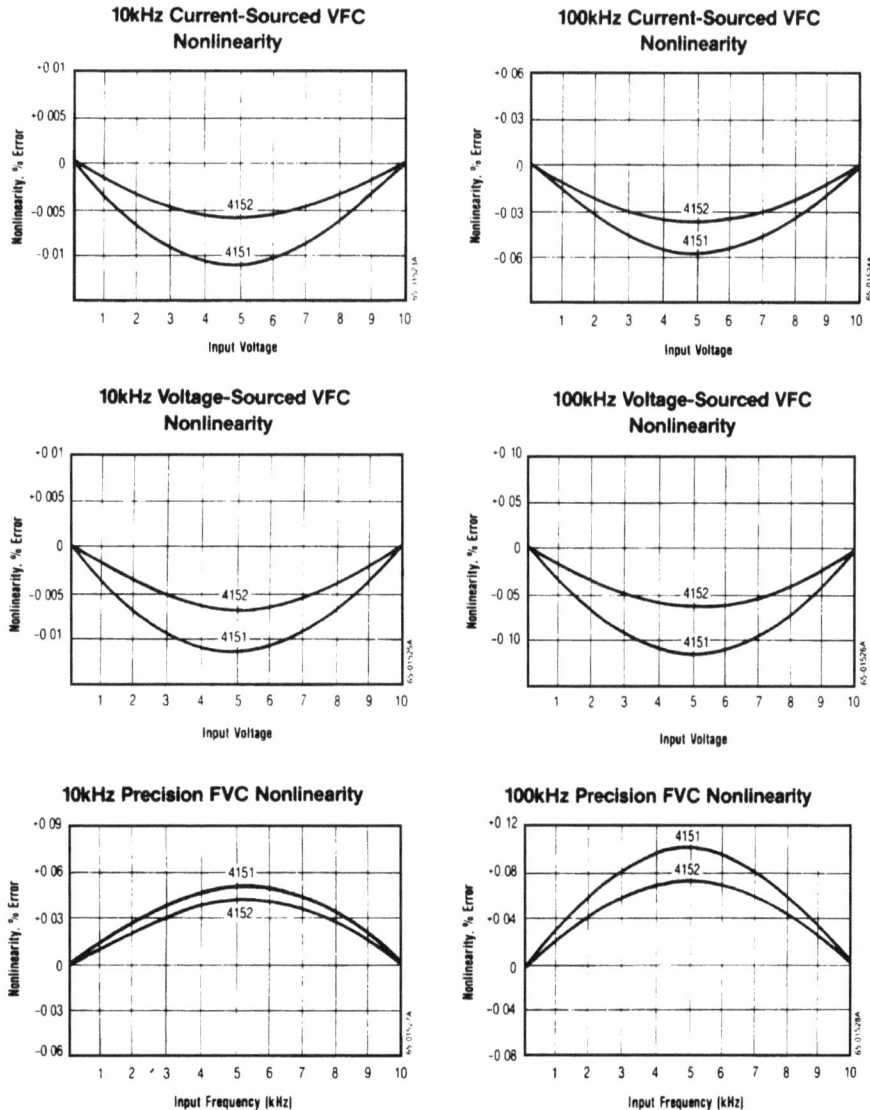

**Figure 11.8**  Typical VFC performance characteristics (*Raytheon Semiconductor Data Book,* 1994, p. 7-5).

During the timing period, the current source, the switched reference, and the output transistor are switched on. For troubleshooting, the switched-current source (pin 1) delivers a current proportional to both the reference voltage and the external resistor ($R_S$). The switched reference (pin 2) supplies an output voltage equal to the internal reference voltage (4151 = 1.9 V, and 4152 = 2.25 V, as shown in Fig. 11.7). The output transistor is turned on, forcing the logic output (pin 3) to a low state. At the end of the timing period, all of these outputs are turned off. As a result, the switched voltage reference has produced an off-on-off voltage pulse, the switched current source has emitted a given charge to the comparator and integrator, and the transistor has produced an output logic pulse.

### 11.2.3  Voltage-frequency converter definitions

The remaining paragraphs of this section describe characteristics that are generally most important when troubleshooting voltage-frequency converters.

### 11.2.4  Compliance

For troubleshooting, *compliance* is the measure of output impedance of a switched-current source, given as a maximum current for a specified voltage change. As shown in Fig. 11.7, compliance for the 4151/4152 goes from a minimum of 1 μA to a typical 2.5 μA, with a pin-1 voltage from 0 V to +10 V.

### 11.2.5  Full-scale frequency

A VFC can operate up to the guaranteed full-scale frequency without violating any of the performance specifications of this frequency range. Although no frequency limit is given for the 4151/4152, the performance characteristics of Fig. 11.8 show operation at a frequency of 100 kHz.

### 11.2.6  Nonlinearity error

On a plot of input voltage versus output frequency (Fig. 11.8), a straight line is drawn from the origin to the full-scale point, which is defined by the intersection of the maximum input voltage and maximum output frequency. The actual plot of output frequency $F_o$ versus input voltage should not deviate from this straight line by more than the increment $dF_o$(max), where $dF_o$ is the difference or deviation of the output frequency.

Nonlinearity is defined here as $(dF_o/dF_s) \times 100\%$, where $F_s$ is the maximum frequency for the range in question. For example, when specifying nonlinearity error for the 0.1-Hz to 10-kHz range, $F_s = 10$ kHz. When specifying nonlinearity error for a FVC, nonlinearity error is defined as $(dV/VF_s) \times 100\%$.

### 11.2.7  Leakage current

Leakage current is the current that flows into the open-collector output transistor when the transistor is in the OFF state, as a result of maximum supply voltage being supplied at the output. As shown in Fig. 11.7, leakage current for the 4151/4152 goes from a typical 1 nA to a maximum of 50 nA, in the OFF state. Leakage current is primarily a design problem, unless a particular circuit shows excessive leakage during troubleshooting. This usually indicates a defect in the IC, but be certain that the maximum supply voltage is not exceeded.

### 11.2.8  Reference voltage

Reference voltage ($V_{REF}$) is the output of the internal voltage reference. This cannot be measured directly in the 4151/4152, but it can be measured as the switched voltage at pin 2 (Fig. 11.5). In some VFCs, the internal reference voltage is accessible at one of the IC pins (such as at pin 5 of the 4153, covered later in this chapter).

When $V_{REF}$ can be measured externally, another characteristic (reference current) is sometimes given on the data sheet. *Reference current* is the current produced (and usually returned back to the circuits in the IC) when the internal voltage reference is at some exact value (such as 7.3 V for the 4153). As shown in Fig. 11.7, $V_{REF}$ for the 4151 goes from a minimum of 1.7 V to a maximum of 2.08 V. $V_{REF}$ for the 4152 goes from 2.0 V to 2.5 V.

### 11.2.9  Scale factor

Scale factor $K$ is the ratio of output frequency divided by a voltage in ($F_o/V_{IN}$). No scale factor is given directly for the 4151/4152. However, a scale-factor tolerance is given from the 4153, and involves $V_{REF}$, resistance-in ($R_{IN}$), and capacitance-out ($C_o$), where $K = \frac{1}{2}V_{REF} \times R_{IN} \times C_o$. The scale-factor tolerance for the 4153 is typically ±0.4% at a frequency of 10 kHz. From a troubleshooting standpoint, a change in scale factor occurs with changes in supply voltage and/or temperature.

### 11.2.10  Absolute maximum ratings and thermal characteristics

The maximum ratings and thermal characteristics shown in Fig. 11.6 deal primarily with the IC, rather than the internal circuits. These characteristics are important to design when selecting heatsinks (if any), but not generally to troubleshooting. A possible exception is when troubleshooting a circuit (experimental or final) that fails at extreme voltages or temperatures. If you are not familiar with such IC characteristics, read the author's *Simplified Design of Linear Power Supplies* (1994) and *Simplified Design of Switching Power Supplies* (1995), both published by Butterworth-Heinemann.

## 11.3  Basic VFC Tests

In practical troubleshooting, the obvious test for any VFC is to vary the input voltage over the range and check that the output frequency varies accordingly.

Use a digital meter at the input and a frequency counter at the output. Using the charge-pump circuit of Fig. 11.9 as an example, the output frequency should vary between 0 and 30 kHz when the input voltage is varied between 0 and 3 V (a 1-V input produces a 10-kHz output, 2 V produces a 20-kHz output, etc.). If practical, the circuit can be subjected to temperature changes and the output frequency monitored for drift. With the circuit of Fig. 11.9, the drift is supposed to be about 20 ppm/°C.

Using the ratio circuit of Fig. 11.10 as an example, two inputs are required for testing, and the output should be the ratio of the two inputs. That is, if $V_1$ is $-10$ V and $V_2$ is $-8$ V, the ratio is 10/8 = 1.25, and the output should be 12.5 kHz. Notice that full scale for the circuit of Fig. 11.10 is 15 kHz, so ratios beyond 1.5 cannot be realized.

## 11.4  Basic FVC Tests

In practical troubleshooting, the test for an FVC is the reverse of that for a VFC. That is, you vary the input frequency over the range and check that the

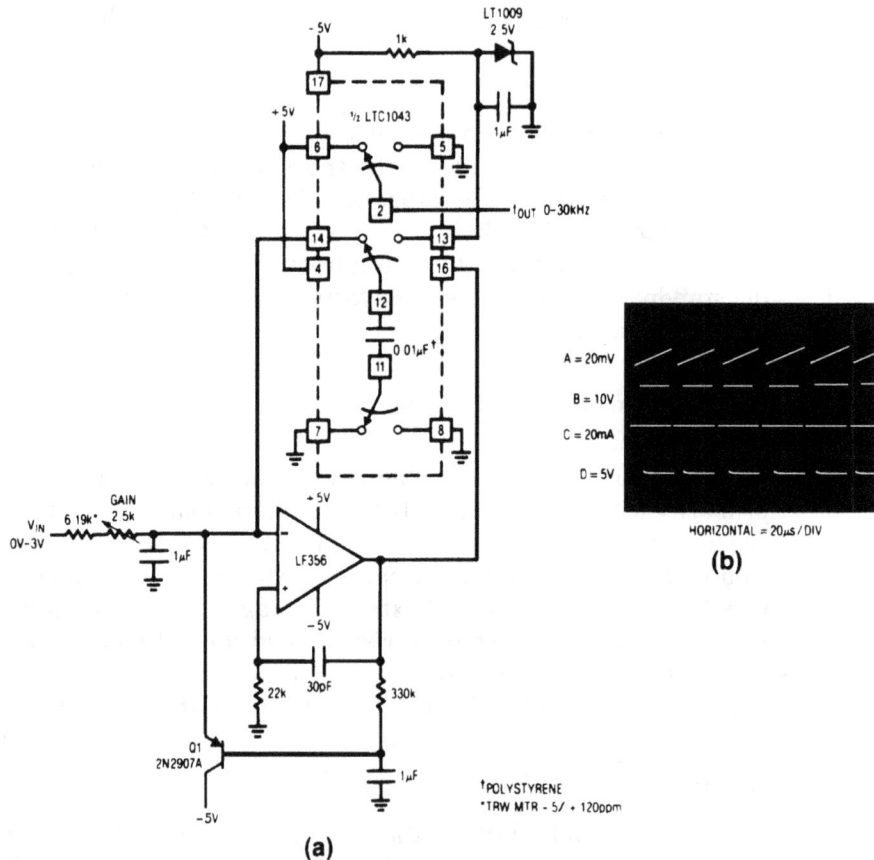

**Figure 11.9**  Charge-pump VFC (Linear Technology, Application Note 3, p. 11).

*Stable components with low tempco

A1, A2 should have low offset and low bias
current: LM351B, LM358B, LF353B, or similar
Q1, Q2: 2N3565, 2N2484, or similar high $\beta$

$$f_{OUT} = \frac{V1}{V2} \times \frac{R_B}{R_A} \times \frac{1}{1.1\,R_T C_T}$$

$$= \frac{V1}{V2} \times 10\ kHz$$

$R_T$

11*

SCALE
FACTOR
TRIM

4.02k*

$C_T$
0.01 $\mu$F

$V_L$

LM331

22k

OUTPUT

$C_L$
0.1 $\mu$F

$V_S$

150

0.1 $\mu$F

A2

47k

$I_2$

Q2

$R_B$
48.9k*

V2
0.1V TO -10V

A1

100k

$I_1$

Q1

$R_A$
100k*

V1
0V TO -10V

SIGNAL
INPUTS

TL/H/8742-4

CURRENT
OUTPUT
100 $\mu$A
FULL-SCALE

$V_{IN}$
0V TO 10V

R1
90k

R2
10k

R3
10k

22M

500 TO 1M
OFFSET ADJ

$-V_S$

$V_S$

a

TL/H/8742-5

22M

$V_S$

500 TO 1M
OFFSET ADJ

b

TL/H/8742-6

R1, R2, R3: Stable components
with low tempco
Q1: $\beta \geq 330$

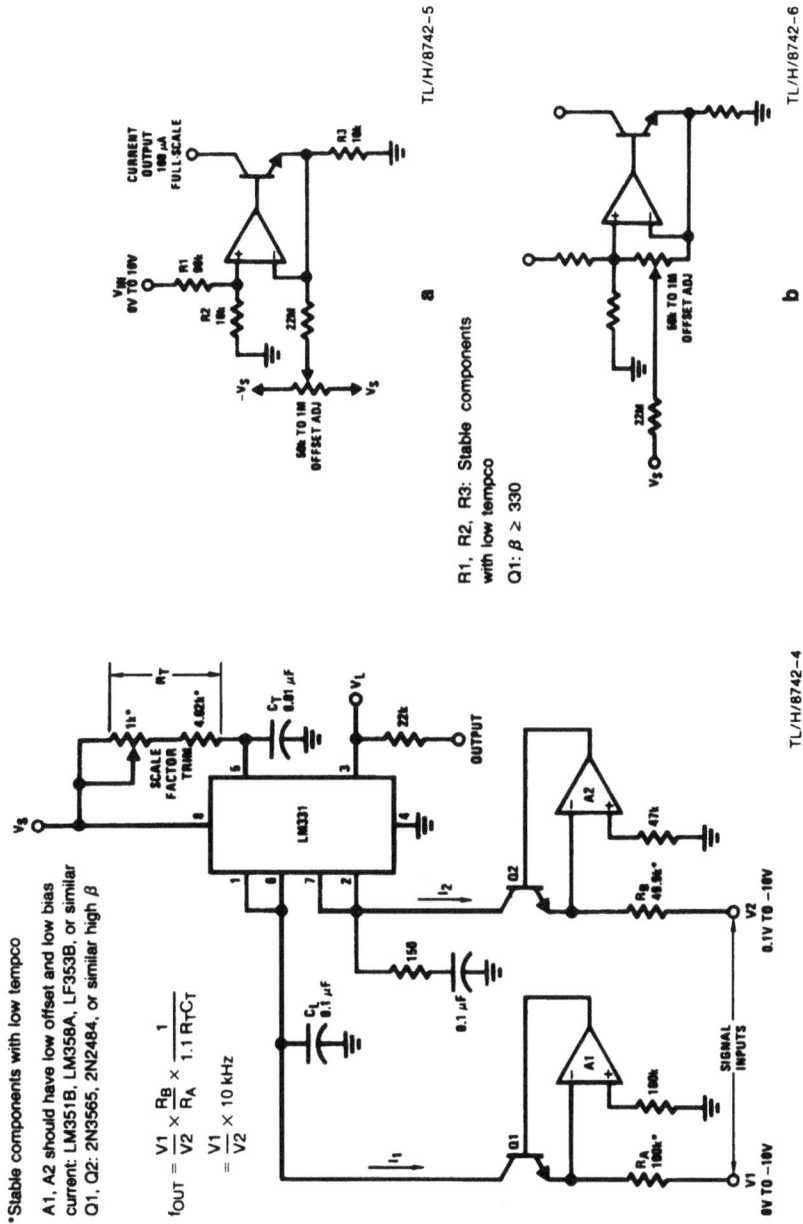

Figure 11.10   Ratio VFC (National Semiconductor, *Linear Applications Handbook*, 1994, p. 1249).

output voltages vary accordingly. Generally, a pulse generator or possibly a square-wave generator is recommended for the input.

In a practical testing/troubleshooting circuit, you can use a simple RC differentiator to convert from sine waves to pulses at the input, if you do not have a square-wave or pulse generator. However, this can disturb some FVC circuits. As a general troubleshooting guideline, use the sine-wave and RC-differentiator combination for the input only when a pulse or square-wave generator is not available, or if a sine wave is specifically recommended in the data sheet.

## 11.5    Basic VFC Troubleshooting

The first step in troubleshooting VFC circuits involves checking that the desired output frequency is produced by a given input voltage. If the output is not correct, try correcting the problem with adjustment. If the problem cannot be corrected by adjustment, trace the signals using a meter or scope from the input (typically a dc voltage) to the output (typically pulses). From that point on, you must make voltage measurements and/or point-to-point resistance measurements. The following are typical examples of practical VFC troubleshooting.

### 11.5.1    Troubleshooting the charge-pump VFC

In the circuit of Fig. 11.9, begin by checking for pulses at pin 2 of the LTC1043. Pin 2 switches between pins 5 (ground) and 6 (+5 V) at a frequency determined by the signal at clock pin 16. If no signal is at pin 16, the LTC switches are at a frequency rate near 200 kHz. In this circuit, the LTC1043 clock is synchronized with the signal at pin 16.

If pin 2 is not switching at any frequency, suspect a problem with the LTC1043. If pin 2 is switching at a fixed frequency, with a variable $V_{IN}$, suspect a problem with LF356 and its associated parts. The output of the LF356, trace B in Fig. 11.9B, should be a series of negative pulses. The inverting input (summing point) is a series of positive ramps (trace A). Current flowing from the LF356 summing point into the 0.01-$\mu$F capacitor at the end of the ramp should produce a series of negative spikes (trace C). Simultaneously, a series of pulses (trace D) should be at the noninverting input of the LF356.

Notice that Q1 prevents the LF356 from going to the negative rail (and staying there) by pulling the summing point negative if the output stays low long enough to charge the 1-$\mu$F/330-k$\Omega$ RC during startup. Also, if the circuit shows excessive drift, or nonlinearity in the output frequency (for a given input voltage), suspect that the 0.01-$\mu$F capacitor has failed.

To trim the circuit of Fig. 11.9, apply 3.0 V and adjust gain trim for a 30-kHz output. The output frequency is directly related to input voltage, with a supposed transfer linearity of 0.005%.

### 11.5.2    Troubleshooting the ramp VFC

In the circuit of Fig. 11.10, start by checking for pulses at the 22-k$\Omega$ resistor (pin 3 of LM331). Pulses should be at this output, regardless of what voltages are applied at the input (including zero input). With both $V_1$ and $V_2$ at zero, the scale-factor trim is adjusted so that the output is 10 kHz.

The circuit of Fig. 11.10 converts the ratio of two voltages to an equivalent frequency. The two op amps convert the inputs to proportional currents. The 1-k$\Omega$ scale-factor trim is adjusted so that the frequency output equals the ratio of $V_1/V_2 \times 10$ kHz. The full-scale output is 15 kHz.

The circuit can accept positive inputs when the op amps are rearranged, as shown in Figs. 11.10A and 11.10B. Trimming out the offset in the op amp provides the ratio converter with better linearity and accuracy. The trim circuit in Fig. 11.10A needs stable positive and negative supplies for the offset trimmer, but the trim in Fig. 11.10B needs only a stable positive supply. The unmarked components in Fig. 11.10B are the same as those in Fig. 11.10A.

If no pulses are at pin 3 of the LM331, suspect a fault with the LM331 or possibly the timing capacitor ($C_T$). The same is true if there are pulses, but the frequency cannot be brought within the desired range, or if the frequency does not change with a changing voltage at the input (pins 1/6 and/or 2/7).

If there are pulses, but the pulse frequency is not controlled by voltages $V_1$ and $V_2$, suspect a problem with Q1, Q2, A1, A2, and the associated parts. Be certain that the voltages at pins 1/6 and 1/7 vary when $V_1$ and $V_2$ are varied.

## 11.6   Basic FVC Troubleshooting

The first step in troubleshooting FVC circuits involves checking that the desired output voltage is produced by a given input frequency. If the output is not correct, try correcting the problem with adjustment. If the problem cannot be corrected by adjustment, trace the signals using a meter or scope from the input (typically pulse or square-wave signals) to the output (typically a dc voltage). After tracing, make voltage measurements and/or a point-to-point resistance measurement. The following are typical examples of practical FVC troubleshooting.

### 11.6.1   Troubleshooting the charge-pump FVC

In the circuit of Fig. 11.11, the frequency input is applied to the clock input of an LTC1043 switched-capacitor IC. The 1000-pF capacitor is switched between a fixed voltage and the inverting input of the LF356. The 1-$\mu$F feedback capacitor averages this action over several cycles, and the circuit output is a dc level that is linearly related to frequency. The feedback resistors set the LF356 gain. Notice that the input pulse must be low for at least 100 ns to allow complete discharge of the 1000-$\mu$F capacitor. To trim the circuit, apply 30 kHz to pin 16 of the LTC1043 and set the gain trim for exactly 3.00 V.

To troubleshoot the Fig. 11.11 circuit, start by checking for pulses at pin 14 of the LTC1043. Pin 12 switches between pins 13 and 14 at a frequency that is determined by the signal at clock pin 16 (trace A of Fig. 11.11B), which is the FVC-circuit input in this case.

If pin 14 is not switching at any frequency, suspect that LTC1043 has failed (of course, check for proper voltages at pins 4, 13, and 17, as well as

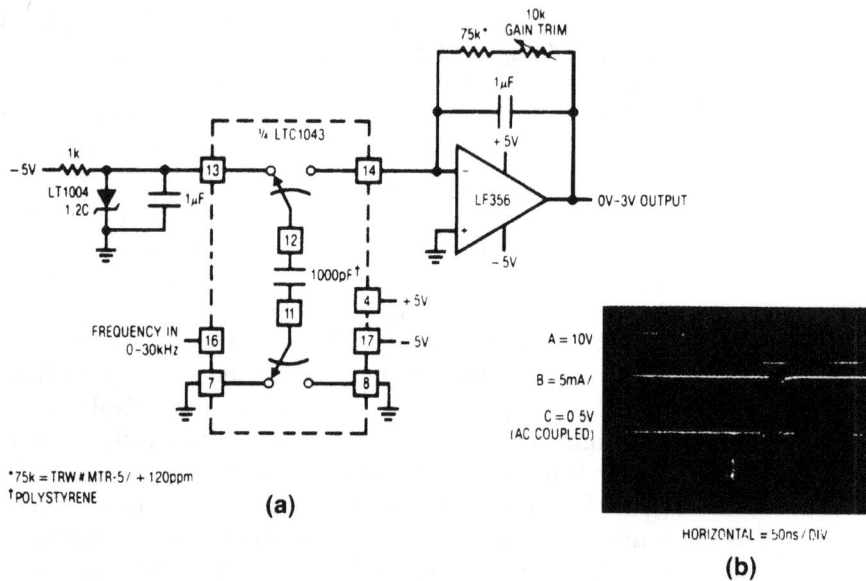

**Figure 11.11**   Charge-pump VFC (Linear Technology, Application Note 3, p. 12).

ground, at pins 7 and 8). If the LTC1043 and voltages are good, but no pulses are at pin 14, the 1000-µF capacitor at pins 11 and 12 is the prime suspect (trace B shows the capacitor signal). The capacitor might be shorted or badly leaking.

If pulses are at the inverting input of the LF356, but there is no output voltage, or if the output voltage does not vary with changes in pulse frequency at the input, suspect a fault with the LF356 or associated parts. The feedback resistors determine the LF356 gain and the 1-µF feedback capacitor averages the pulse input to a dc output (trace C shows the negative and positive swing of the LF356 output). Notice that if the circuit shows excessive drift or nonlinearity in output voltage (for a given input frequency) suspect the 1000-pF capacitor at pins 11 and 12.

## 11.6.2   Troubleshooting the VFC IC connected as an FVC

To troubleshoot the circuit of Fig. 11.2 (Section 11.1.2), start by checking for a dc voltage at pin 1 of the LM331, with a signal at the input. (If the input to the LM331 is zero, the voltage at pin 1 is zero.) If the voltage does not vary at pin 1, when the frequency of the signal at pin 6 is varied (between 0 and 10 kHz), suspect that the LM331 has failed. It is also possible that timing capacitor $C_T$ is shorted or badly leaking.

Try monitoring the output voltage (with the input frequency steady) while varying the gain-adjust control. If the output voltage does not change, check the voltage and resistance from pin 2 of the LM331 to ground. If the connection from pin 2 is good, suspect a problem with the LM331.

## 11.7   Troubleshooting a Typical VFC/FVC IC

Figure 11.12 shows the connection information, maximum ratings, and thermal characteristics for the Raytheon RC4153 converter. This IC can be used as both a VFC and an FVC. Figures 11.13 and 11.14 show the electrical characteristics and typical performance characteristics, respectively. Figures 11.15,

### Absolute Maximum Ratings(1)

Supply Voltage .......................................................±18V
Internal Power Dissipation ...............................500 mW
Input Voltage ...............................................-$V_S$ to +$V_S$
Output Sink Current
(Frequency Output).........................................20 mA
Storage Temperature Range ...............-65°C to +150°C
Operating Temperature Range
RM4153 .........................................-55°C to +125°C
RC4153 ...............................................0°C to +70°C

Note:
1.   "Absolute maximum ratings" are those beyond which the
      safety of the device cannot be guaranteed. They are not meant
      to imply that the device should be operated at these limits. If
      the device is subjected to the limits in the absolute maximum
      ratings for extended periods, its reliability may be impaired.
      The tables of Electrical Characteristics provide conditions for
      actual device operation.

### Ordering Information

| Part Number | Package | Operating Temperature Range |
|---|---|---|
| RC4153D | D | 0°C to +70°C |
| RM4153D | D | -55°C to +125°C |

Notes:
D = 14-lead ceramic DIP

### Connection Information

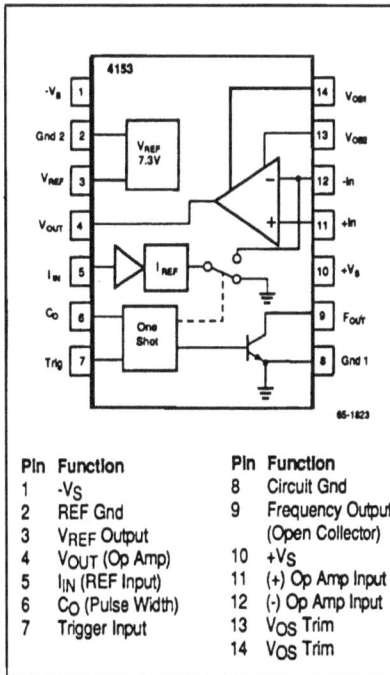

| Pin | Function | Pin | Function |
|---|---|---|---|
| 1 | -$V_S$ | 8 | Circuit Gnd |
| 2 | REF Gnd | 9 | Frequency Output |
| 3 | $V_{REF}$ Output | | (Open Collector) |
| 4 | $V_{OUT}$ (Op Amp) | 10 | +$V_S$ |
| 5 | $I_{IN}$ (REF Input) | 11 | (+) Op Amp Input |
| 6 | $C_O$ (Pulse Width) | 12 | (-) Op Amp Input |
| 7 | Trigger Input | 13 | $V_{OS}$ Trim |
| | | 14 | $V_{OS}$ Trim |

### Thermal Characteristics

| | 14-Lead Ceramic DIP |
|---|---|
| Max. Junction Temp. | +175°C |
| Max. $P_D$ $T_A$ <50°C | 1042 mW |
| Therm. Res $\theta_{JC}$ | 60° C/W |
| Therm. Res. $\theta_{JA}$ | 120°C/W |
| For $T_A$ >50°C Derate at | 8.33 mW/°C |

Figure 11.12   Connection information, maximum ratings, and thermal characteristics for RC4153 (*Raytheon Semiconductor Data Book,* 1994, p. 3-810).

11.16, and 11.17 show the IC connected as a minimum-circuit VFC, FVC, and VFC with offset and gain adjustments, respectively.

### 11.7.1 Operating principles of an IC connected as a VFC

Figure 11.18 shows a block diagram of the IC connected for VFC operation. Figure 11.19 shows the corresponding waveforms and timing.

Both capacitors shown in Fig. 11.18 are discharged when power is first applied. The input current, produced by $V_{IN}/R_{IN}$, causes C1 to charge, and a

## Electrical Characteristics

($V_S$ = ±15V and $T_A$ = +25°C unless otherwise noted)

| Parameters | Min | Typ | Max | Units |
|---|---|---|---|---|
| Power Supply Requirements | | | | |
| Supply Voltage | ±12 | ±15 | ±18 | V |
| Supply Current   (+V$_S$, I$_{OUT}$ = 0) | | +4.2 | +7.5 | mA |
| (-V$_S$, I$_{OUT}$ = 0) | | -7 | -10 | |
| Full Scale Frequency | 250 | 500 | | kHz |
| Transfer Characteristics | | | | |
| Nonlinearity Error Voltage-to-Frequency[1] | | | | |
| 0.1 Hz ≤ F$_{OUT}$ ≤ 10 kHz | | 0.002 | 0.01 | %FS |
| 1.0 Hz ≤ F$_{OUT}$ ≤ 100 kHz | | 0.025 | 0.05 | %FS |
| 5.0 Hz ≤ F$_{OUT}$ ≤ 250 kHz | | 0.06 | 0.1 | %FS |
| Nonlinearity Error Frequency-to-Voltage[1] | | | | |
| 0.1 Hz ≤ F$_{IN}$ ≤ 10 kHz | | 0.002 | 0.01 | %FS |
| 1.0 Hz ≤ F$_{IN}$ ≤ 100 kHz | | 0.05 | 0.1 | %FS |
| 5.0 Hz ≤ F$_{IN}$ ≤ 250 kHz | | 0.07 | 0.12 | %FS |
| Scale Factor Tolerance, F = 10 kHz | | | | |
| $K = \dfrac{1}{2V_{REF}\,R_{IN}\,C_O}$ | | ±0.5 | | % |
| Change of Scale Factor with Supply | | 0.008 | | %/V |
| Reference Voltage (V$_{REF}$) | | 7.3 | | V |
| Temperature Stability (0°C to 70°C) [1, 2, 3] | | | | |
| Scale Factor 10 KHz Nominal | | ±75 | ±150 | ppm/°C |
| Reference Voltage | | ±50 | ±100 | ppm/°C |
| Scale Factor (External Ref) 10 KHz FS | | ±25 | ±50 | ppm/°C |
| Scale Factor (External Ref) 100 KHz FS | | ±50 | ±100 | ppm/°C |
| Scale Factor (External Ref) 250 KHz FS | | ±100 | ±150 | ppm/°C |

Notes:
1.  Guaranteed but not tested.
2.  V$_{REF}$ Range: 6.6V ≤ V$_{REF}$ ≤ 8.0V.
3.  Over the specified operating temperature range.

**Figure 11.13**  Electrical characteristics for RC4153 (*Raytheon Semiconductor Data Book*, 1994, pp. 3-811, 3-812).

ramp to be developed at point C. The trigger threshold of the one shot is about +1.3 V and, if the integrator output (point C) is less than +1.3 V, the one shot will fire and pulse the open-collector output (point E) and the switched current-source output (point A).

Because point C is less than +1.3 V, the one shot fires and the switched current source delivers a negative current pulse to the integrator. This causes C1 to charge in the opposite direction, and point C ramps up until the end of the one-shot pulse. At that time, the positive current ($V_{IN}/R_{IN}$) again makes point C ramp down until the trigger threshold is reached.

The one shot continuously fires until the integrator output exceeds the trigger threshold. When this point is reached, the one-shot fires (as needed) to keep the integrator output above the trigger threshold. If $V_{IN}$ is increased, the slope of the down ramp increases, and the one shot fires more often to keep the integrator output high.

Because the one-shot firing frequency is the same as the open-collector output frequency, any increase in $V_{IN}$ causes an increase in $F_{OUT}$ (as shown in Figs. 11.15 through 11.17). This relationship is almost linear because the amount of charge in each output current pulse is carefully defined, both in magnitude and duration. The pulse duration is set by timing capacitor $C_o$ (point D). This feedback system is called a *charge-balanced loop*.

| Parameters | Min | Typ | Max | Units |
|---|---|---|---|---|
| Op Amp | | | | |
| Open Loop Output Resistance | | 230 | | Ω |
| Short Circuit Current | | 25 | | mA |
| Gain Bandwidth Product [1] | 2.5 | 3.0 | | MHz |
| Slew Rate | 0.5 | 2.0 | | V/μS |
| Output Voltage Swing ($R_L \geq 2K$) | 0 to +10 | -0.5 to +14.3 | | V |
| Input Bias Current | | 70 | 400 | nA |
| Input Offset Voltage (Adjustable to 0) | | 0.5 | 5.0 | mV |
| Input Offset Current | | 30 | 60 | nA |
| Input Resistance (Differential Mode) | | 1.0 | | MΩ |
| Common Mode Rejection Ratio | 75 | 100 | | dB |
| Power Supply Rejection Ratio | 70 | 106 | | dB |
| Large Signal Voltage Gain | 25 | 350 | | V/mV |
| Switched Current Source | | | | |
| Reference Current (External Reference) | | 1.0 | | mA |
| Digital Input (Frequency-to-Voltage, Pin 7) | | | | |
| Logic "0" | | | 0.5 | V |
| Logic "1" | 2.0 | | | V |
| Trigger Current | | -50 | | μA |
| Logic Output (Open Collector) | | | | |
| Saturation Voltage (Pin 9) | | | | |
| $I_{SINK}$ = 4 mA | | 0.15 | 0.4 | V |
| $I_{SINK}$ = 10 mA | | 0.4 | 1.0 | V |
| $I_{LEAK}$ (Off State) | | 150 | | nA |

Notes:
1. Guaranteed but not tested.

**Figure 11.13**  (Continued)

The scale factor, $K$ (Fig. 11.13), which is the number of pulses per second for a specified $V_{IN}$, is adjusted by changing either $R_{IN}$ (and thus the input current) or by changing the amount of charge in each $I_{OUT}$ pulse. Because the magnitude of $I_{OUT}$ is fixed at 1 mA, the amount of charge is adjusted by changing $C_o$ to alter the one-shot duration. ($I_{OUT}$ can also be adjusted by changing $V_{REF}$, but this requires altering the fixed 7.4-V reference.)

## Typical Performance Characteristics

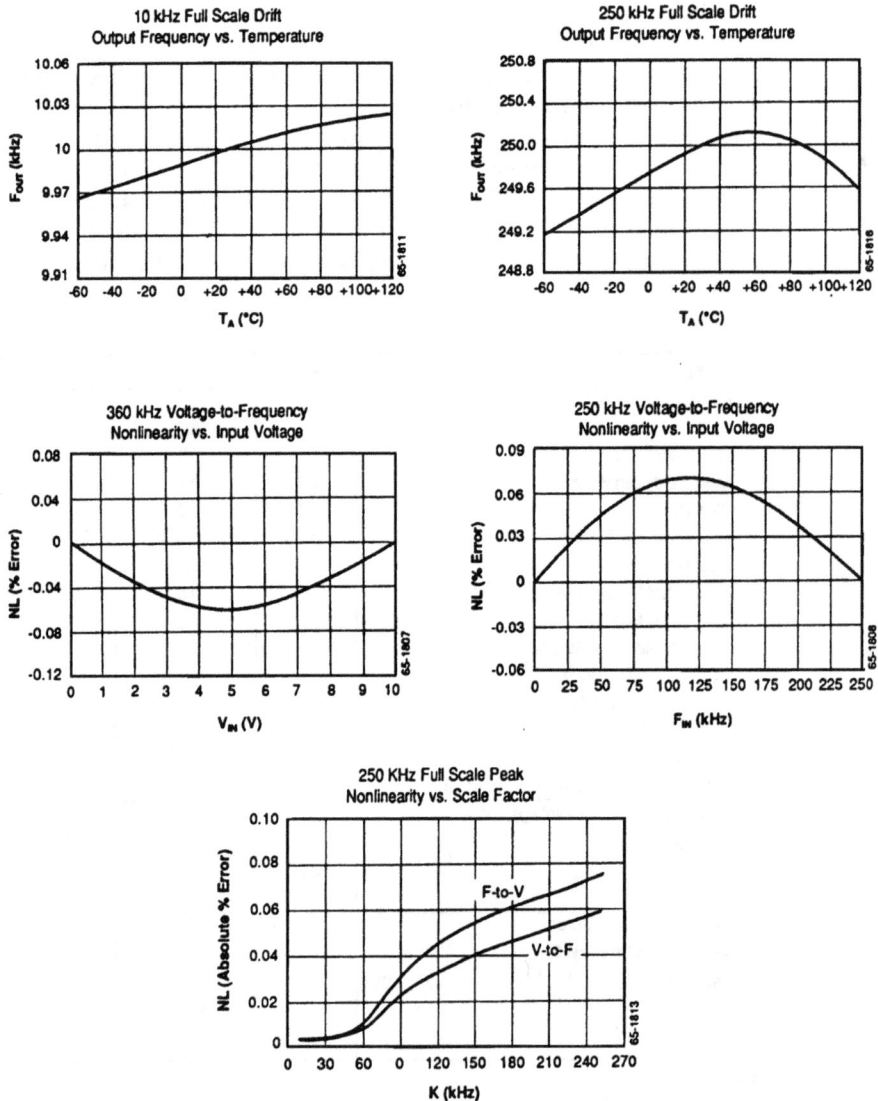

**Figure 11.14**  Typical performance characteristics for RC4153 (*Raytheon Semiconductor Data Book*, 1994, pp. 3-813, 3-814).

In troubleshooting a VFC circuit, notice that the accuracy between $V_{IN}$ and $F_{OUT}$ is affected by three major sources of error: temperature drift, nonlinearity, and offset.

The greatest source of drift in a typical application is the timing capacitor $(C_o)$. Low $T_C$ capacitors, such as silver mica and polystyrene, should be measured for drift using a capacitance meter. (The manufacturer recommends wiring parallel capacitors composed of 70% silver mica and 30% polystyrene.) Troubleshooting temperature-coefficient $(T_C)$ problems is covered further in Section 11.8.

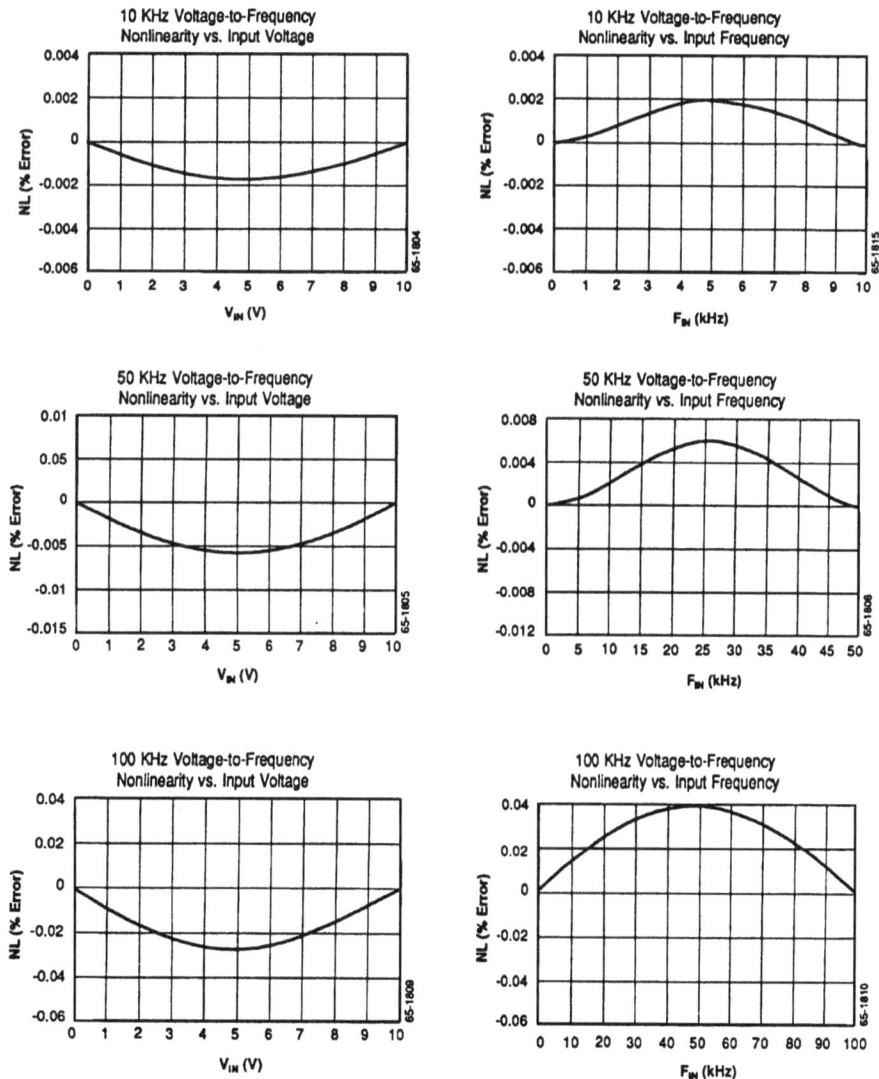

**10 KHz Voltage-to-Frequency Nonlinearity vs. Input Voltage**
NL (% Error) vs. $V_{IN}$ (V) — 65-1804

**10 KHz Voltage-to-Frequency Nonlinearity vs. Input Frequency**
NL (% Error) vs. $F_{IN}$ (kHz) — 65-1815

**50 KHz Voltage-to-Frequency Nonlinearity vs. Input Voltage**
NL (% Error) vs. $V_{IN}$ (V) — 65-1805

**50 KHz Voltage-to-Frequency Nonlinearity vs. Input Frequency**
NL (% Error) vs. $F_{IN}$ (kHz) — 65-1808

**100 KHz Voltage-to-Frequency Nonlinearity vs. Input Voltage**
NL (% Error) vs. $V_{IN}$ (V) — 65-1809

**100 KHz Voltage-to-Frequency Nonlinearity vs. Input Frequency**
NL (% Error) vs. $F_{IN}$ (kHz) — 65-1810

**Figure 11.14**  (Continued)

| Full Scale | $C_I$ | $C_o$ | $R_{IN}$ |
|---|---|---|---|
| 10 kHz | 0.1 µF | 3300 pF | 20K |
| 50 kHz | 0.02 µF | 680 pF | 20K |
| 100 kHz | 4300 pF | 330 pF | 20K |
| 250 kHz | 1000 pF | 130 pF | 20K |

65-1825

$$T = \frac{1}{F_{OUT}}$$

$$t = 1.5 \times 10^{-4} C_o$$

$$C_o \le \frac{5 \times 10^{-5}}{F_{OUT} \ (\text{Max})}$$

$V_{OS1}$ (14)  
$V_{OS2}$ (13)  
$-In$ (12)  
$+In$ (11)  
(10)  
$F_{OUT}$ (9)  
Gnd 1 (8)

$+V_S^*$  
$R_L$ 5.1K  
$F_{OUT}$

4153  
$V_{REF}$ 7.3V  
$I_{REF}$  
One Shot

(1) $R_S^{**}$ $-V_S$  
(2) Gnd 2  
(3) $V_{REF}$ $V_{OUT}$  
(4)  
(5) $I_{IN}$  
(6) $C_o$  
(7) $C_o$ Trig

$R_{IN}$  
$V_{IN}$

$C_I = 30 \ C_o$

$$F_{OUT} = \frac{V_{IN}}{2 V_{REF} \ R_{IN} \ C_o}$$

$(V_{REF} = 7.3V)$

\* ±V s must be thoroughly decoupled.  
\*\* For bipolar input.  
Resistance in Ohms unless otherwise specified.

\*\* For Bipolar Input

$$F_{OUT} = \frac{V_{IN} \ R_{REF} + V_{REF} \ R_{IN}}{2 R_{IN} \ R_S \ V_{REF} \ C_o}$$

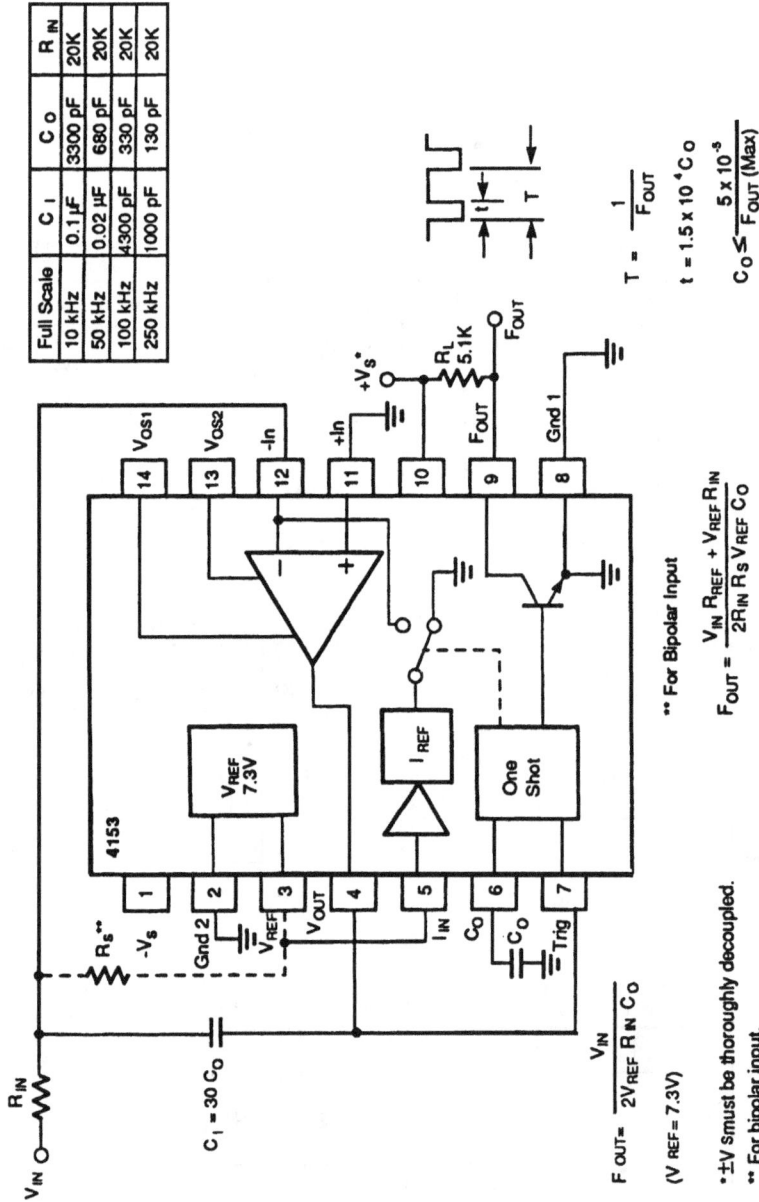

Figure 11.15 Minimum-circuit VFC (*Raytheon Semiconductor Data Book*, 1994, p. 3-815).

348

| Full Scale | $C_1$ | $C_0$ | $R_B$ |
|------------|-------|-------|-------|
| 10 kHz | 10 μF | 3300 pF | 20K |
| 50 kHz | 2 μF | 330 pF | 40K |
| 100 kHz | 1 μF | 150 pF | 43K |
| 250 kHz | 0.2μF | 60 pF | 39K |

$$V_{RIPPLE} = \frac{2V_{REF}C_0(1-1.5 \times 10^4 C_0 F_{IN})}{C_1}$$

$$T_{RECOVERY} = 1.36 \times 10^4 C_1 C_0 R_B \, \Delta F_{IN}$$

$$V_{OUT} = 2V_{REF} R_B C_0 F_{IN} \quad C_0 \leq \frac{5 \times 10^{-4}}{F_{IN} \, (Max)}$$

* ±$V_s$ must be thoroughly decoupled.
** Optional.
  Resistance in Ohms unless otherwise specified.

4153

Voltage Output

Full Scale Adjust

$V_{OS1}$  $R_{OS}$ 10K  $V_{OS2}$  -In  +In

$R_{IF}$ 20K

$C_{IF}$ 0.01 F (Cer Disk)

$R_L = 5.1K$**

5K  18.7K  $R_B$

$C_1$

$F_{OUT}$  Gnd 1  $F_{OUT}$

+$V_s$  +$V_s$  $V_s$

14  13  12  11  10  9  8

$V_{REF}$ 7.3V  $I_{REF}$  One Shot

1  2  3  4  5  6  7

-$V_s$*  Gnd 2  $V_{REF}$  $V_{OUT}$  $I_{IN}$  $C_0$  Trig

$C_0$ 3.3 nF

+$V_s$*  R1 10K  R2 5.1K

$F_{IN}$ (0-10 kHz)  $C_{IN}$ 0.002 pF

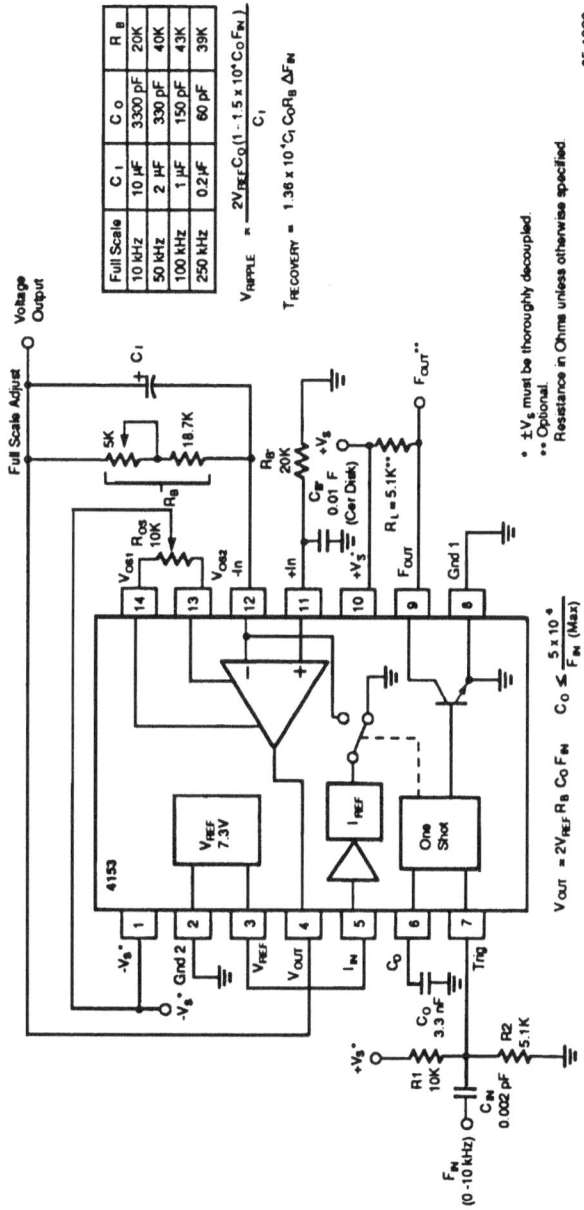

**Figure 11.16** Minimum-circuit FVC (*Raytheon Semiconductor Data Book*, 1994, p. 3-815).

349

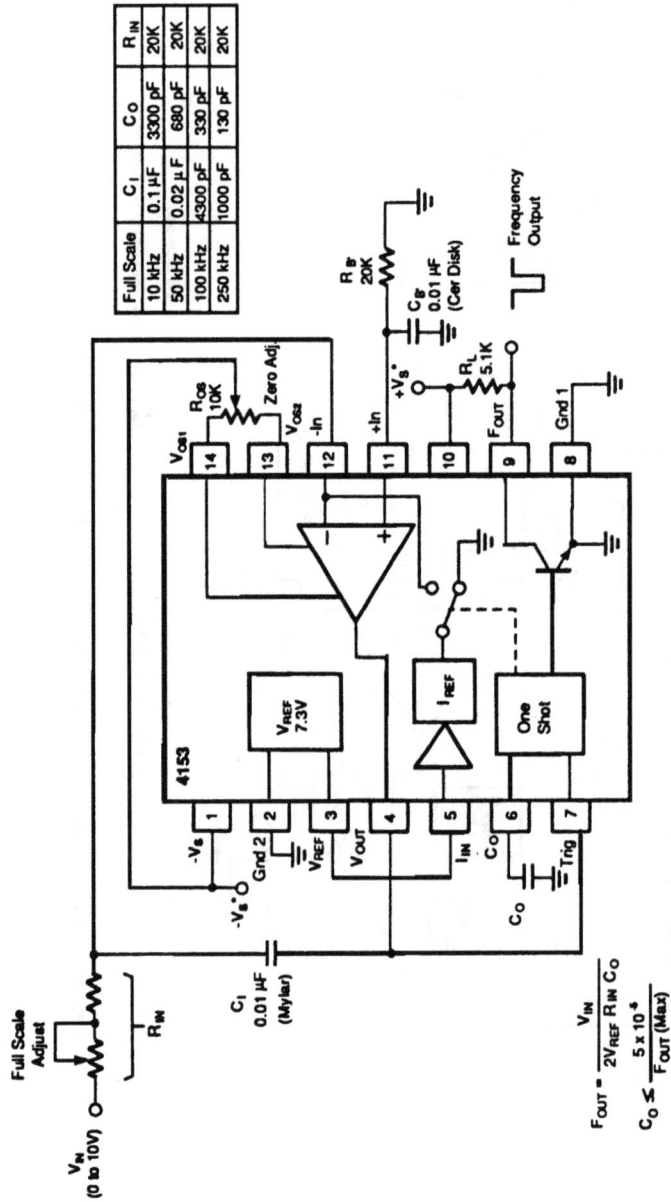

| Full Scale | $C_1$ | $C_O$ | $R_{IN}$ |
|---|---|---|---|
| 10 kHz | 0.1 μF | 3300 pF | 20K |
| 50 kHz | 0.02 μF | 680 pF | 20K |
| 100 kHz | 4300 pF | 330 pF | 20K |
| 250 kHz | 1000 pF | 130 pF | 20K |

65-1828

$$F_{OUT} = \frac{V_{IN}}{2V_{REF}\, R_{IN}\, C_O}$$

$$C_O \leq \frac{5 \times 10^{-5}}{F_{OUT}\ (Max)}$$

* ±$V_S$ must be thoroughly decoupled.
Resistance in Ohms unless otherwise specified.

Figure 11.17  VFC with offset and gain adjustments (*Raytheon Semiconductor Data Book*, 1994, p. 3-816).

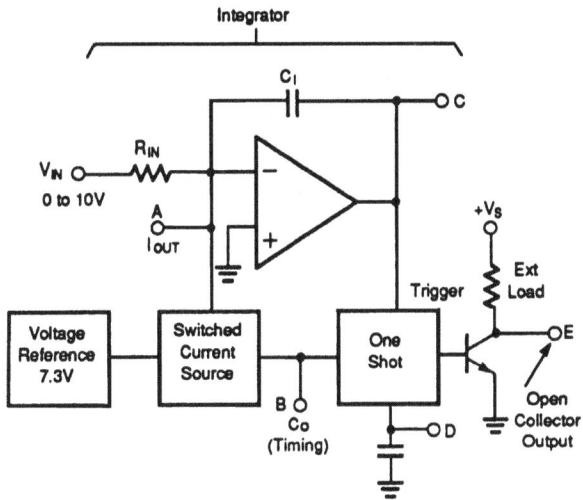

**Figure 11.18** 4153 connected for VFC operation (*Raytheon Semiconductor Data Book,* 1994, p. 3-817).

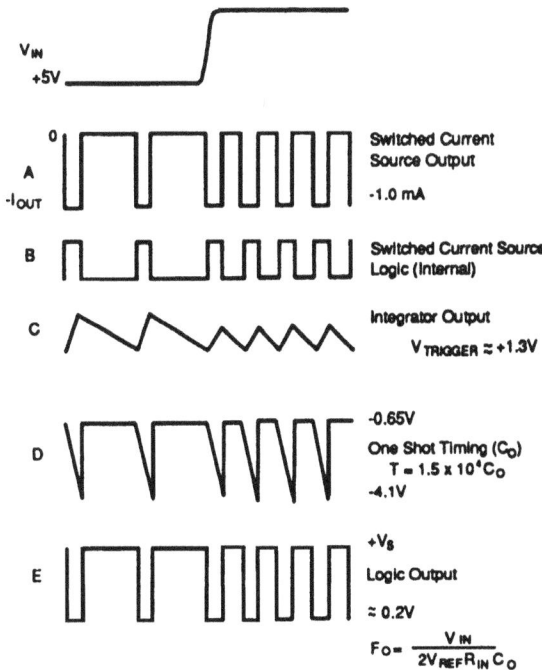

**Figure 11.19** 4153 VFC timing waveforms (*Raytheon Semiconductor Data Book,* 1994, p. 3-817).

$$F_O = \frac{V_{IN}}{2V_{REF}R_{IN}C_O}$$

65-1818

The built-in voltage reference (7.3 V) can also be replaced with a fixed external reference, if temperature/drift is critical. The manufacturer recommends an LM199 that has an output of 6.9 V with less than 10 ppm/°C drift.

The major cause of nonlinearity in this VFC (and most others) is that there is some change in the exact amount of charge in each $I_{OUT}$ pulse. When the frequency increases, internal stray capacitances and switching problems change the width and amplitude of the $F_{OUT}$ pulses, causing a nonlinear relationship between $V_{IN}$ and $F_{OUT}$. For this reason, the scale factor should be below 1 kHz/V or as low as the acquisition time of the system will permit.

Notice that the circuits of Figs. 11.16 and 11.17 have an offset adjustment (between pins 13 and 14). This makes it possible to offset any drift in the internal op amp. As always, offset problems are most crucial when the input voltages are low.

### 11.7.2   Calibrating the VFC

The circuit of Fig. 11.17 is calibrated as follows. First, apply a measured full-scale input voltage and adjust $R_{IN}$ for the desired scale factor. Then apply a small (but precise) input voltage (typically 10 mV) and adjust the op-amp offset until the output frequency equals the input, multiplied by the scale factor.

For example, assume that the input voltage range is 0 V to 10 V, and that the desired full-scale output frequency is 250 kHz. Using the values for input and output capacitances shown in Fig. 11.17 (1000 pF for input and 130 pF for output), adjust $R_{IN}$ for a 250-kHz output ($R_{IN}$ should be about 20 kΩ). Then reduce the input voltage to 10 mV and adjust the offset for a 250-Hz output. Recheck the calibration at the full-scale input of 10 V.

For applications in which precision is crucial, the manufacturer recommends that trimpots (for $R_{IN}$ and offset) be replaced by metal-film resistors soldered in parallel. Even the best trimpots have bad temperature coefficients, and are easily taken out of adjustment by mechanical shock.

In any VFC/FVC IC, the reference, switched-current source, external resistor, and external capacitor temperature coefficients all contribute to temperature drift. In an experimental circuit, when temperature-drift specifications are fanatical, use an external reference, such as an LM199. For best results, heat the 4153 with an external heater, measure temperature drift, then select resistors and capacitors for the best temperature performance.

### 11.7.3   Operating principles of IC connected as an FVC

Figure 11.20 shows the 4153 connected as a precision FVC. Figure 11.21 shows the corresponding waveforms and timing.

This circuit converts the input frequency to a proportional voltage by integrating the switched current-source output. When the input frequency increases, the number of output pulses taken from the inverter increases, thus increasing the average output voltage. Some ripple is at the output, with the ripple amplitude dependent on the integrator time constant. The output can be further filtered, but this reduces the response time. In general, a second-order filter will decrease ripple and improve response time, as covered in Section 11.7.4.

When troubleshooting the Fig. 11.20 circuit in experimental form, remember that the input waveform must meet three conditions. First, the input must

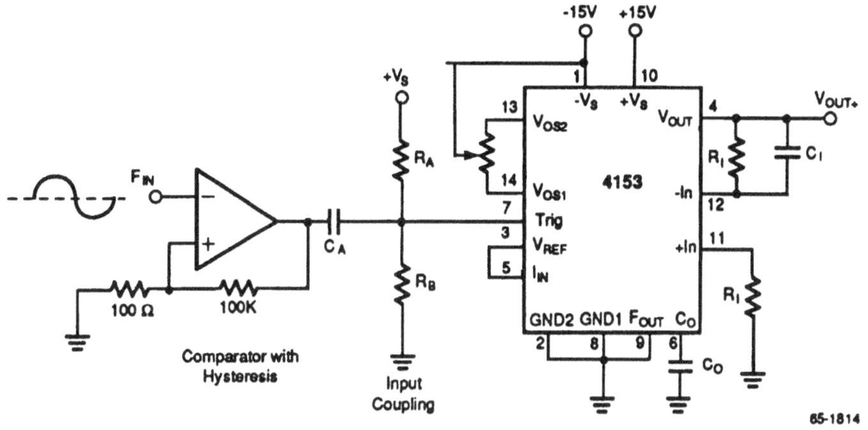

**Figure 11.20**   4153 connected for precision VFC operation (*Raytheon Semiconductor Data Book,* 1994, p. 3-819).

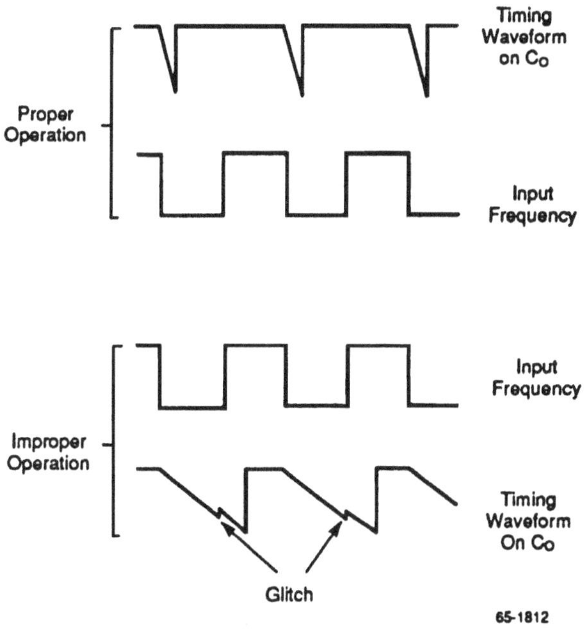

**Figure 11.21**   4153 FVC timing waveforms (*Raytheon Semiconductor Data Book,* 1994, p. 3-819).

have sufficient amplitude and offset to swing above and below the 1.3-V trigger threshold. (The values of $R_A$ and $R_B$ must be selected to provide the correct offset for a given value of $+V_S$.)

Second, the input must be a fast-slewing waveform with a quick rise time. This can be done with the comparator (to square up the input) and with ac coupling (capacitor $C_A$).

Third, the input pulse width must not exceed the one-shot time (to avoid triggering the one shot). Capacitive coupling between the trigger input (pin 7) and the timing-capacitor input (pin 6) might occur if the input waveform is a square wave or if the input has a short time period. This can cause excessive nonlinearity because of changes in the one-shot timing waveform (as shown in Fig. 11.21). The problem can be avoided by keeping $C_o$ small. This will keep the timing period less than the input waveform. See Fig. 11.16 for some typical component values.

### 11.7.4   Troubleshooting FVC ripple problems

As mentioned, any FVC circuit has some ripple problem. The dc output contains a ripple component equal in frequency to the input pulses. This can be minimized with filtering, but with a decrease in response time. Consequently, there is always a tradeoff between response time and ripple. When troubleshooting an experimental circuit, use the largest amount of filtering that an acceptable response time will allow.

Figure 11.22 shows an FVC with a single-pole integrator at the switched current-source output. Figure 11.23 shows a second-order (double-pole) filter that will improve both response time and output ripple. The ratio of the time constants $R_1C_1$ and $R_2C_2$ determines the response to a step change in input frequency. The response is critically damped if $R_1C_1 = 4(R_2C_2)$. Best results are obtained when $R_1C_1 = R_2C_2$, which provides a damping factor of one half.

Choose filter capacitors C1 and C2 (Fig. 11.23) as well as the one-shot timing capacitor (Fig. 11.22) for minimum ripple over the desired range of operation. As a troubleshooting guideline for the 4153, peak ripple is less than 100

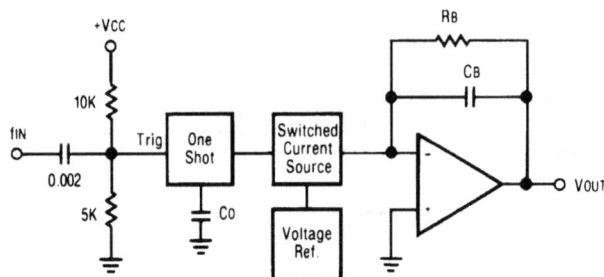

**Figure 11.22**   FVC with single-pole integrator (*Raytheon Semiconductor Data Book*, 1994, p. 7-9).

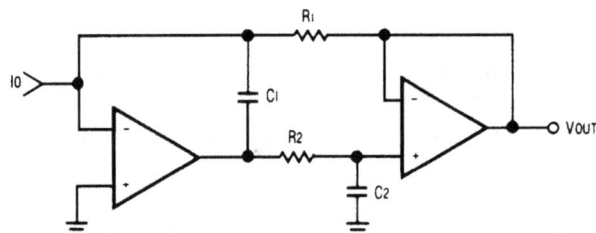

**Figure 11.23**   Second-order (double-pole) active filter (*Raytheon Semiconductor Data Book*, 1994, p. 7-9).

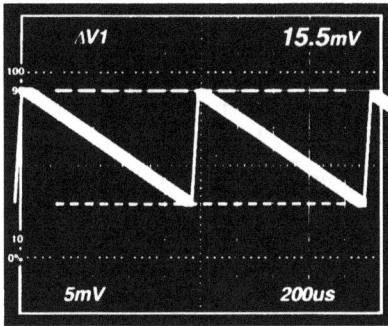

**Figure 11.24** Typical ripple output (Analog Devices, *Applications Reference Manual,* 1993, p. 23-62).

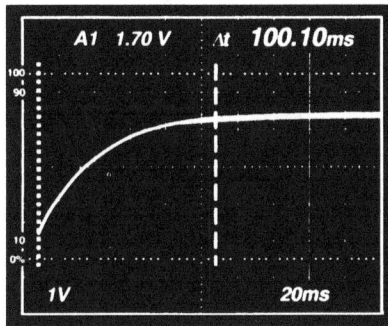

**Figure 11.25** Response to step change in frequency (Analog Devices, *Applications Reference Manual,* 1993, p. 23-62).

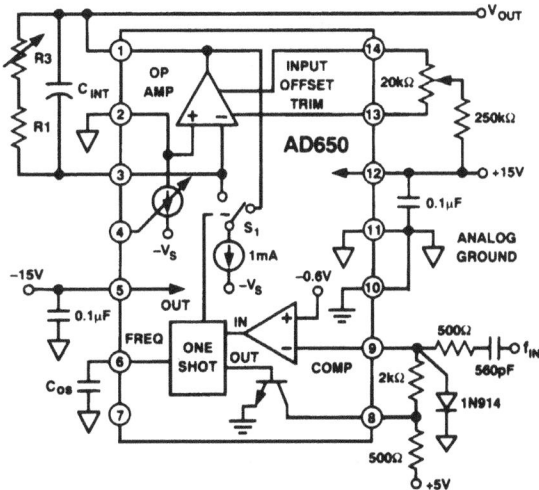

**Figure 11.26** Practical FVC circuit (Analog Devices, *Applications Reference Manual,* 1993, p. 23-63).

mV (10 Hz to 10 kHz) when $R_1$ = 100 kΩ and $C_1$ = 0.01 μF. Notice that the values for the input capacitor and offset-bias resistors in Fig. 11.22 correspond to those shown in Fig. 11.16.

Figures 11.24 and 11.25 show typical ripple output and response time, respectively, for another VFC (the Analog Devices AD650) used as an FVC (as shown in Fig. 11.26). In this particular circuit, the response time versus ripple can be controlled by the external integrating capacitor, $C_{INT}$ (connected between pins 1 and 3).

If response time is of primary importance, the value of $C_{INT}$ can be lowered, but at the expense of increased ripple. On the other hand, if ripple is paramount, $C_{INT}$ can be increased. However, this results in slower response. In Figs. 11.24 and 11.25, the value of $C_{INT}$ is about 1.5 μF. This produces a ripple of about 12 mV to 15.5 mV.

The equation for determining ripple is long and complex (far beyond the scope of this book). However, the graph of Fig. 11.27 shows the peak-to-peak ripple versus frequency for the circuit of Fig. 11.26 in a typical application. In troubleshooting, the important point to remember is that ripple amplitude changes with frequency and is largest at the lowest frequency.

To sum up the ripple problem for a basic FVC, such as shown in Fig. 11.28, the circuit has two related problems that must be approached with a compromise in

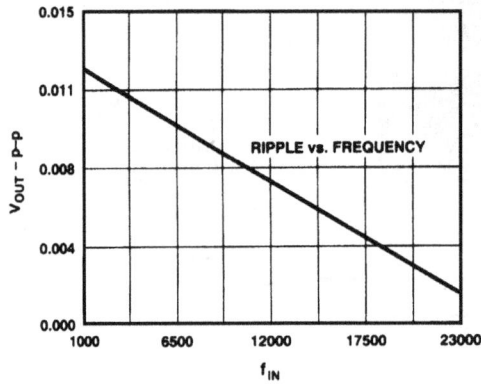

**Figure 11.27**  Peak-to-peak ripple versus frequency (Analog Devices, *Applications Reference Manual,* 1993, p. 23-61).

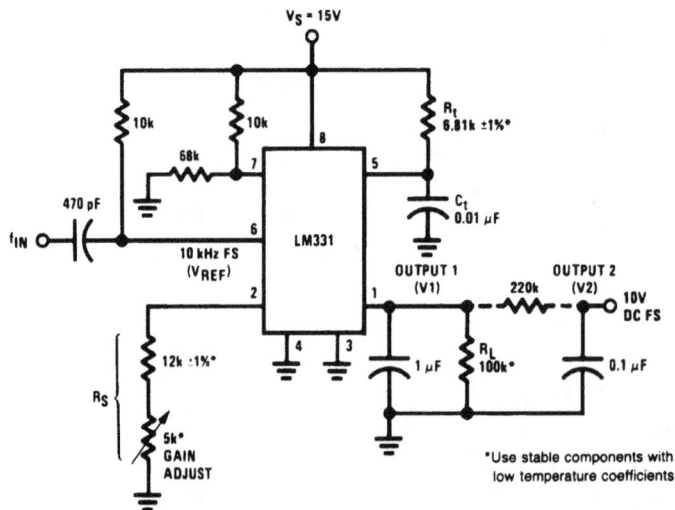

TL/H/8741–1

**Figure 11.28**  Stand-alone FVC (National Semiconductor, *Linear Applications Handbook,* 1994, p. 1241).

mind. The circuit has about 13 mVp-p of ripple and lags 0.1 second behind an input-frequency step change, settling to within 0.1% of full scale in about 0.6 second. Both of these conditions can be corrected (but with a tradeoff) as follows.

Increasing the filter-capacitor value reduces ripple, but also increases response time. Conversely, lowering the filter-capacitor value improves response time at the expense of larger ripple. In most cases, the problem is solved by adding an active filter. This results in faster response time, and less ripple, for high input frequencies.

In cases where ripple is crucial, but response time is not, the addition of a passive filter can produce satisfactory results. For example, adding a 220-$k\Omega/0.1$-$\mu$F post filter (as shown in the dotted lines on Fig. 11.28) slows the response slightly, but also reduces ripple to less than 1 mVp-p for frequencies from 200 Hz to 10 kHz.

## 11.8  Troubleshooting Temperature Problems

FVCs often have temperature-related problems that are the result of passive-component temperature coefficients. Use the following guidelines to keep such problems to a minimum.

Teflon or polystyrene capacitors usually show a $T_C$ of $-110 \pm 30$ ppm/°C. If such capacitors are used as timing capacitors (such as $C_t$ in Fig. 11.29), the output voltage (or gain, in terms of volts per kHz) also shows a corresponding $T_C$. The effect of the $C_t$ temperature coefficient is offset by the addition of a resistor-diode network at pin 2. When $R_x = 240$ k$\Omega$, the current flowing through pin 1 will have an overall $T_C$ of 110 ppm/°C, effectively canceling the $T_C$ of the timing capacitor. This eliminates the need for a zero-$T_C$ timing capacitor. However, $C_t$ should have a stable $T_C$. The resistor-diode network also compensates (almost) for the $T_C$ of the remaining circuit components.

After the circuit has been assembled, and checked out at room temperature, a brief oven test will indicate the sign and size of the $T_C$ for the complete FVC.

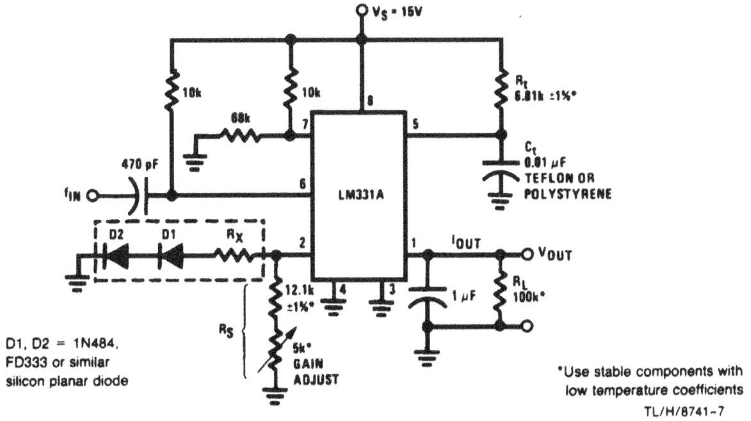

**Figure 11.29**  FVC with $T_C$-control circuit (National Semiconductor, *Linear Applications Handbook,* 1994, p. 1244).

You can then add resistance in series with $R_x$, or add conductance in parallel with $T_C$, to reduce the $T_C$ to a minimum.

For example, if the circuit increases the full-scale output by 0.1% per 30°C (33 ppm/°C) during the oven test, adding 120 kΩ in series with an $R_x$ of 240 kΩ cancels the temperature-caused deviation. Or, if the full-scale output decreases by −0.04% per 20°C (−20 ppm/°C), add 1.2 MΩ in parallel with $R_x$.

To allow trimming in both directions, start with a finite fixed $T_C$ (such as the −110 ppm/°C of $C_t$). Then cancel the fixed $T_C$ with an adjustable $T_C$. This procedure makes it possible to compensate for whatever polarity of $T_C$ is found by the oven test, and get $T_C$s as low as 20 ppm/°C, possibly even 10 ppm/°C. Consider the following guidelines for best results in $T_C$ trimming.

Use a good capacitor for $C_t$. The cheapest polystyrene capacitors can shift value by 0.05% or more per temperature cycle. Using such a capacitor makes it impossible to distinguish the actual temperature sensitivity from hysteresis.

After soldering, bake or temperature-cycle the circuit (at a temperature not exceeding 75°C in the case of polystyrene) for a few hours to stabilize all components and to relieve the strains of soldering.

Do not rush the trimming. Recheck the room-temperature value before and after the high-temperature information is taken. This procedure will ensure a reasonably low hysteresis per cycle.

Do not expect a perfect $T_C$ at −25°C if you trim for a ±5 ppm/°C at temperatures from +25°C to 60°C. None of the components in the circuit of Fig. 11.29 offer linearity much better than 5 ppm/°C or 10 ppm/°C cold, if trimmed for a zero $T_C$ at warm temperatures. Even so, it is still possible to get a data-converter circuit with an 0.02% accuracy and 0.003% linearity, for a ±20°C range near room temperature.

Start the trimming with $R_x$ installed, at a value near the design-center value (240 kΩ or 270 kΩ). Such values should produce a $T_C$ near zero. Do not start trimming without an $R_x$ installed.

If you change $R_x$ (for example, from 240 kΩ to 220 kΩ), do not pull out the 240-kΩ resistor and put in a new 220-kΩ part. The results will be more consistent if a 2.4-MΩ resistor is added in parallel. This is also true when adding resistances in series with $R_x$.

Use reasonably stable components. (This is a good idea when troubleshooting any experimental circuit, especially VFC and FVC circuits!) If you use an LM331A (±50 ppm/°C maximum) and RN55D film resistors (each ±100 ppm/°C) for $R_L$, $R_t$, and $R_s$, it will generally be impossible to trim out the worst-case ±350 ppm/°C $T_C$. Resistors with a specification of 25 ppm/°C usually work well.

Use the same resistor for both $R_s$ and $R_t$. When these resistors come from the same manufacturer's batch, their $T_C$ tracking will usually be better than 20 ppm/°C.

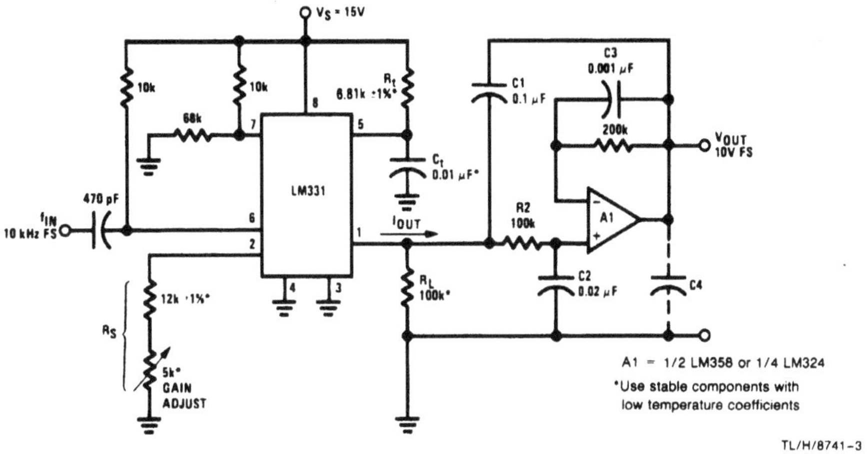

**Figure 11.30**  FVC with output buffer/filter (National Semiconductor, *Linear Applications Handbook,* 1994, p. 1242).

Whenever an op amp is used as a buffer (such as in Fig. 11.30), the offset voltage and current (±7.5 mV maximum and ±100 nA, respectively, for most inexpensive IC op amps) can cause a ±17.5-mV worst-case output offset. However, if both plus and minus supplies are available, a symmetrical offset adjustment can be added. With only one supply, a small positive current can be added to each op-amp input and one input can be trimmed.

# Oscillator/Generator Circuit Troubleshooting

This chapter is devoted to troubleshooting for oscillator/generator circuits. It starts with a review of basic oscillator/generator testing and troubleshooting problems. This is followed by a section of practical techniques for troubleshooting specific circuit components, as well as for specific trouble symptoms.

Before getting started, let's resolve certain differences in terms. In this book, a circuit is considered to be an *oscillator* if the primary function is to produce sine waves (such as the circuit of Fig. 12.1). If the circuit output is a pulse, square wave, triangle wave, ramp, etc. (such as the circuit of Fig. 12.2), the circuit is considered a *generator*.

## 12.1 Basic Testing/Troubleshooting Approach

No matter how complex or simple the circuits of this chapter appear, they are essentially oscillators (or generators) and can be treated as such for practical testing and troubleshooting. For example, each circuit produces output signals, possibly crystal-controlled, but often where frequency depends on RC (resistance-capacitance) or LC (inductance-capacitance). The output signals must have a given amplitude and must be at a given frequency (or must be capable of tuning across a given frequency range). For generators, the output must be of a given shape (square, pulse, triangular, etc.). If you measure the signals and find them to be of the correct frequency, amplitude, and shape, the oscillators or generators are good from a troubleshooting standpoint. If not, the test results provide a good starting point for troubleshooting.

A scope is the logical instrument for both testing and troubleshooting generators because the shape of the circuit output is often critical. However, it is convenient to monitor output amplitude with a meter. A frequency counter is

L2–L5 #1891

LOW FREQ (<50Hz)
LOW DISTORTION MODE

L1 #327

NORMAL MODE

100Ω

430Ω

LT1037

OUTPUT

*1% FILM RESISTOR.
10k DUAL POTENTIOMETER—
MATCH TRACKING 0.1%.
MATCH ALL LIKE CAPACITOR
VALUES 0.1%.

20Hz–200Hz    200Hz → 2kHz    2kHz → 20kHz

0.82    0.082    0.0082

0.82    0.082    0.0082

953*

10k

953*

10k

(a)

A = 10V/DIV

B = 0.01V/DIV
(0.003% DISTORTION)

HORIZONTAL = 20μs/DIV

(b)

0.050
0.045
0.040
0.035
0.030
0.025
0.020
0.015
0.010
0.005
0

PERCENT DISTORTION

NORMAL MODE

LOW FREQUENCY
LOW DISTORTION
MODE

0    20    200    2k    20k
FREQUENCY (Hz)

(c)

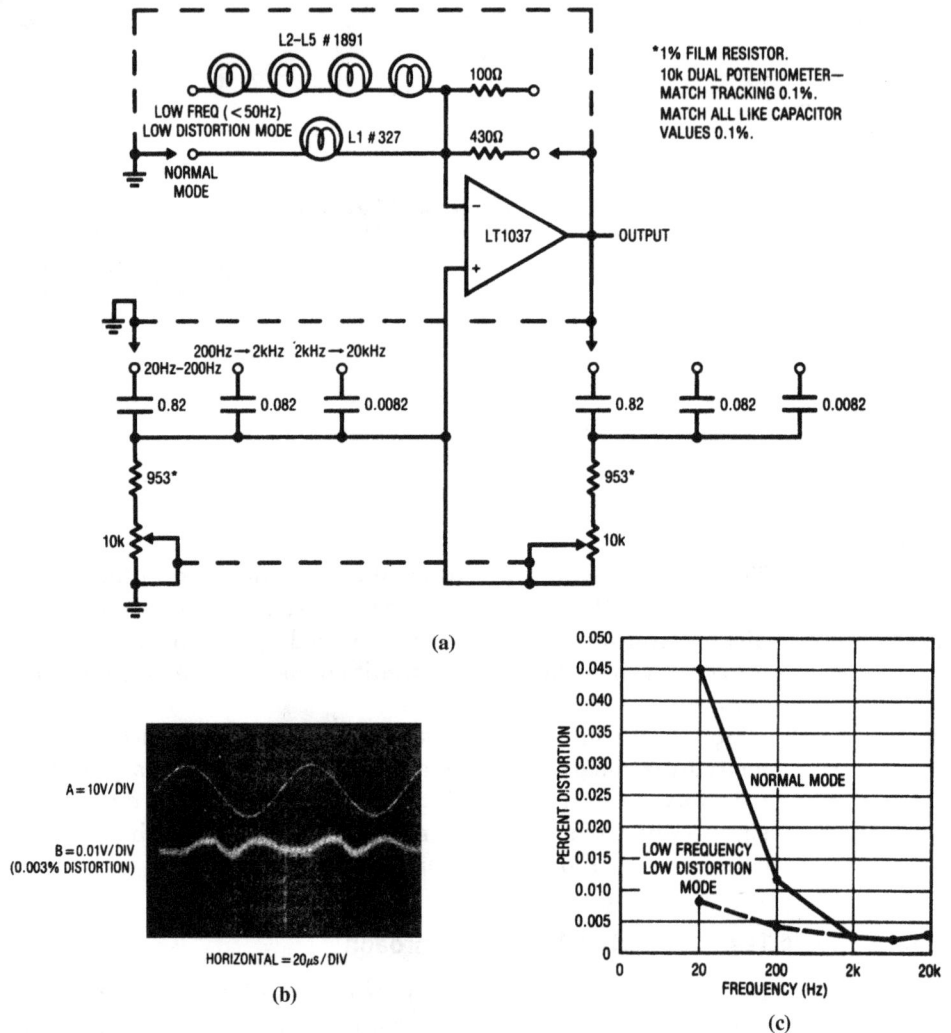

**Figure 12.1** Low-distortion sine-wave oscillator (Linear Technology, Application Note 5, p. 8).

much easier to use when measuring output frequency (although you can measure both amplitude and frequency with a scope).

## 12.2  Basic Oscillator/Generator Tests

The first step in troubleshooting oscillator/generator circuits is to measure the amplitude, frequency, and shape of the output signals. Most oscillator/generator circuits have a built-in test point. For example, the sine and cosine output of the Fig. 12.3 circuits are available at the op-amp outputs. Likewise, triangular and square-wave outputs are available at $V_4$ and $V_1$, respectively, in Fig. 12.4.

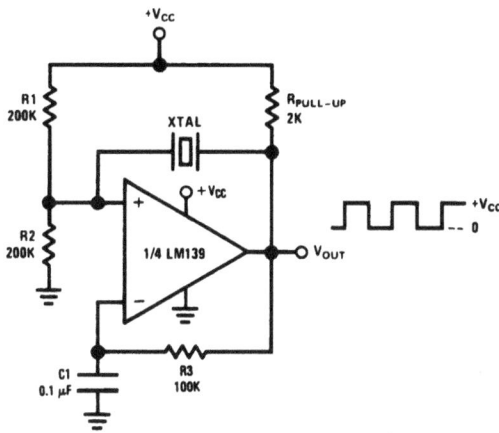

Figure 12.2  Crystal-controlled square-wave generator (National Semiconductor, *Linear Applications Handbook*, 1991, p. 262).

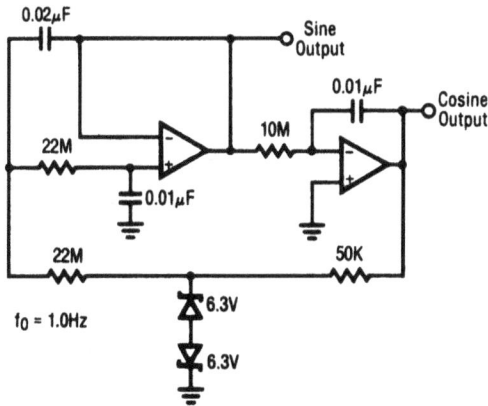

(a)

8-Lead Metal Can TO-99
(Top View)

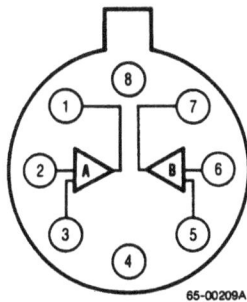

65-00209A

8-Lead Dual In-Line Package
(Top View)

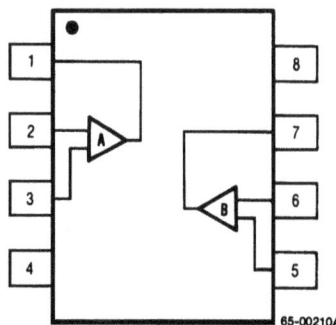

65-00210A

| Pin | Function |
|-----|----------|
| 1 | Output (A) |
| 2 | -Input (A) |
| 3 | +Input (A) |
| 4 | -V$_S$ |
| 5 | +Input (B) |
| 6 | -Input (B) |
| 7 | Output (B) |
| 8 | +V$_S$ |

(b)

Figure 12.3  Low-frequency sine-wave generator (*Raytheon Linear Integrated Circuits,* 1989, p. 4-204).

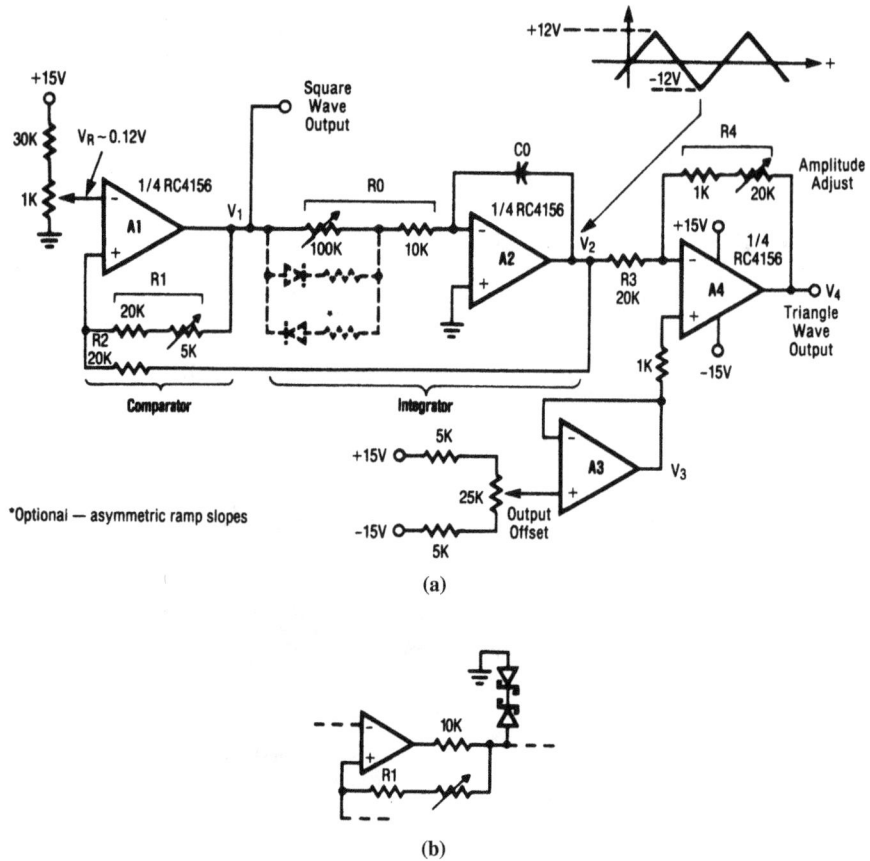

**Figure 12.4** Triangle and square-wave generator (*Raytheon Linear Integrated Circuits, 1989*, p. 4-188).

For an RF (radio-frequency) oscillator, the signal can be monitored at the collector or emitter as shown in Fig. 12.5. In this illustration, signal amplitude is monitored with a meter or scope using an RF probe. If you are interested only in the frequency, use a frequency counter.

## 12.3  Oscillator/Generator Frequency Problems

When you measure an oscillator/generator signal, the frequency can be (1) right on, (2) slightly off, or (3) way off. If the frequency is slightly off, you can sometimes correct the problem with adjustment. Some oscillator/generator circuits are adjustable (even those with crystal control). The most precise adjustment is made by monitoring the oscillator/generator signal with a frequency counter and adjusting the circuit for exact frequency. However, it is possible to adjust a crystal oscillator with a meter or scope.

Generally, when a crystal-controlled circuit is adjusted for maximum signal amplitude, the circuit is at the exact crystal frequency. However, it is possible that the circuit is being tuned for a harmonic (multiple or submultiple) of the crystal frequency. A frequency counter shows this, but a meter or scope does not.

If the circuit frequency is way off, look for a defect, rather than improper adjustment. For example, a coil or transformer might have shorted turns, a transistor or capacitor might be leaking badly, and an IC or crystal might be defective, or you might have wired the circuit incorrectly (impossible?).

## 12.4   Measuring the Frequency with a Scope

If you do not have a frequency counter and must measure frequency with a scope, use the following procedure. Adjust the scope controls so you can measure the duration of one complete cycle (along the horizontal trace). For example, if one complete cycle occupies two horizontal divisions, and each division

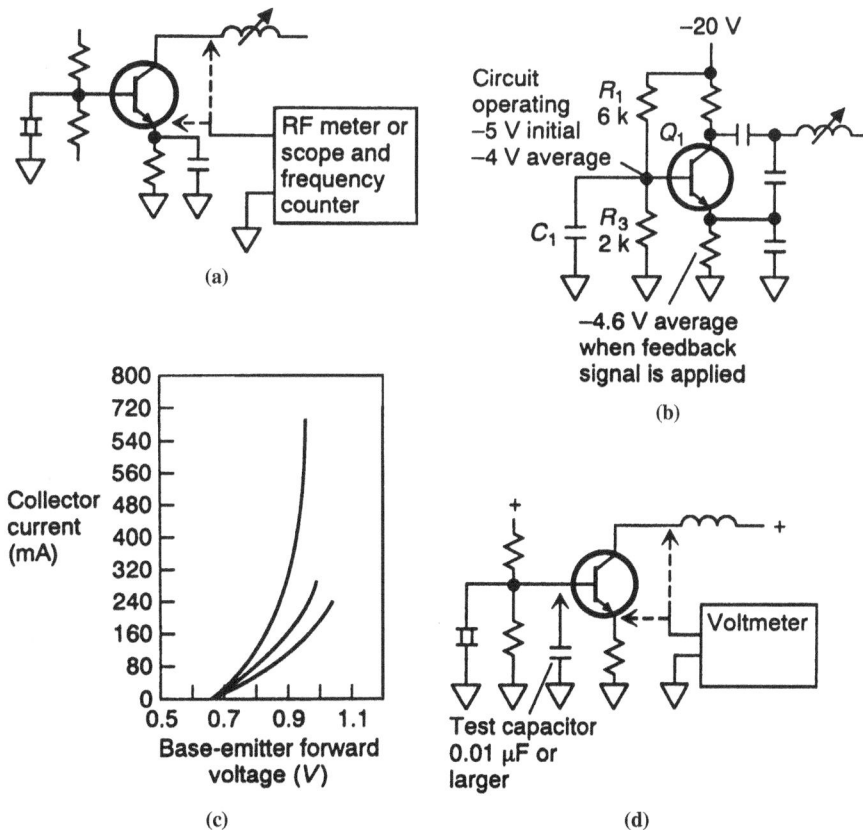

**Figure 12.5**   Oscillator testing and troubleshooting.

is 10 ns, as shown in Fig. 12.6, each cycle is 20 ns ($20^{-9}$ s) in duration. Find the reciprocal of 20 ns:

$$1/20 = 0.05 \times 10^9 = 50^6 = 50 \text{ MHz}$$

## 12.5 Oscillator/Generator Signal-Amplitude Problems

When you measure an oscillator/generator signal, the amplitude is (1) right on, (2) slightly low, or (3) very low. If the amplitude is right on, leave the circuit alone (unless the frequency is off). If the amplitude is slightly low, it is sometimes possible to correct the problem with adjustment. Monitor the signal with a meter or scope, and adjust the circuit for maximum signal (or for the desired output signal level).

For example, in Fig. 12.4, adjust $R_1$ and the $V_R$ voltage (at A1) for a triangle at $V_2$ with a ±12-V amplitude. Then adjust R4 for the desired triangular-output level.

Notice that with an adjustable crystal-oscillator circuit, such as shown in Fig. 12.5, adjusting for maximum amplitude also locks the oscillator at the correct frequency. In any case, if the adjustment does not correct the problem, look for leaking transistors or capacitors, or a possible defective IC.

If the amplitude is very low, look for defects, such as lower power-supply voltages, badly leaking transistors and/or capacitors, defective ICs, and shorted coil or transformer turns (for RF oscillators). Usually, if the signal output is very low, there are other indications, such as abnormal voltages and resistance values.

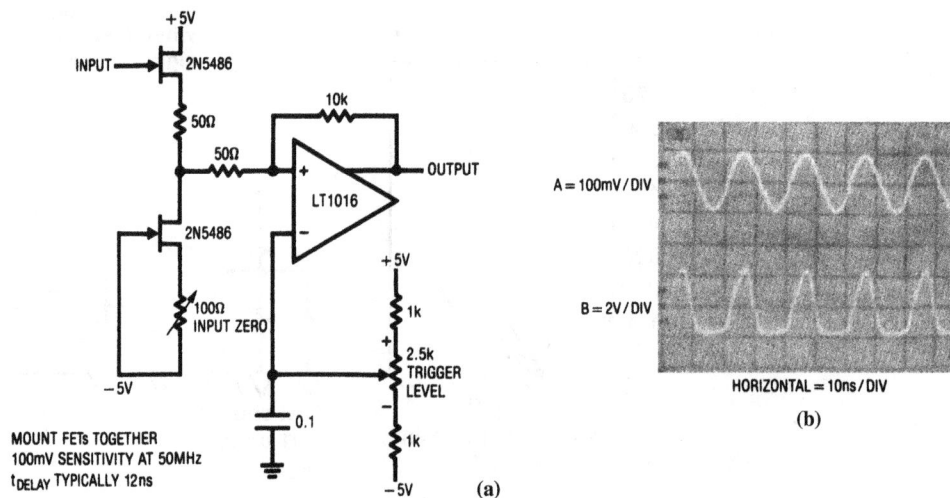

**Figure 12.6**  50-MHz trigger (Linear Technology, Application Note 13, p. 24).

## 12.6   Determining the Output Voltage Amplitude

When troubleshooting an experimental circuit, notice that the output voltage should be slightly lower than the supply (in most cases). For example, in the circuit of Fig. 12.4, the square-wave and triangular-wave outputs can be no greater than the ±15-V supply. (The amplitude is arbitrarily adjusted for ±12 V.) The same is usually true for the RF oscillator in Fig. 12.5D. However, in the oscillator of Fig. 12.5B, the output voltage will be less than 20 V because of the drop across the collector resistor of Q1.

## 12.7   Measuring Oscillator/Generator Amplitude
## with a Scope

If you want to measure oscillator/generator signal amplitude with a scope, use the following procedure. Adjust the scope controls so that you can measure the amplitude of several cycles (along the vertical scale). For example, Fig. 12.6 shows two traces. Trace A is a sine wave taken at the input, with an amplitude of two vertical divisions. Each division is 100 mV, so the input is 200 mV. Trace B is a pulse or trigger response taken at the output with an amplitude of about two vertical divisions. Each division is 2 V, so the output is 4 V (adjustable by the 2.5-kΩ pot).

## 12.8   Generator Waveshape Problems

When troubleshooting a generator circuit where the frequency and amplitude are good, but the waveshape is not correct, suspect leakage. Usually, leaking capacitors and/or a leaking transistor (in discrete circuits) are the culprits. For example, if either or both the triangular-wave and square-wave outputs in Fig. 12.4 are of the correct amplitude and frequency, but are distorted (square-wave sides not straight, tops not flat, triangular-wave ramps bending in or out, etc.), suspect a problem with capacitor CO.

## 12.9   Capacitor Checks During Troubleshooting

During the troubleshooting process (for oscillator/generator circuits, as well as most other electronic circuits), suspected capacitors can be removed from the circuit and tested on bridge-type checkers. This test establishes that the capacitor value is correct. With a correct value, it is reasonable to assume that the capacitor is not open, shorted, or leaking. From the opposite standpoint, if the capacitor shows no shorts, opens, or leaks, it is also fair to assume that the capacitor is good. So, from a practical troubleshooting standpoint, simple tests for shorts, opens, or leaks are usually sufficient. The following paragraphs describe such tests, which can be applied to most electronic circuits that contain capacitors.

### 12.9.1   Checking capacitors with circuit voltages

As shown in Fig. 12.7, using circuit voltages involves disconnecting one lead of the capacitor (the ground or cold end) and connecting a voltmeter between the disconnected lead and ground. A good capacitor should have a momentary voltage indication (or surge) as the capacitor charges up to the voltage at the hot end.

If the voltage indication remains high, the capacitor is probably shorted. If the voltage indication is steady, but not necessarily high, the capacitor is probably leaking. If it has no voltage indication whatsoever, the capacitor is probably open. Notice that this test is good only where one end of the capacitor is connected to a point in the circuit with a measurable voltage above ground (such as the capacitor in Fig. 12.6).

### 12.9.2   Checking capacitors with an ohmmeter

As shown in Fig. 12.7B, using an ohmmeter involves disconnecting one lead of the capacitor (usually the hot end) and connecting an ohmmeter across the capacitor. Be certain that all power is removed from the circuit. As a precaution, short across the capacitor (after the power is removed) to be sure that no charge is retained. A good capacitor should have a momentary resistance indication (or surge) when the capacitor charges up to the voltage of the ohmmeter battery.

If the resistance indication is near zero and remains so, the capacitor is probably shorted. If the resistance indication is steady at some high value, the capacitor is probably leaking. If there is no resistance indication (or surge) whatsoever, the capacitor is probably open.

### 12.10   Troubleshooting Generator and VCO ICs

The circuit of Fig. 12.8A uses a Harris ICL8038 to provide simultaneous sine, square, and triangle outputs. These outputs can be swept in frequency or fre-

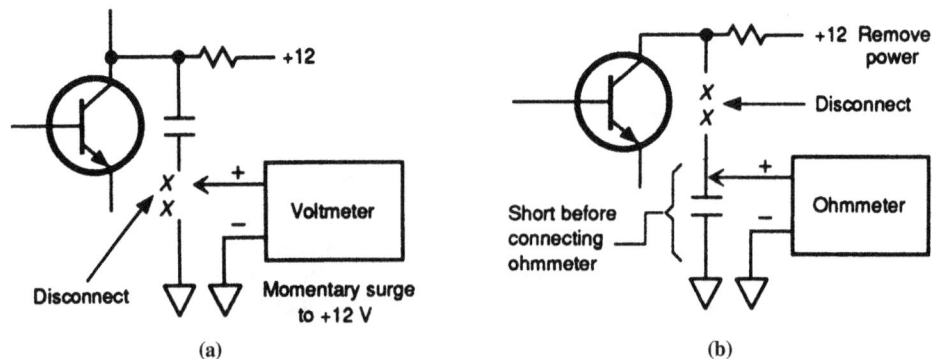

Figure 12.7   In-circuit capacitor tests.

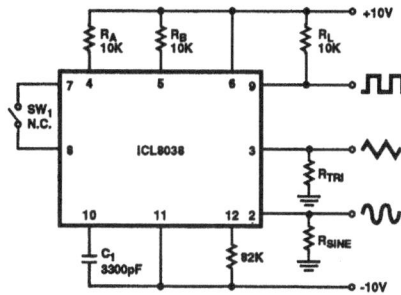

(a)

### Absolute Maximum Ratings

Supply Voltage (V- to V+)............................. 36V
Power Dissipation (Note 1)........................ 750mW
Input Voltage (Any Pin)........................... V- to V+
Input Current (Pins 4 and 5)........................ 25mA
Output Sink Current (Pins 3 and 9) ................ 25mA
Lead Temperature (Soldering 10 Sec.)................ +300°C

### Operating Conditions

Operating Temperature Range
ICL8038AM, ICL8038BM................... -55°C to +125°C
ICL8038AC, ICL8038BC, ICL8038CC.......... 0°C to +70°C
Storage Temperature Range................. -65°C to +150°C

CAUTION: Stresses above those listed in "Absolute Maximum Ratings" may cause permanent damage to the device. This is a stress only rating and operation of the device at these or any other conditions above those indicated in the operational sections of this specification is not implied.

### Electrical Specifications   $V_{SUPPLY} = \pm10V$ or +20V, $T_A$ = +25°C, $R_L$ = 10kΩ, Test Circuit Unless Otherwise Specified

| PARAMETERS | SYMBOL | TEST CONDITIONS | ICL8038CC | | | ICL8038BC(BM) | | | ICL8038AC(AM) | | | UNITS |
|---|---|---|---|---|---|---|---|---|---|---|---|---|
| | | | MIN | TYP | MAX | MIN | TYP | MAX | MIN | TYP | MAX | |
| Supply Voltage Operating Range | $V_{SUPPLY}$ | | | | | | | | | | | |
| Single Supply | V+ | | +10 | | +30 | +10 | | 30 | +10 | | 30 | V |
| Dual Supplies | V+, V- | | ±5 | | ±15 | ±5 | | ±15 | ±5 | | ±5 | V |
| Supply Current | $I_{SUPPLY}$ | $V_{SUPPLY} = \pm10V$ (Note 2) | | | | | | | | | | |
| 8038AM 8038BM | | | | | | | 12 | 15 | | 12 | 15 | mA |
| 8038AC, 8038BC, 8038CC | | | | 12 | 20 | | 12 | 20 | | 12 | 20 | mA |
| FREQUENCY CHARACTERISTICS (ALL WAVEFORMS) | | | | | | | | | | | | |
| Max. Frequency of Oscillation | $f_{MAX}$ | | 100 | | | 100 | | | 100 | | | kHz |
| Sweep Frequency of FM Input | $f_{SWEEP}$ | | | 10 | | | 10 | | | 10 | | kHz |
| Sweep FM Range (Note 3) | | | | 35:1 | | | 35:1 | | | 35:1 | | |
| FM Linearity | | 10:1 Ratio | | 0.5 | | | 0.2 | | | 0.2 | | % |
| Frequency Drift with Temperature (Note 5) | $\Delta f/\Delta T$ | | | | | | | | | | | |
| 8038 AC, BC, CC | | 0°C to +70°C | | 250 | | | 180 | | | 120 | | ppm/°C |
| 8038 AM, BM | | -55°C to +125°C | | | | | | 350 | | | 250 | ppm/°C |
| Frequency Drift with Supply Voltage | $\Delta f/\Delta V$ | Over Supply Voltage Range | | 0.05 | | | 0.05 | | | 0.05 | | %/V |
| OUTPUT CHARACTERISTICS | | | | | | | | | | | | |
| Square Wave | | | | | | | | | | | | |
| Leakage Current | $I_{OLK}$ | $V_9$ = 30V | | | 1 | | | 1 | | | 1 | μA |
| Saturation Voltage | $V_{SAT}$ | $I_{SINK}$ = 2mA | | 0.2 | 0.5 | | 0.2 | 0.4 | | 0.2 | 0.4 | V |
| Rise Time | $t_R$ | $R_L$ = 4.7kΩ | | 180 | | | 180 | | | 180 | | ns |
| Fall Time | $t_F$ | $R_L$ = 4.7kΩ | | 40 | | | 40 | | | 40 | | ns |
| Typical Duty Cycle Adjust (Note 6) | $\Delta D$ | | 2 | | 98 | 2 | | 98 | 2 | | 98 | % |
| Triangle/Sawtooth/Ramp | | | | | | | | | | | | |
| Amplitude | $V_{TRIANGLE}$ | $R_{TRI}$ = 100kΩ | 0.30 | 0.33 | | 0.30 | 0.33 | | 0.30 | 0.33 | | $xV_{SUPPLY}$ |
| Linearity | | | | 0.1 | | | 0.05 | | | 0.05 | | % |
| Output Impedance | $Z_{OUT}$ | $I_{OUT}$ = 5mA | | 200 | | | 200 | | | 200 | | Ω |
| Sine Wave | | | | | | | | | | | | |
| Amplitude | $V_{SINE}$ | $R_{SINE}$ = 100kΩ | 0.2 | 0.22 | | 0.2 | 0.22 | | 0.2 | 0.22 | | $xV_{SUPPLY}$ |
| THD | THD | $R_S$ = 1MΩ (Note 4) | | 2.0 | 5 | | 1.5 | 3 | | 1.0 | 1.5 | % |
| THD Adjusted | THD | Use Figure 14 | | 1.5 | | | 1.0 | | | 0.8 | | % |

NOTES:

1. Derate ceramic package at 12.5mW/°C for ambient temperatures above 100°C.
2. $R_A$ and $R_B$ currents not included.
3. $V_{SUPPLY}$ = 20V; $R_A$ and $R_B$ = 10kΩ, f ≅ 10kHz nominal; can be extended 1000 to 1. See Figures 15A and 15B.
4. 82kΩ connected between pins 11 and 12, Triangle Duty Cycle set at 50%. (Use $R_A$ and $R_B$.)
5. Figure 1, pins 7 and 8 connected, $V_{SUPPLY}$ = ±10V. See Typical Curves for T.C. vs $V_{SUPPLY}$.
6. Not tested, typical value for design purposes only.

(b)

**Figure 12.8**   Waveform generator and VCO (Harris Semiconductors, *Linear & Telecom ICs,* 1994, pp. 7-121, 7-123, 7-125).

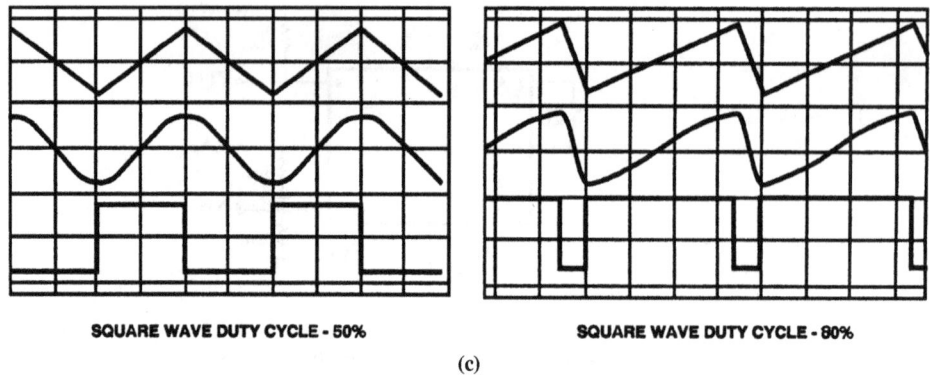

SQUARE WAVE DUTY CYCLE - 50%                SQUARE WAVE DUTY CYCLE - 80%

(c)

**Figure 12.8**   (Continued)

quency modulated. The output frequency range is from 0.001 Hz to 300 kHz, and is set by $R_A$, $R_B$, and $C_1$, using the equation:

$$\text{frequency} = 0.33/RC$$

where $R = R_A = R_B$

The characteristics are shown in Fig. 12.8B. Figure 12.8C shows the phase relationship of the waveforms.

The most likely cause of trouble in this circuit is capacitor C1. If the capacitor is open or shorted the IC will probably not oscillate. If C1 is leaking (even a little), the waveforms will be distorted.

Also notice that all of the output waveforms change when the square-wave duty cycle is changed (with different values for $R_A$ and $R_B$). This is common for generator ICs with multiple outputs.

## 12.11   Troubleshooting Generator/VCO ICs with Variable Duty Cycle

The circuit of Fig. 12.9 is similar to that of Fig. 12.8, except that the Fig. 12.9 circuit has alternate timing configurations and a duty-cycle control. If the circuit is experimental, use Fig. 12.9A where the timing resistors are separate.

In troubleshooting this circuit, remember that $R_A$ controls the rising portion of the triangle and sine wave, as well as the 1-state of the square wave. When $R_A = R_B$, the duty cycle is 50%. If the duty cycle is to be varied over a small range about the 50% point, the connections of Fig. 12.9B are more convenient. Adjust the 1-k$\Omega$ pot for a minimum sine-wave distortion of about 1%.

Again, capacitor C is the most likely cause of trouble. However, remember that it is normal for the triangle and sine waves to be distorted if the square wave is not set at 50% duty cycle. As shown in Fig. 12.8C, the triangular wave turns into a linear ramp, and the sine wave appears as a nonlinear ramp (with a hump).

## 12.12   Troubleshooting Generator with Sweep and Frequency Modulation

The circuit of Fig. 12.10 is similar to that of Fig. 12.8, except that the circuits of Fig. 12.10 provide for FM (frequency modulation) or sweep modulation. Use the circuit of Fig. 12.10A for FM (small deviations, about ±10%). Use the circuit of Fig. 12.10B for a sweep range of 1000:1. The external resistor between pins 7 and 8 is not absolutely necessary, but it can be used to increase input impedance from about 8 kΩ (pins 7 and 8 connected together) to about $(R+8$ kΩ). The capacitor at pin 8 provides decoupling to the FM source.

As a guideline for troubleshooting, notice that the sweep frequency approaches 0 Hz when the voltage at pin 8 equals the supply voltage ($V+$). The frequency reaches maximum at the lower pin-8 voltage limit of $V + -(1/3\ V_{\text{SUPPLY}} - 2)$. Waveform symmetry variations can be minimized when a 10-MΩ resistor is added between pins 5 and 11.

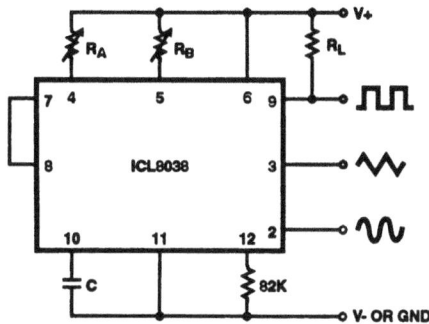

(a)

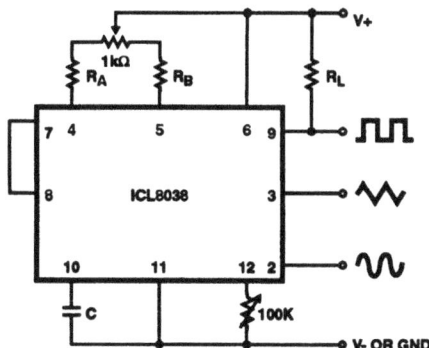

(b)

Figure 12.9   Waveform generator with variable duty cycle (Harris Semiconductors, *Linear & Telecom ICs*, 1994, p. 7-126).

## 12.13   Troubleshooting Low-Frequency Reference Oscillator

The circuit of Fig. 12.11 uses two timers (Harris ICM7242) connected in cascade to produce a low-frequency reference signal. Figure 12.11B shows the basic timer connections and equations for operating frequency.

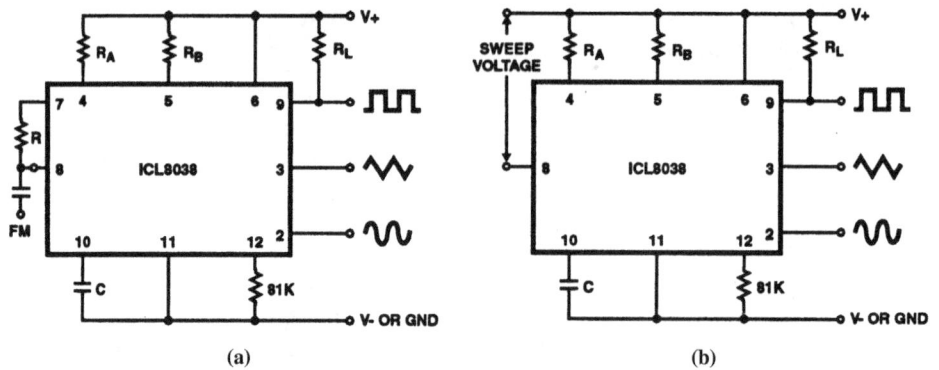

**Figure 12.10**  Waveform generator with sweep and frequency modulation (Harris Semiconductors, *Linear & Telecom ICs,* 1994, p. 7-127).

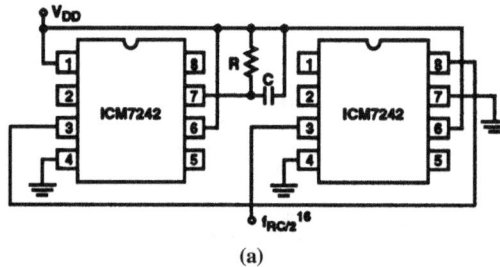

(a)

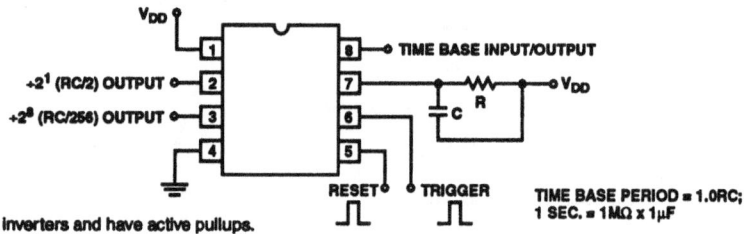

NOTE: +2¹ and +2⁸ outputs are inverters and have active pullups.

(b)

**Figure 12.11**  Low-frequency reference oscillator (timer) (Harris Semiconductors, *Linear & Telecom ICs,* 1994, pp. 7-141, 7-144).

In troubleshooting this or any similar circuit, remember that the frequency can be changed by a change in either (or both) $R$ and $C$. (The frequency is the reciprocal of the time-base period.) The minimum value for timing capacitor C is 10 pF. The range for timing resistor R is from 1 kΩ to 22 MΩ. $V_{DD}$ can be from 2 V to 16 V. As usual, C is the most likely cause of circuit problems.

## 12.14    Troubleshooting Timer Oscillators

Figures 12.12, 12.13, and 12.14 show general-purpose timers (Harris ICM7555/7556) connected as oscillators. The circuits of Figs. 12.12 and 12.13

are astable oscillators, whereas the circuit of Fig. 12.14 provides monostable (or one-shot) operation.

In troubleshooting any of these circuits, capacitor C is the key component. This applies to the frequency of operation and to the circuit performance. A capacitor C of incorrect value will produce an incorrect operating frequency. A leaking, shorted or open capacitor C will cause total circuit failure, or distorted waveforms. The supply can range from 5 V to 15 V, and the output will be rail to rail. However, the duty cycle of the Fig. 12.13 circuit is set by the ratio of $R_A$ and $R_B$, using the equation:

$$\text{duty cycle } D = (R_A + R_B)/(R_A + 2R_B)$$

The duty cycle for the circuit of Fig. 12.12 should be 50%, with the output frequency set by $R$ and $C$ using the equation:

$$\text{output frequency} = 1/(1.4\,RC)$$

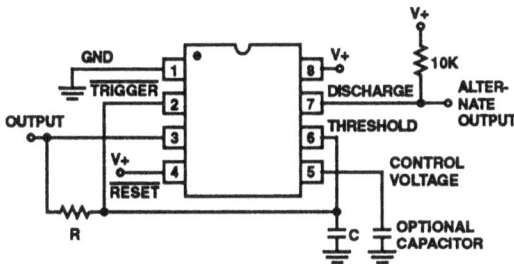

Figure 12.12   Astable oscillator (timer) (Harris Semiconductors, *Linear & Telecom ICs*, 1994, p. 7-152).

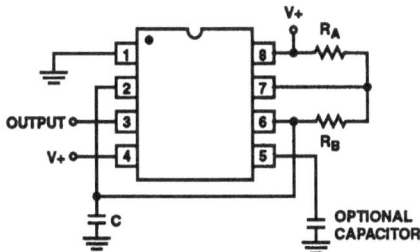

Figure 12.13   Alternate astable oscillator (timer) (Harris Semiconductors, *Linear & Telecom ICs*, 1994, p. 7-152).

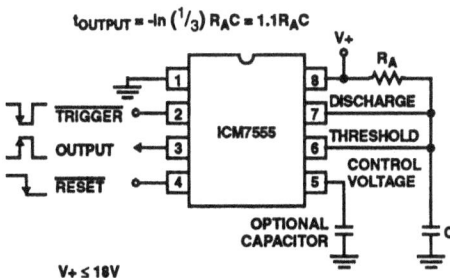

Figure 12.14   Monostable oscillator (timer) (Harris Semiconductors, *Linear & Telecom ICs*, 1994, p. 7-152).

The output frequency for the circuit of Fig. 12.13 is set by $C$, $R_A$, and $R_B$ using the equation:

$$\text{output frequency} = \frac{1.44}{(R_A + 2R_B)} C$$

The output timing for the circuit of Fig. 12.14 is set by the equation shown.

## 12.15    Troubleshooting Programmable Square-Wave Generators

Figure 12.15 shows an 8-bit (255-frequency) programmable oscillator with a TTL-compatible square-wave output (so it could be called a *programmable generator*). The digital input produces a corresponding current from A1. The current is applied to A2, which produces a square-wave output with a frequency that is proportional to the numerical value of the digital input word.

When troubleshooting an experimental circuit, with the values shown, the circuit has a nominal full-scale frequency of 10 kHz (9961 Hz for all 1s). To calibrate the circuit, apply all 1s to A1 and adjust $R_1$ for an output frequency of 9961 Hz. Worst-case nonlinearity is 0.16%. Capacitor C1 is the crucial component in this circuit.

## 12.16    Troubleshooting Programmable Sine-Wave Oscillators

Figure 12.16 shows a sine-wave oscillator with linear control using a state-variable filter. Figure 12.16A shows the circuit configuration where the fre-

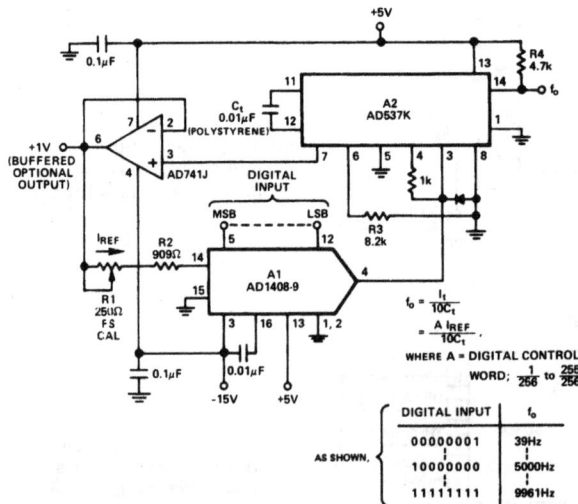

**Figure 12.15** Programmable square-wave generator (Analog Devices, *Applications Reference Manual*, 1993, p. 8-72).

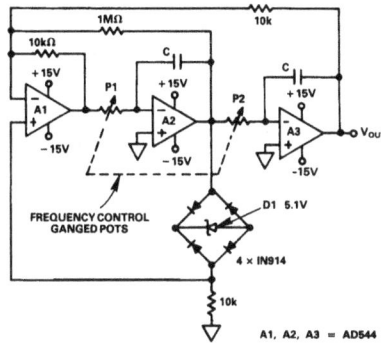

(a)

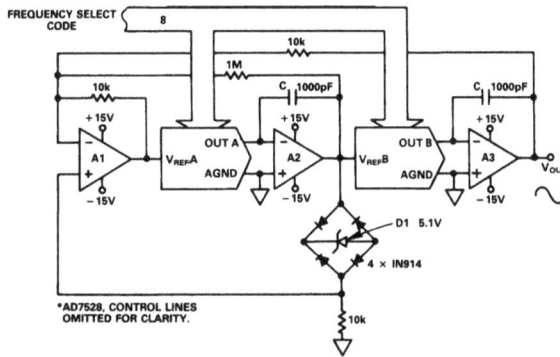

(b)

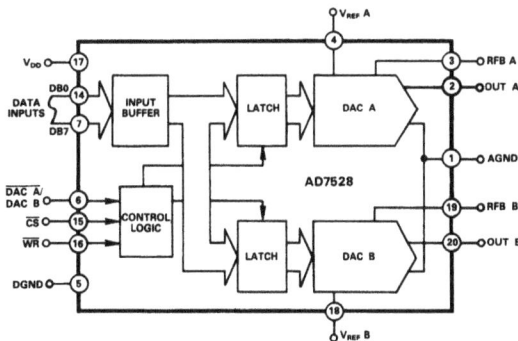

(c)

Figure 12.16  Programmable sine-wave oscillator (Analog Devices, *Applications Reference Manual*, 1993, pp. 8-113, 8-115).

quency is set by ganged potentiometers P1 and P2. Figure 12.16B shows the same circuit with P1 and P2 replaced by a matched pair of AD752s (Analog Devices). Figure 12.16C shows the internal functions and pin configuration for the AD7528.

In troubleshooting either version of this circuit, notice that frequency depends on *R* and *C*, using the equation:

$$\text{frequency (in Hz)} = \frac{N}{256\ (6.28\ RC)}$$

where $R$ = DAC ladder resistance ($V_{REF}$ input resistance), $C$ is as shown in Fig. 12.16B, $N$ = decimal representation of the digital input code. For example, $N$ = 128 for input code 10000000. For the values given in Fig. 12.16B, output frequency is variable from 0 to 15 kHz. The output amplitude is set by D1. The THD is −53 dB at 1 kHz and −43 dB at 14 kHz. A cosine output is also available at the output of A2.

The crucial components in this circuit are the feedback resistors and capacitors for A1, A2, and A3. However, any leakage in D1 or the four 1N914 bridge diodes can produce erratic results.

# Index

# Index

www.ingramcontent.com/pod-product-compliance
Lightning Source LLC
Chambersburg PA
CBHW082104220326
41598CB00066BA/5218